Handbook on
FEDERATED LEARNING
Advances, Applications and Opportunities

Editors

Saravanan Krishnan
Associate Professor, Department of Computer Science and Engineering
College of Engineering, Guindy, Anna University, Chennai, India

A. Jose Anand
Professor, Department of Electronics and Communication Engineering
KCG College of Technology, Karapakkam, Chennai, India

R. Srinivasan
Professor, Department of Computer Science and Engineering
Vel Tech Rangarajan Dr. Sagunthala R & D Institute of Science and
Technology, Chennai, India

R. Kavitha
Professor, Department of Computer Science and Engineering
Vel Tech Rangarajan Dr. Sagunthala R & D Institute of Science and
Technology, Chennai, India

S. Suresh
Professor, Department of Computer Science and Engineering
KPR Institute of Engineering and Technology, Coimbatore, India

CRC Press
Taylor & Francis Group
Boca Raton London New York

CRC Press is an imprint of the
Taylor & Francis Group, an **informa** business

A SCIENCE PUBLISHERS BOOK

First edition published 2024
by CRC Press
2385 NW Executive Center Drive, Suite 320, Boca Raton FL 33431

and by CRC Press
4 Park Square, Milton Park, Abingdon, Oxon, OX14 4RN

CRC Press is an imprint of Taylor & Francis Group, LLC

Library of Congress Cataloging-in-Publication Data (applied for)

ISBN: 978-1-032-47162-4 (hbk)
ISBN: 978-1-032-47163-1 (pbk)
ISBN: 978-1-003-38485-4 (ebk)

DOI: 10.1201/9781003384854

Typeset in Palatino Linotype
by Radiant Productions

Preface

Federated Learning (FL) is a Distributed Machine Learning model that has been used in many applications today. It has been pioneered by companies like Google Inc and Apple Inc as a technique to create machine learning models with improved performance on distributed datasets without sacrificing privacy. Currently, federated learning is used by Apple Inc to enhance the accuracy of Face ID and Siri, and by Google to power keyboard predictions in Gboard. Its ability to handle the privacy issues raised by conventional ML models contributed to its appeal. As a result, federated learning is currently one of the most intriguing and promising areas in AI. It has lately altered how we gather, examine, and use data. Federated Learning has widespread applications in Industry 4.0, Healthcare and Robotics, etc. This book provides a comprehensive approach in FL for various aspects.

The book comprises four sections namely Concepts and Types, Learning Process and Optimization, Tools and Techniques and Applications and use cases. The purpose of this edited book is to present a detailed explanation about concepts, tools and techniques in various applications with real time use cases.

Section 1 discusses federated learning definitions, architectures, and applications for the federated-learning framework, and provides a comprehensive survey on existing works.

Section 2 provides chapters with advantages in data security, data diversity, real-time continuous learning, hardware efficiency, etc., in terms of FL. However, when certain participants are not believed to be completely honest, it poses new privacy risks such membership inference attacks and data poisoning assaults. Furthermore, self-centered participants may receive the collective data of others while withholding or even providing false information about their own local area. Consequently, modern privacy and fairness mechanisms have been included into FL systems like blockchain concepts were also discussed in chapters. It addresses the challenges in data privacy, data security, data access, etc. In Section 3,

Chapters investigate the evolution of Federated Reinforcement Learning from a technical perspective, and highlight its advantages over previous Reinforcement Learning algorithms.

Section 4 discusses a learning paradigm seeking to address the problem of data governance and privacy by training algorithms collaboratively without exchanging the data itself. It recently gained attraction for healthcare applications which were discussed in a few chapters. FL enables gaining insights collaboratively, e.g., in the form of a consensus model, without moving patient data beyond the firewalls of the institutions in which they reside. This book provides a comprehensive approach in federated learning in various aspects.

Contents

Preface   iii

1. **Introduction to Federated Learning: Methods and Classifications** 1
   Shashikiran Venkatesha and *Ramanathan Lakshmanan*

2. **Federated Data Model - Go Local, Go Global and**   32
   Go Fusion - In an Industry 4.0 Context
   Daniel Einarson and *Charlotte Sennersten*

3. **Federated Learning Architectures, Opportunities,**   61
   and Applications
   Pradipta kumar Mishra, Rabinarayan Satapathy and *Debashreet Das*

4. **Secure and Private Federated Learning through Encrypted**   80
   Parameter Aggregation
   K Vijayalakshmi, PM Sitharselvam, I Thamarai, J Ashok,
   Goski Sathish and *S Mayakannan*

5. **Navigating Privacy Concerns in Federated Learning:**   106
   A GDPR-Focused Analysis
   G Anitha and *A Jegatheesan*

6. **A Federated Learning Approach for Resource-Constrained**   131
   IoT Security Monitoring
   P Sakthibalan, M Saravanan, V Ansal, Amuthakkannan Rajakannu,
   K Vijayalakshmi and *K Divya Vani*

7. **Efficient Federated Learning Techniques for Data Loss**   155
   Prevention in Cloud Environment
   A Peter Soosai Anandaraj, S Sridevi, R Vaishnavi, M Meenalakshimi
   and *RV Chandrashekhar*

8. **Maximizing Fog Computing Efficiency with Federated**   174
   Multi-Agent Deep Reinforcement Learning
   G Anitha and *A Jegatheesan*

9. **Future of Medical Research with a Data-driven Federated Learning Approach** — 202
G Arun Sampaul Thomas, S Muthukaruppasamy, S Sathish Kumar and K Saravanan

10. **Collaborative Federated Learning in Healthcare Systems** — 226
Bini M Issac and SN Kumar

11. **Federated Learning for Efficient Cardiac Disease Prediction based on Hyper Spectral Feature Selection using Deep Spectral Convolution Neural Network** — 245
B Dhiyanesh, G Kiruthiga, P Saraswathi, Gomathi S and R Radha

12. **A Federated Learning based Alzheimer's Disease Prediction** — 264
S Suchitra, N Senthamarai, M Jeyaselvi and RJ Poovaraghan

13. **Detecting Device Sensors of Luxury Hotels using Blockchain-based Federated Learning to Increase Customer Satisfaction** — 283
Moyeenudin HM, Shaik Javed Parvez, Jose Anand A, Anandan R and Sam Goundar

14. **Navigating the Complexity of Macro-Tasks: Federated Learning as a Catalyst for Effective Crowd Coordination** — 308
S Mayakannan, N Krishnamurthy, K Vimala Devi, R Deepalakshmi, Sandya Rani and Jose Anand A

15. **Stock Market Prediction via Twitter Sentiment Analysis using BERT: A Federated Learning Approach** — 333
M Rajeev Kumar, S Ramkumar, S Saravanan, R Balakrishnan and M Swathi

Index — 355

Chapter 1

Introduction to Federated Learning
Methods and Classifications

Shashikiran Venkatesha[1],* and Ramanathan Lakshmanan[2]

1. Motivation and Introduction

Advancement in the applications and tools related to different realms such as recommender systems, natural language processing, computer vision, and automatic speech recognition is a result of rapid progress in machine learning (ML) algorithms, as observed in the recent past [1–3]. The formidable and favourable outcomes of ML platforms tuned by Deep Learning (DL) are supported by massive volumes of legitimate data or "Big Data" accessible from the reservoirs of data, also known as "data repositories" or "data centres" [2–4]. The ultimate performance is achievable in various tasks executed by DL platforms which are sometimes on par with human cognitive excellence. A face recognition system tuned by DL trained with billions of images can produce outcomes that are effective, and marketable. For Example, 3.5 billion images are taken from Instagram [5] for training in an object detection system namely Detectron, a classic example of a scalable distributed system framework and an integral part of Facebook. Software architects, ML Researchers

[1] Assistant Professor Senior Grade-1, SITE, Vellore Institute of Technology, Vellore India.
[2] Associate Professor, SCOPE, Vellore Institute of Technology, Vellore India.
Email: lramanathan@vit.ac.in
* Corresponding author: shashikiran@vit.ac.in

and developers, face many challenges such as creating, developing, operationalizing, and maintaining a distributed computing framework. The challenges are enumerated and presented below.

- Data ownership rights: In the 21st Century, people are aware of their ownership rights when it comes to their data and its usage in constructing ML platforms. In modern ML driven products like recommendation services, there exists an ambiguity in claiming ownership rights over consumer buying patterns and remittance modes, unlikely in the case of product data and supplier transactions, its ownership rights are claimed by the service provider.

- Compliance with Data protection bills: In 2018, European Union adopted a bill named "General Data Protection Regulation" which is a prominent example of data protection law and compliance [6]. In the year 2020, the California Consumer Privacy Act (CCPA) enacted in the state of California, United States provides a legal framework for data protection and usage [7]. The basic formulations adopted in GDPR for data protection provide legal remedies in case of breach of private data.

- Data privacy and confidentiality: In distributed systems, consumer privacy and confidentiality of data have become the primary issue in designing ML applications. Weak governments and commercial establishments manipulate or use private data to fulfil their objectives. Internet based multi-national companies have failed to protect users' private data, and have been punished by higher jurisprudence in the form of "fines". Internet based companies have been punished in courts for (a) sending unsolicited messages, (b) phishing, and (c) unethical data exchanges.

- Data in the independent warehouse: Delicate or susceptible information in the documents (records in finance and medical sections) that can influence decision making in government institutions and commercial establishments forces data to remain in isolation or as independent data silos managed by data center owners [8].

- Lack of understanding of benefits of collaborative models: Another pertinent reason for ML organizations suffering from data scattering is the lack of understanding in quantifying the gains from collaborative models. For example, two organizations try to construct a collaborative ML model by training using medical images. In legacy systems, data is transferred from one organization to another and the owner of the data loses control over the transferred data when it leaves the sphere of influence of the organization. The non-existence of a tangible measure to allocate gains among the participants who contributed towards building better joint ML models is a major challenge faced by ML researchers.

However, in recent times, the computational abilities, and the storage capacity of the Internet of Things (IoT) devices have grown over a period. Software developers and architects have leveraged the computational abilities of millions of devices to build collaborative models that mitigate the challenges mentioned above.

1.1 Federated Learning is the Answer

Considering the above imperatives, the distributed learning platforms (a) that leverage enhanced computational abilities of devices, (b) connecting devices executing local training models, and (c) connecting devices co-operate to build consensus global models of learning, form a Federated Learning (FL). In the year 2016, McMahan et al., advocated federated learning in an edge-server architecture for aggregating the language models deployed over mobile phones [9–12]. Many IOT devices or edge-compute devices maintain the private data of the owners. Google's keyboard system known as the Gboard system uses a federated learning algorithm named Federated averaging (FedAvg).

The Gboard performs programmed word completion, an application that uses FedAvg for aggregating the model parameters from various mobile phones and informs the word-prediction model about the user's choice and selection of words. No data is moved from mobile phones to any central server forms the basis for federated learning. The model parameters or the model itself gets encrypted on any device and transferred to cloud. Participating edge devices will not be able to identify the data of other participating edge devices because model parameters (or the model itself) are encrypted at the server and shipped to the cloud. All encrypted models are consolidated to build a global model in the cloud and servers in the network will not be able to identify the edge-compute devices [9–14]. The global model under encryption is apprised and posted to all participating devices on the edge of the cloud system [12, 13, 15, 16]. The customer's data present in edge devices is not disclosed to others, or to servers in the network that participates in the entire communication process.

FL algorithms find usage or deployment in various applications related to different realms of science and engineering. Recommendation systems, keyword spotting, next-word prediction, smart healthcare, smart object detection, and training of Unmanned aerial vehicles (UAV) – enabled IOT devices are a few examples of federated learning applications. The FL is also widely used in developing applications that could detect

Table 1. Federated Learning applications.

Method name	Ref. No.	Concept description	Learning model
Gboard	[21]	Linear regression	Linear model
FL-ranking browser history suggestions	[22]	Support vector machine	
Fedner	[23]	Reinforcement learning	Neural model
FedMF	[24]	Matrix Factorization	Linear model
Smart Robotics	[25]	Reinforcement learning	Neural model
FCF	[26]	Collaborative filter	Linear model
Smart object detection	[27]	Kullback-Leibler divergence	
FL predicting emoji	[28]	Natural language processing	Neural model
FL next word prediction	[29]	Natural language processing	
UAV-enabled IOT devices	[30]	CNN-based Long Short-Term Memory (LSTM) model	
FL OOV	[31]	Natural language Processing	
Fedrec	[32]	Reinforcement learning	

money laundering fraud [17] or fight financial fraud [18] executed by non-state actors, brokers, and banking institutions. Significant usage of FL in drug discovery [19], forecasting the oxygen needs for COVID-19 patients [20] has drawn the attention of translational researchers worldwide. In Table 1, a few existing FL applications are listed with contributors, concept descriptions, and learning models adopted. An article is being proposed to facilitate research on federated learning methods.

2. Fundamentals of Federated Learning

2.1 Machine Learning

Machine learning (ML) can be accomplished in two phases, and they are (a) the training phase, and (b) the prediction phase as shown in Fig. 1. In the training phase, input data vectors are fed into the ML model and the model learns by incrementally updating the function that maps the input vectors to output class labels. The learning in the ML model can be classified into four classes, and they are (a) Supervised learning, (b) Unsupervised learning, (c) Semi-supervised learning, and (d) Reinforcement learning. Supervised learning, tries to figure out a mapping function between the input data vectors and the known class labels or anticipated output values. Then, the mapping function is subjected to new input data vectors to determine the class label. In the unsupervised learning, the class labels or anticipated output values are not known in advance. The mapping

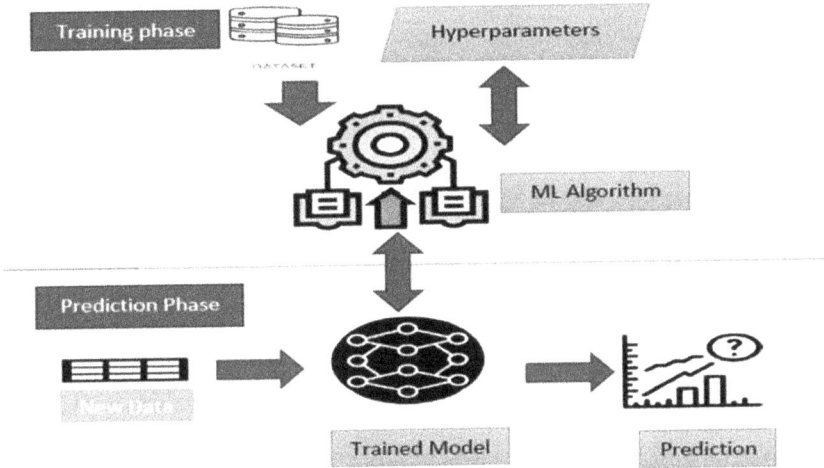

Figure 1. Training and prediction phase of machine learning.

function tries to identify the similarity and dissimilarity among the input data vectors and groups the data input vectors using similarity measures. In semi-supervised learning, the majority of the input data vectors in the dataset are not labelled, the remaining being minority data vectors possessing class labels. Class labels for the majority of data vectors are deduced using clustering algorithms. Choice of actions and the reward that return over a period that leads to goals or fulfils the objectives, forms the basis for formulating the reward (or mapping) function in reinforcement learning. In the prediction phase of machine learning, the output of the training phase called the trained model is put to exercise. The trained model admits input data vectors whose class label is yet to determined and produces the output known as the prediction (or class label). Accuracy in the prediction determines the efficiency of the machine learning algorithms.

2.1.1 Formal Definition of Machine Learning

The Function mapping F with the input $X(x_1, x_2 ..., x_n)$ and output $Y(y_1, y_2, ..., y_n)$ and w_L being the learning parameter, we express the gap between the prediction and the desired value in the Equation (1).

$$\arg \frac{min}{w_L} Loss(X, Y, w_L) = \left\| F_{w_L}(X) - Y \right\| \tag{1}$$

where *Loss* is the loss function at the points X and Y. The objective is to minimize the loss between the prediction and desired values, it forms the basis for all machine learning algorithms.

2.1.2 *Classes of Machine Learning algorithms*

Machine learning can be broadly classified into five different classes:

- Matrix factorization algorithms: Data elements presented in the form of matrices may have concealed factors or missing elements in them, and are identified by Factorization. It finds application in the recommender systems in the form of a "User item rating matrix" and "Drug target protein matrix," used to determine user preferences by knowing their ratings on the items [33], and helpful in innovative drug discovery [34] respectively.

- Topic models or statistical models: The Latent Dirichlet Allocation is a statistical method that produces a mapping between unstructured text documents and topics set with probabilistic values [35]. This approach uses a few key topics or words to learn about unstructured text documents. The Naïve Bayes classifier is a probabilistic approach that uses supervised learning for training. This approach has very less accuracy when compared to other complex approaches. Matrix based mapping between the topics and text documents is constructed in Latent Semantic Analysis/Latent Semantic Indexing approaches to classify the text documents. Gaussian distribution for inputs is assumed in this approach. Probabilistic Latent Semantic Analysis/ Probabilistic Latent Semantic Indexing adopts Poisson distribution for inputs such as topics and text document [36].

- Rule based machine learning (RBML) algorithms: The RBML algorithms construct "If <condition> Then" rules for the given Input dataset. Each rule appropriates a portion of the problem [37]. The rules are simple and easy to interpret when compared to neural network models. Decision Trees and Association Rule Learning are typical examples of rule-based machine learning methods. In Association Rule Learning, it builds rules by identifying the association between the items or variables. In decision trees, it follows a hierarchical representation of rules with branches that guides to the leaf which denotes the decision or class label for the given input data vector.

- Stochastic gradient descent (SGD) based algorithms: The loss function is a minimization multi-variable objective function that moves towards the optimal point guided by the negative gradient. The negative gradient is a first order partial derivative of the loss function for the given problem. The negative gradient is used to adjust the model parameters along its direction. The movement in the search space toward the optimal point is governed by the learning rate. Since the gradient is computed over randomly selected input data vectors, it is called a stochastic gradient. This is an iterative approach to find the minimal point for the given loss function. The SGD is used in popular ML models like Support Vector Machines (SVMs), Perceptrons [38], and Artificial Neural Networks (ANNs). In SVMs, input data is mapped to n-dimensional space, and an n-1 dimensional hyperplane that appropriately groups the two sets of data points separated by a larger distance can be used for classification. The Perceptron model performs binary classification. All the input data are given weights by the perceptrons. The sum of the products of the respective weights with inputs is compared with threshold levels to determine the class label of the input data. In ANN, multiple layers of neurons form a network and they are connected in a specific manner to recognize complex features in the large data. Each layer of neurons is connected by edges assigned with weights and an activation function introduced on the outcome of the layer generates non-linearities. Generative Adversarial Networks (GAN), Convolutional Neural Networks (CNNs), Deep Neural Networks (DNN), and Recurrent Neural Networks are a few examples of specific models of ANN.

- Evolutionary algorithms: The Genetic algorithm is designed based on the evolution of human beings [39]. It represents a decision variable in the form of genotype (or string of ones and zeros). Every genotype is evaluated using a fitness function and the one that satisfies the fitness function undergoes crossover and mutation to produce a new genotype. New genotypes are again subjected to quality tests using a fitness function. Only better genotypes retained in the population are considered for the reproduction of new genotypes. It is iterative and has a slow convergence rate.

2.2 Distributed Machine Learning

Training over larger datasets on a single machine increases the runtime of ML algorithms. Complex applications of terabytes order may need a higher degree of parallelism and larger I/O bandwidth in distributed systems [40]. The parallelism can be conceived at the data level (or creating data fragments that are disjoint or not overlapping), or at a model level where identical copies of the dataset are trained by different portions of the training model. Compute node topology is important when parallelism at the data and model level is implemented. Parallel execution and inter-compute node communication should go in hand for faster model convergence with faster updates.

2.2.1 *Data-parallelism and Model-parallelism*

In the Data-parallelism method, the dataset is divided into disjoint subsets equalling the number of compute (or worker) nodes in the system. The same machine learning algorithm is used for training the disjoint subsets of the dataset associated with different compute nodes. All the compute nodes execute the same instructions over different data assuming the independent and identical distribution of data over disjoint subsets of the dataset. This model produces only one comprehensible output. In the Model-parallelism method, all compute nodes receive identical datasets for training. But compute nodes execute different portions of the training model. Data-parallelism and Model Parallelism are illustrated in Fig. 2. At the program execution level, different instruction executes parallelly on different compute nodes. All machine learning algorithms do not offer themselves to this model. Alternatively, different models of training are allowed to run on the same dataset on multiple compute nodes. Finally, global aggregates are constructed using the outcomes from individual compute nodes, and this format of learning is called Ensemble Learning [41].

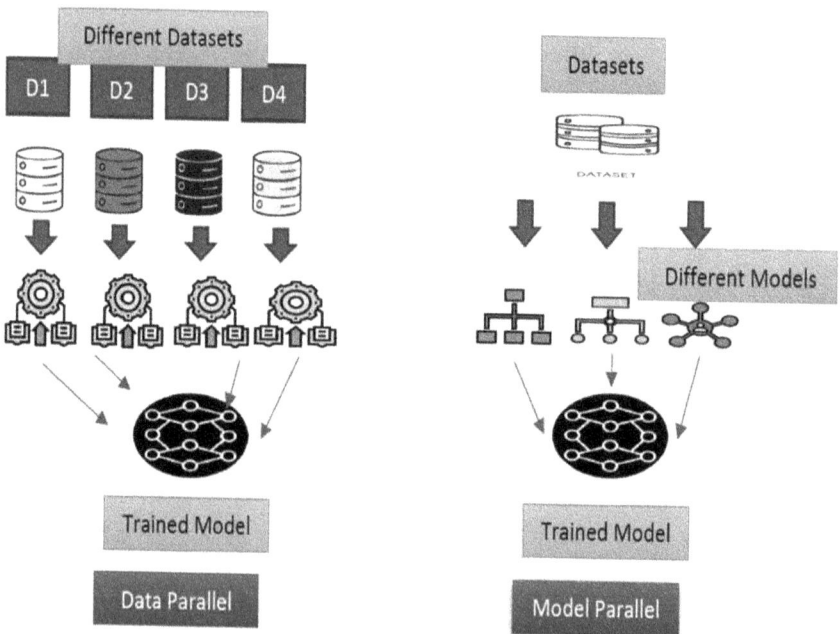

Figure 2. Data parallelism and model parallelism.

2.2.2 Topologies for Distributed Machine Learning

Fully distributed systems, are a network of autonomous compute nodes that collect the final solutions together. Compute nodes do not have any specific part to play in deciding the outcomes. In the Centralized systems, a systematic approach is followed in the central server for result aggregation of results. Result aggregation is performed hierarchically. Tree structure is followed in the Decentralized systems, which allows halfway aggregations of results (as shown in Figs. 3 and 4) for a replicated model or apportioned model that is sharded over multiple Parameter servers. Tree topology is popular in distributed machine learning frameworks, used in ALLReduce [42] to compute local gradients in child nodes and pass on the value to their parents. Ring topology is popular in systems that require minimal communication latency or poor broadcasting backing. Parameter servers have a set of centralized master compute nodes with shared memory and a set of decentralized worker compute nodes. The shard on every parameter server stores the model parameters. Key-value store approach is adopted by all clients for read and write. Parameter servers can be a single point of failure since it handles all communication.

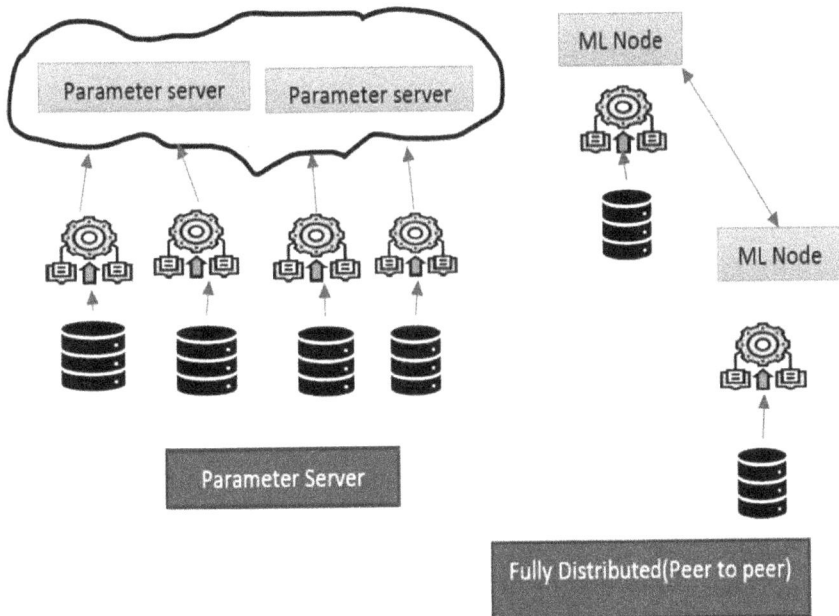

Figure 3. Topologies for distributed Machine Learning.

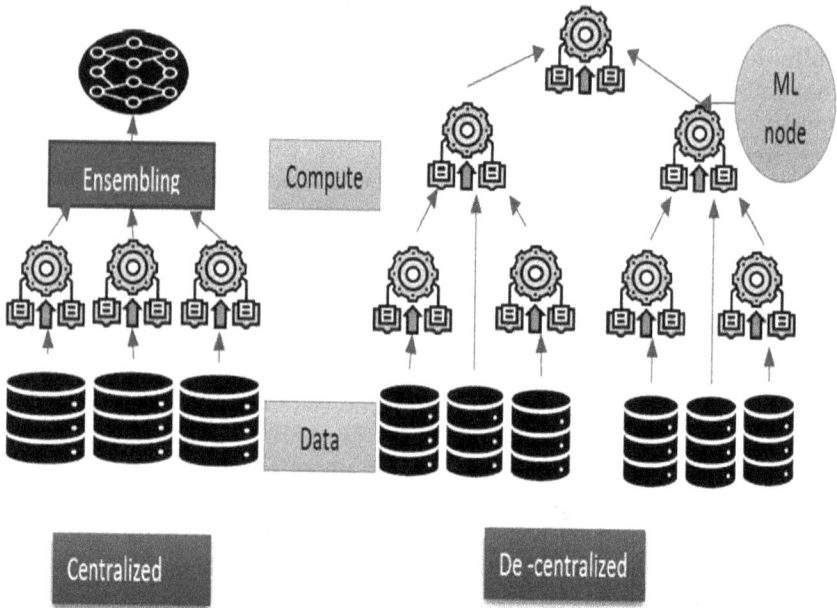

Figure 4. (Contd. from Figure 3) Topologies for distributed Machine Learning.

2.2.3 Communication Models for Distributed Machine Learning

The synchronization between the communication and computation phase of compute (or worker nodes) implies accuracy in updates performed, communication latency, and compute time. Different communication models are identified because of the above imperatives. Accurate model convergence at a quicker pace and newer updates at a faster pace can be placed at two extreme points in the spectrum of communication models. Bulk Synchronous Parallel (BSP) [43], Stale Synchronous Parallel (SSP), Approximate Synchronous Parallel (ASP) [44], and Barrierless Asynchronous Parallel [45]/Total Asynchronous Parallel [44] are the four communication models adopted in distributed machine learning ecosystem. Applications such as Baidu AllReduce [46], Horovod [47], Caffe2 [48], and CNTK or Microsoft cognitive toolkit [49] are examples of parallel-synchronous model based distributed ML systems. DistBelief [50], DIANNE [51], Tensorflow [52], MXNet [53], Petuum [54], and MXNet-MPI [55] are examples of parallel-asynchronous and parallel-asynchronous/synchronous distributed ML systems respectively.

2.2.4 Debrief on Challenges

Privacy, Performance, Fault tolerance, and portability are the bottlenecks for deploying ML algorithms over scalable computing architectures. The communication models like parameter servers do not ensure privacy in distributed systems. The distributed system does not have any form of compliance with Data protection laws and is being legitimized in various countries.

2.3 Federated Learning

Federated learning is a collaborative ML model trained and built over large datasets placed at many locations. It has two phases and they are (a) Model training phase, and (b) model inference phase. In the model training phase, one or many participants collaborate to construct an ML model by contributing towards training the model without sharing the data. In this phase, the transfer of data among the participants does not expose the private part of the data at the location. The instances of the trained model can be sent from one participant to other participants by encryption thereby privacy of the data is maintained. The performance of the final model constructed from multiple participants should be comparable to a model built on a single location or by a single participant located at a site. In the model inference phase, the trained model admits new data input and deduces outcomes for the same. A sustainable federated model ensures reasonable appropriation of gains among the participants who contributed to the construction of a joint ML model. In this section, formal definitions of federated learning, client-server architecture, and peer to peer architecture for federated learning are presented.

2.3.1 Formal Definition of Federated Learning

A separable linear objective function denotes the algebraic summation of accumulated Loss for number of P participants in federated learning given in the Equation 2.

$$\arg \frac{min}{w_L} Loss(X,Y,w_L) = \Sigma_1^P U_P \, Loss_p \, (X,Y,w_p) \tag{2}$$

In the Equation (2), U_p denotes the weight of the Pth participant in the FL; w_p denotes learning rate of the participants; Summation of Lp loss functions in the objective function makes it a separable objective function.

2.3.2 Federating Learning Architectures

The application built using FL systems might include or might not include a single point co-ordinating compute node. The single point co-ordinating compute node in an FL ecosystem is shown in Fig. 5. The co-ordinating

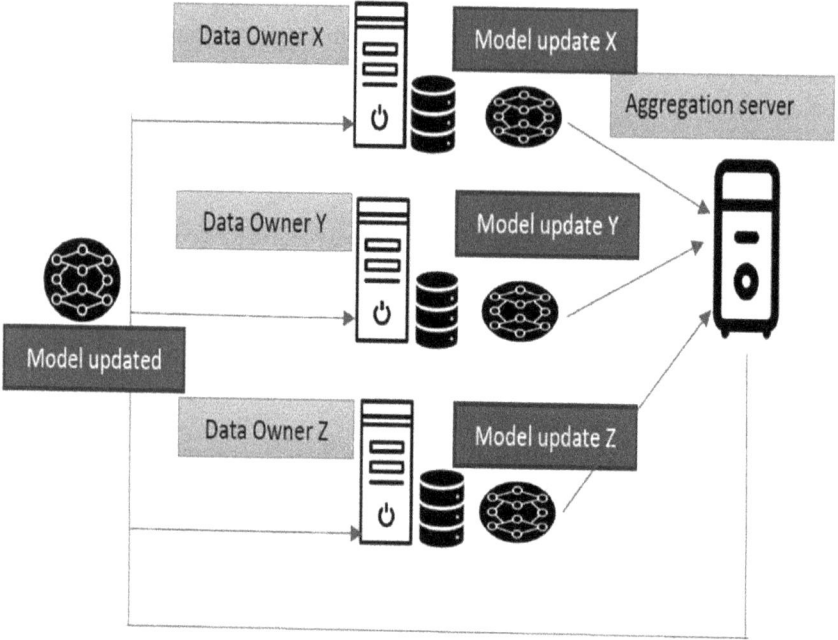

Figure 5. Client server model for Federated Learning.

compute node is designated as the parameter server. The parameter server sends the primary untrained instance of the model to all the local participants. The local participants train the model using their local datasets and post model weights to the parameter server. The parameter server performs aggregation from all the participants and updates the model. Until convergence criteria are met, the FL system repeats local training and global model updates iteratively. The original data located at the local participant's site on no occasion leaves the site. To avoid leakage of model updates, homomorphic encryption is used for encrypting the data exchange parameter server and local participants [8]. Peer-to-peer approach is also a viable design architecture for FL systems as shown in Fig. 6. The computation increases due to the absence of a parameter server, and is the demerit of this approach. This approach provides an enhanced security level for peer-to-peer communication between participants in the absence of third-party intervention.

2.3.3 Open-source Projects in Federated Learning: A Debrief

Many open-source projects are being developed at a faster pace in federated learning. A debrief is presented for selected open-source frameworks that were launched recently in the public domain and are itemized in Table 2.

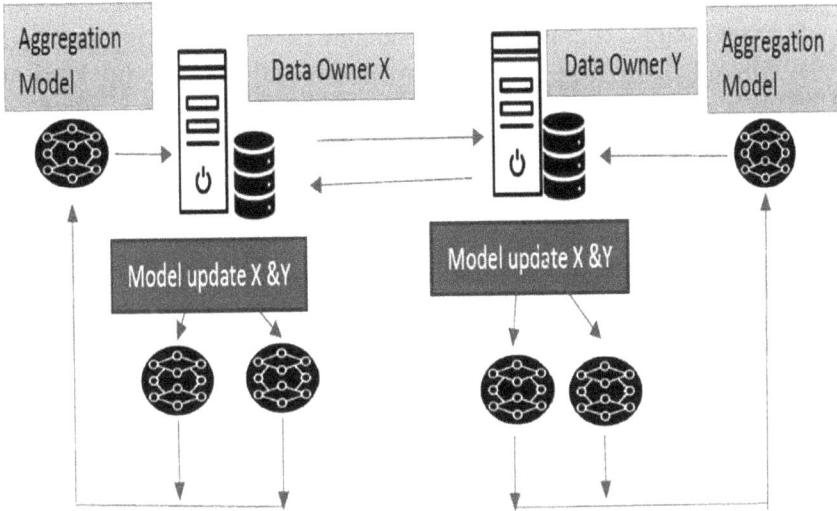

Figure 6. Peer to peer model for Federated Learning.

Table 2. Open-source projects in FL..

Open-source projects	Algorithms/Library	Description
coMind [56] and [57]	Fedavg algorithm	TensorFlow is the bottom layer for coMind
TensorFlow Federated (TFF) [58] and TensorFlow-Encrypted [59]	Python library	It used for evaluating privacy preserving ML algorithms, and benchmarks in high performance computing
Horovod [47, 60]	TensorFlow and PyTorch frameworks	It uses message passing interface to realize FL
OpenMined and PySyft [61–64]	PySyft library	Suitable for multi-party computation and homomorphic encryption
LEAF benchmark [65, 66]	–	Suitable framework for high end researchers and product developers in FL domains

3. Classification of Federated Learning based on Data Samples

Let us consider the dataset or sample set represented in the form of a Matrix M_i for the ith data owner. The columns of the dataset represent features and the rows represent actual data or sample sets. Class labels are assigned to some samples. The sample space is identified by *sample ID*.

A and B are used to denote features and labels of the sample set. The sample set used for training is represented by *(A, B, sample ID)*. Federated Learning can be classified into three classes and they are (a) Vertical Federated Learning, (b) Horizontal Federated Learning, and (c) Federated Transfer Learning. Classification has the basis of different approaches adopted for data sample partitioning concerning feature and sample space.

3.1 Vertical Federated Learning (VFL)

It is a feature-based partitioning of the samples in the dataset. Two sets of data may have a common *sample ID* but different features associated with them, as shown in Fig. 7. Consider state owned banks and commercial enterprises like e-commerce companies. Both entities may have common customers as their business clients but attributes are different. Business clients to banks have features like credit ratings, capital and revenue expenditures, and the same clients have different features like purchase behaviours when they get associated with e-commerce enterprises. The VFL trains the model over samples with different features and class label space by computing the loss and gradients aggregated finally. Formally VFL is stated in the following Equation (3).

$$A_i \neq A_j, B_i \neq B_j, Sample\ ID_i = Sample\ ID_j, \forall M_i M_j, i \neq j \qquad (3)$$

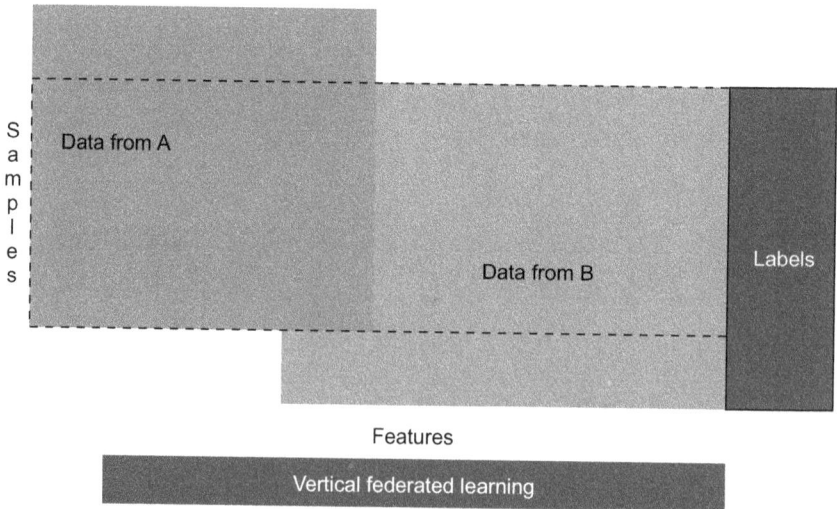

Figure 7. Vertical Federated Learning.

3.2 *Horizontal Federated Learning (HFL)*

It is sample-based partitioning with overlapping attributes or features as shown in Fig. 8. Let us consider two sets of data with common feature space but different samples. Consider two banks owned by the state, which may have different customers or sample spaces. But the business behaviour of the two banks has many features in common like performing assets, liabilities, profits from bonds and non-performing assets, etc. The HFL builds a model over heterogeneous samples with overlapping features. Formally HFL is stated in the following Equation (4).

$$A_i = A_j, B_i = B_j, Sample\ ID_i \neq Sample\ ID_j, \forall M_i M_j, i \neq j \qquad (4)$$

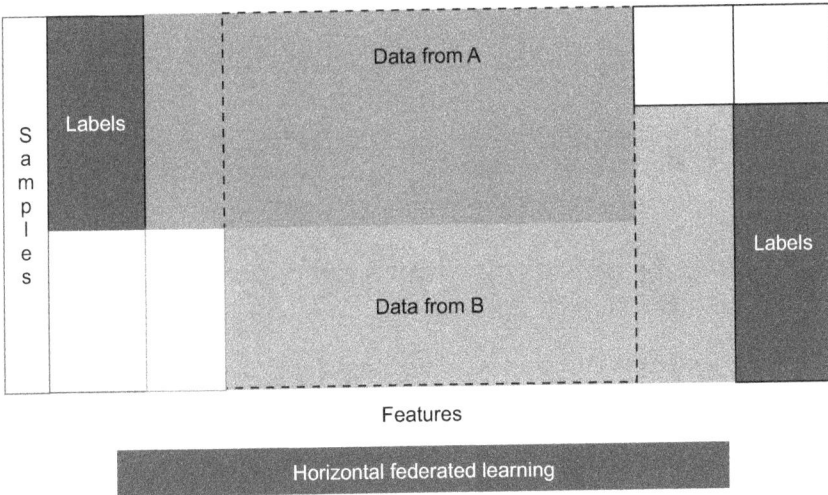

Figure 8. Horizontal Federated Learning.

3.3 *Federated Transfer Learning (FTL)*

It is a dataset that has a highly heterogeneous sample space with differing features (see Fig. 9). Consider two datasets from two independent institutions such as the Reserve Bank of India and the Federal Bank in the US. Both the Central banks may have very few features in common, but with different objectives and models of financial governance rooted in their respective countries. FTL algorithms are applied to a small sample space with two feature spaces of both entities and a shared format is learned. Formally FTL is stated in the following Equation (5).

$$A_i \neq A_j, B_i \neq B_j, Sample\ ID_i \neq Sample\ ID_j, \forall M_i M_j, i \neq j \qquad (5)$$

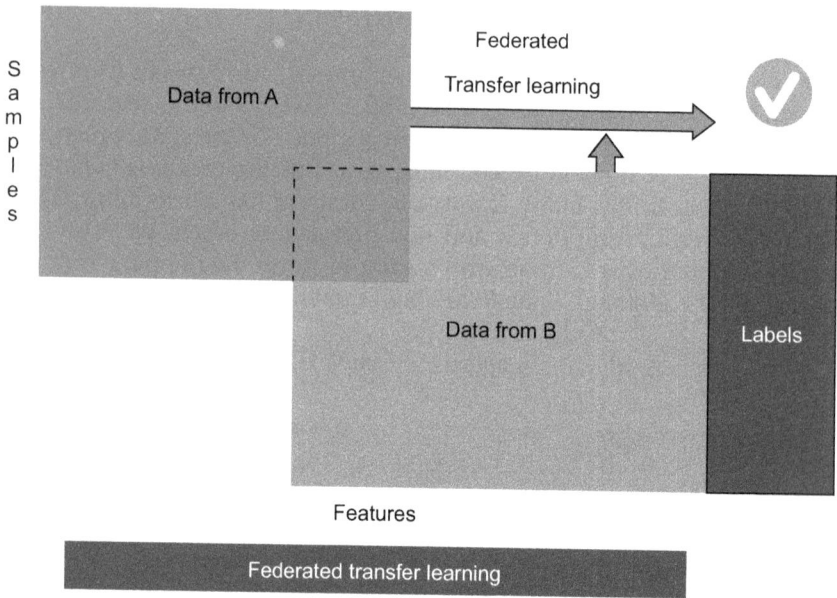

Figure 9. Federated Transfer Learning.

4. Machine Learning Algorithms in Federated Learning: A Classification

The Federated Learning Average (FedAvg) [9] is the most popular and classical algorithm proposed by McMahan et al., in the year 2017. The FedAvg has been tested over image classification benchmarks like MNIST [67] and CIFAR-10 [68]. The results of testing (FedAvg) meet the desired results in terms of performance, and classification accuracy. Further, the federated learning algorithms can be classified as (1) Supervised learning in FL, (2) Unsupervised learning in FL, and (3) Semi-supervised learning in FL. Each one of them is briefly presented below.

4.1 Supervised Learning in FL

4.1.1 Decision Tree Algorithm in Federated Learning

Yang Liu et al., proposed a Federated decision tree (FDT) [69] in the year 2022. It uses a random forest approach for the VFL framework. Master communicates to clients a few selected features and sample IDs. Clients will not know all the existing features and sample IDs. The split criteria are estimated to construct trees locally by the clients. The clients share the split criteria to the master node privately using encryption. Master decrypts the split criteria and computes global best split conditions and the same is notified to the clients. The client got hold of the best split will construct left and right subtrees using the samples. The above steps are repeated recursively. Finally, the master gathers all the local optimal split features from the clients thereby paving the way for computing the global best split features. Another similar model named Secure Boost Model [70] proposed by Kewei Cheng et al., in the year 2021, is a regression tree method realised in a decentralized VFL framework. The objective function value for loss for augmenting g_t a tree is stated in Equation (6).

$$\arg\min Loss^t = \Sigma_{i=1}^{n} \, Loss(B_i, \hat{B}_i^{t-1}) + G(Sample\ ID_i) \tag{6}$$

In the Equation (6), class labels are denoted by B; $\widehat{B}_i = \Sigma_1^K g_K(Sample\ ID_i)$ denotes the output for the class label prediction using K regression trees; iteration is denoted by t; first and second order derivative of loss function is denoted by $G(Sample\ ID_i)$.

4.1.2 Support Vector Machine in Federated Learning

Hartmann et al., advocated a linear support vector machine in federated learning in the year 2019. The local clients are updated in blocks. In the exchange of model parameters between the central server and the clients, the private feature is not revealed to the server such as the labels of the user. Hence, the name secret vector machine (SecVM) [71], performs splitting of the local gradients, hashing to conceal the labels of the users, and transmitting updates to the server complying with communication protocols. For a binary classifier, the linear objective function is stated in the following Equation (7).

$$G(z) = \frac{1}{T} \Sigma_{i=1}^{T} \, Loss(z; A_i, B_i) + \theta \; regular(z) \tag{7}$$

In Equation (7), features of the *ith* user and labels are denoted by $A_i \in R^d$, & $B_i \in [-1,1]$ respectively; the number of samples is denoted by T; regularization of parameters is performed using θ & $regular(z)$; Loss is the loss function of every user or samples; model parameter is denoted by z. Model parameter z is updated by computing the sum of the partial derivatives of the loss function and regularization function, which is given in Equation (8).

$$z_j = z_j - \eta \left(\frac{1}{T} \Sigma_{i=1}^{T} \frac{\partial Loss(z; A_i, B_i)}{\partial z_j} + \theta \frac{\partial regular(z)}{\partial z_j} \right) \tag{8}$$

In Equation (8), the learning rate is denoted by η for this model; The loss function is defined as $loss\,(z; A_j, B_j) = \max(0, 1 - B_i\, z^T A_i)$ for linear SVM. The gradient computation for Linear SVM is given in the Equation (9).

$$\frac{\partial Loss(z; a_i, b_i)}{\partial z_j} = \vartheta(1 - B_i\, z^T A_i)B_i\, A_{ij}, \ \vartheta(A) = 1, \ if \ A > 0, \ otherwise \ \vartheta(A) = 0 \ (9)$$

In the linear SVM model, the server sends the present model to all the clients. Every client computes the local gradient $\nabla w Loss\,(z; a_i, b_i)$ and is posted to the server. The server computes the global gradient and sums with the regular function denoted by $regular(z)$. The step length in Equation (8), $\eta = 1/\theta t$, t denotes the iteration number, is estimated using the Pegasos algorithm [72]. The limitations of this approach are (a) failure of the hashing function which is very rare, and (b) SVM may not yield results when one of all gradients from clients is zero. FL model will not be able to reconstruct the features of one client (with gradient zero) from others because of hashing which conceals the labels of the clients.

4.1.3 Federated Linear Algorithm

The Logistic regression method is integrated into a centralized VFL model for choosing the all-out number of workers (or devices) in a wireless network scenario [73]. It is proposed by Yang et al., in the year 2020. The objective function of this VFL model is a linear separable function with minimization as stated in Equation (10).

$$MinG(z) = \frac{1}{D = (x_j, y_j)} \Sigma_{j=1}^{T}\, g_j(z) \tag{10}$$

In Equation (10), the vectorized parameter of the model is denoted by z; number of data points is denoted by T; loss function of every device selected is denoted by $g_j = loss\,(z; x_j, y_j)$; $z \in R^d$, d is the optimization dimension; local dataset is denoted by $D_i \in D$. This model has a higher convergence rate and minimal error in global aggregation, and higher prediction accuracy is observed. The privacy in the information exchange between central server and the data owners is ensured using homomorphic encryption techniques.

4.2 Semi-supervised Learning in FL

The dispersed class labelling in data samples on the client side incurs huge cost which can be reduced by semi-supervised learning in FL. On the other hand, the required amount of data available for training is to be guaranteed when FL is realized for the problem. Jeong et al., proposed

Federated Match (FedMatch) [74] introduces a novel concept called Inter-client consistency loss (ICCL), which is formally stated in Equation (11) given below.

$$ICCL = \frac{1}{H}\Sigma_{j=1}^{H} KLoss(p^*_{\theta^{hj}}(Y \mid U)\|p^*_{\theta^i}(Y \mid U)) \tag{11}$$

In Equation (11), the helper agents are denoted by H; the model comparable to clients chosen on the server side is denoted by $p^*_{\theta^{hj}}(Y \mid U)$, these models do not undergo training on the client side; the number of clients is denoted by K. The ICCL performs disjoint learning for predicting data labels that could reside either at the client or server. To enhance the efficiency of the semi-supervised model, complex models like variable auto-encoding [75], and generative models [76] are recently proposed. The semi-supervised learning is further classified into two classes (a) Horizontal federated semi-supervised learning, and (b) Vertical federated semi-supervised learning. In the horizontal federated semi-supervised learning, the N participants, A denotes feature space which is the same for all samples and sample ID is different for spatial samples, i.e., $A_j = A_k$, *Sample ID$_j$ ≠ Sample ID$_k$*, $k \neq j$ and for every user i, the dataset is denoted by $D_i = D_{il} \cup D_{iU}$, where $D_{iU} = X_p \in A$, $p = lower_i + 1, ... , lower_i + upper_i$; $D_{iL} = X_p \in A$, B_p, $p = 1, ... , lower_i$. In the vertical federated semi-supervised learning, feature space is different, and the sample ID are the same, i.e., $A_k \neq A_j$, *Sample ID$_k$ = Sample ID$_j$*, $k \neq j$ and for every user i, the dataset is denoted by $D_i = D_{il} \cup D_{iU}$, where $D_{iU} = X_p \in A_i$, $p = lower_i + 1, ... , lower_i + upper_i$; $D_{iL} = X_p \in A_i$, B_p, $p = 1, ... , lower_i$.

Yang et al., proposed a logical regression model for a decentralized VFL, where a contract exists between A and B parties who have labelled data and unlabelled data respectively [77]. The regression method is used by either party to get the outcomes for its unlabelled data. Both parties A and B update the model using gradients.

4.3 *Unsupervised Learning in FL*

The unsupervised algorithms used in ML to find constructs or patterns in the dataset which has no class labels. In this framework, clients participating have data formations in a uniform, inter-client dependent manner. Migration of data across the domains exists in this FL framework. With no third-party intervention, data from different domains and knowledge is transmitted to new devices from the decentralized FL framework. Peng et al., advocated an Unsupervised federative domain adaptive model for FL framework [78]. In this framework, demonstrative models learned from different compute nodes are brought in line with the data distribution of the target compute node. The migration of data between ordinary compute nodes and target compute nodes are different from conventional migration that happens between ordinary compute nodes, and in the former case, compute node may tend to contribute unfavourably.

5. Deep Learning Algorithms in Federated Learning: A Classification

A brief conceptual discussion on deep learning algorithms integrated with the FL framework is presented in this section.

5.1 Neural Network in Federated Learning

McMahan et al., proposed a neural network model in an FL framework in the year 2017 [9]. The model is tested on the MNIST dataset for digit recognition. The neural network model has one input layer, two hidden layers, and one output layer, altogether four-layer model is proposed. 200 neuron units are present in each hidden layer. The MNIST dataset is appropriated to clients and they do not intersect with each other. The two different experiments are conducted on two groups: in the first experiment, the same stochastic initial values are assigned to the local model parameters of two clients; in the second experiment, different stochastic initial values are assigned to the local model parameters of two clients. The final model of the neural network is constructed using the model parameters from two groups.

5.2 Convolution Neural Network (CNN) in Federated Learning

Rong et al. [79] advocated an intrusion detection approach that uses CNN for feature extraction in an FL framework in the year 2020. Firstly, data filling method is used to construct two-dimensional structure. The Diffusion CNN is used to construct and learn feature parameters in an FL system. SoftMax classifier is used to determine the class label in the detection phase. The CNN model used in this FL has 4 convolution layers and two fully connected layers. The high detection rate and reduced training time are the important attributes of this model. Three different types of experiments are conducted and their results are presented. It is observed that increasing the number of hidden layers from two to three has no significant implication on performance. It is verified that two dimensional structures reduce the operational costs and improves the accuracy. Application of the FC algorithm in the intrusion detection method, and experiments conducted on NSL-KDD dataset show no change in accuracy, but reduction in the training time is observed. In the FC algorithm-based model, a smaller number of parameters are exchanged between the client and the servers in the training phase thereby reducing the training time.

Sattler et al. [80] proposed an approach named Spare Ternary Compression (STC), which reduces the bandwidth usage when model parameters are exchanged. Uploading the model, training the model, and downloading the model from the server and final aggregation is part of the training process in FL designed with STC. The non-independent and identically distributed (non-IID) data are prevalent among the edge devices thereby reducing the effectiveness of training in the FL model. Obviously, the gradient of the local edge device would be biased towards the local dataset. The mean value of the estimated gradient is expected to be unbiased. Improvements in the accuracy by 30% are observed in experiments carried out using CIFAR – 10 datasets consisting of 5% of common data globally.

5.3 Bayesian Network in Federated Learning

Yurochkin et al., proposed a Bayesian network in an FL framework [81]. Beta Bernoulli process (BBP) is a Bayesian nonparametric (BNP) model that permits parameter matching between the local parameters and the global parameters. New global parameters are reconstructed when current global parameters do not match with the local ones. Two important features of BBP are (a) dissociating local model learning from a blend or combination of local clients agglomerating to a global model in FL thereby different learning algorithms might be used on data sources, and (b) pre-trained models being known, BBP can use its matching arm to combine them to form global models without the support of additional learning algorithms and can construct pretrained models. The objective of BNP is to identify neuron subsets in the K local model and have equivalence with neurons in the remaining local models. These clusters of neurons from local models are put together to form a global model.

5.4 Long Short-Term Memory (LSTM) in Federated Learning

Li et al., proposed federated proximal (FedProx) [82] for character recognition and sentiment analysis. The FedProx has a better convergence when compared to FedAvg. In case of statistical heterogeneity, local clients in FedAvg will not be able complete updates at the same pace when local client changes. To address this challenge, FedProx introduces a stability term that stabilizes the FL framework. The stability term significantly increases the restrictions on the gap between local parameters and global parameters. LSTM classifier consisting of two layers with 100 hidden units and 80 embedded layers are incorporated in FedProx. FedProx has better accuracy (by 22%), and quicker convergence in a heterogeneous ecosystem when compared to FedAvg.

5.5 Reinforcement Learning in Federated Learning

Anwar et al., proposed reinforcement learning for architecting model poisoning attacker model, which can pose a serious challenge to Federated Learning. General attacks on mobile devices are ineffective. A multi-tasked federated reinforcement learning (MT-FedRL) [83] capitalizes on information gain minimization to launch an effective attack on FL systems. In MT-FedRL, 'n' agents are present, each has its environment, and update policy operates within the aegis of the Markov decision process (MDP). Each agent has its own state space and action space. Each agent is characterized by a 5-tuple and they are reward function, discount factor, transition probabilities, state space, and action space. The objective of the MT-FedRL is to construct a co-operative policy that is optimal in all environments of n agents [84]. The framework for Reinforcement Learning in Federated Learning is illustrated in Fig. 10.

5.6 Meta-Learning in Federated Learning

In Meta learning, each agent or device collaborates to train the meta-learner. The parameters from the devices are aggregated in the server, despite independent meta-learning takes place on every device. The global trained model is identical on all devices. Chen et al., advocated Federated Meta Learning (FedMeta) [85], which is an approach that integrates Model Agnostic Meta Learning (MAML) algorithm and Meta-SGD into the framework. In FedMeta, parameters exchange clients and servers, unlike MT-FedRL. Rapid convergence on newer datasets is exhibited by MAML algorithms. It is found that MAML and Meta-SGD together can reduce communication costs and increase the accuracy of the training model. FedMeta has a communication cost less by 2.82–4.33 times and a convergence rate faster by 3% to 14% when compared to FedAvg. Figure 11 illustrates Meta-learning in FL frameworks.

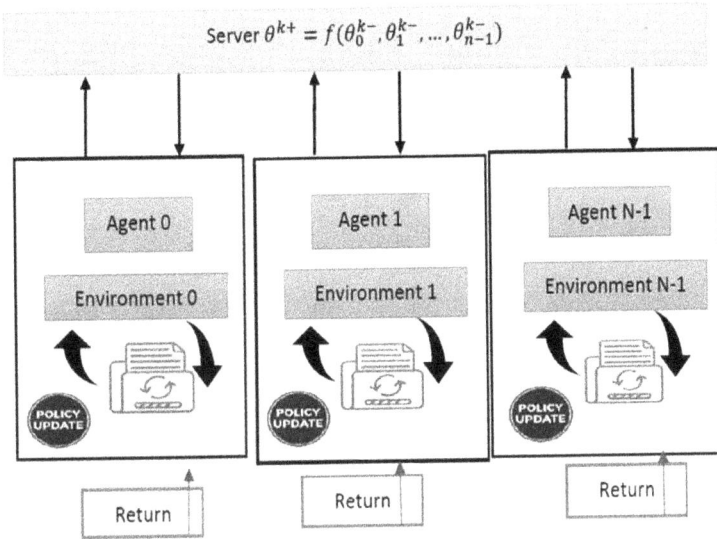

Reinforcement Learning in Federated Learning

Figure 10. Framework for reinforcement learning in Federated Learning.

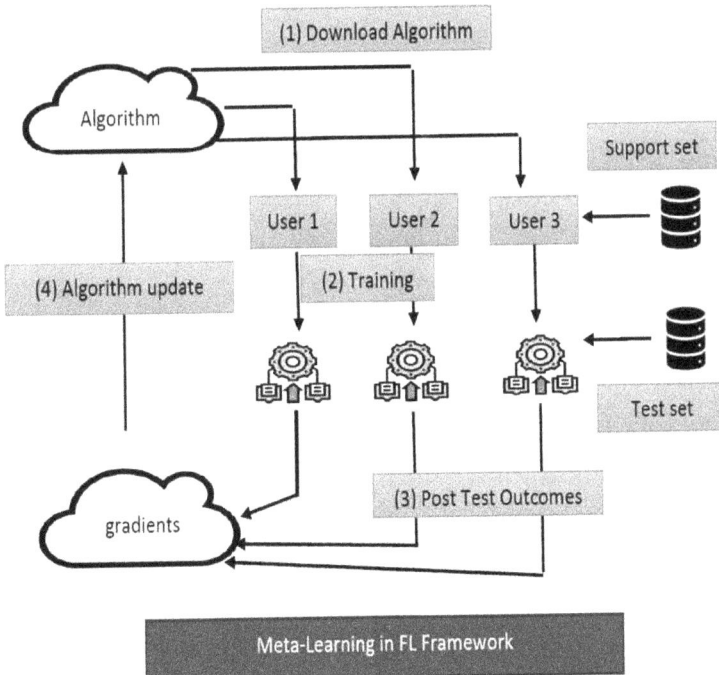

Meta-Learning in FL Framework

Figure 11. Illustrations of meta-learning in FL Framework.

5.7 *Residual networks in Federated Learning*

Huang et al., advocated the use of Residual Pooling Network (RPN) [86] in an FL framework to reduce the size of data transmitted in uploading and downloading to the server. The Tasks in an RPN are scheduled in an order that reduces data transmission with the same levels of performance maintained. Firstly, in RPN, all clients are not permitted to transfer parameters to the server. Client selection is done and selected ones alone perform the model updates. In the second step, every client is initialized with a global update. In the third step, the local model is constructed and trained using local datasets. In the fourth step, the residual network is computed by determining the difference between model updates after and before, thereby changes in parameters are captured. In the fifth step, Spatial mean pooling is performed to compress the model parameters as shown in Fig. 12. In the sixth step, RPN from clients is sent to the server for aggregation. Finally, RPN is sent to selected clients and the process repeats. RPN can be adopted in all CNN based training models to reduce overheads in data communication, and can easily be deployed in real applications.

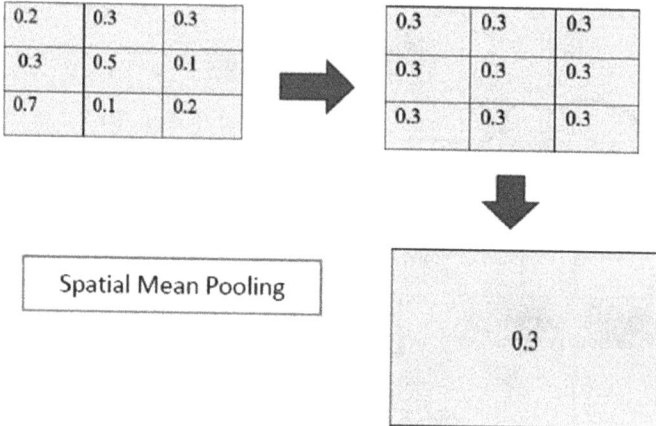

0.2	0.3	0.3
0.3	0.5	0.1
0.7	0.1	0.2

0.3	0.3	0.3
0.3	0.3	0.3
0.3	0.3	0.3

Spatial Mean Pooling

0.3

Figure 12. Spatial pooling by computing Mean.

6. Threat Models for Privacy in FL

Privacy and confidentiality are the building blocks for the evolution of FL frameworks from distributed machine learning computing platforms. Understanding privacy threat models is essential for designing privacy preserving FL frameworks. Three important roles of FL are (a) user of the trained model (or model querier), (b) Model builder (or agent responsible for computation and deducing the result), and (c) data owner (or clients or data holder). FL systems may encounter attacks at any stage or phase of training. Broadly phases of training can be stated as (a) model training, (b) Inferencing, and (c) data dissemination.

A brief on different types of attacks to an FL system is presented below.

(a) Attribute Inference attacks: Attribute Inference attacks in the data dissemination phase cause de-anonymization of data holders or clients for malicious drives. The De-anonymization of the data owners for hateful purposes is the objective of the malicious challengers or attackers. The privacy of clients is safeguarded from attribute inference attacks by avoiding the presence of personal identifiable information (PII) such as user IDs when data is disseminated in the ML model.

(b) Model Poisoning attacks: It is observed from experimental and practical implementation, the FL system is susceptible to model poisoning attacks. It is also called back door attacks. In the case of word prediction application, the book door attacker will introduce words in the sentences so that sentence completes perfectly. Furthering up the weights of the malicious clients will cause a loss in the training thereby accomplishing the attacker's goal.

(c) Reconstruction attack: In reconstruction attacks, malicious challengers or an attacker's objective is to recreate or reformulate the feature vectors of the samples in the model building phase or inference phase. In the centralized FL, clients upload the original data into the model builder, thereby making it susceptible to malicious model builders. In FL frameworks, each client performs ML training over their local dataset. The gradient values or the weights of parameters are given to other clients in the ecosystem. Sometimes, attackers or malicious challengers may capitalize on gradient values to extract additional statistics about the training dataset.

(d) Model Inversion attacks: There are two types of access to the model, they are (a) black box access, and (b) white box access. In case of model inversion attacks, both types of access for malicious challengers (or attackers) are assumed. White box access affords text information about the model to be known to the malicious challengers, but not the feature vectors. In black box access, malicious challengers can gather data by querying the model. The main objective of the malicious challengers is to construct a feature vector from the training data of the model. The query-response duo on models may be used by malicious challengers to learn about the model.

(e) Membership-Inference attacks: In Members-inference attacks, the malicious challenger has block box access to the training model. It is presumed that attackers know the specific sample in the dataset. The attackers try to ascertain the presence of a specific sample in the training dataset by examining the response of the model. Examining the difference between the prediction for known samples in the training data versus other samples, attacks are launched on the model.

7. Challenges in Federated Learning

(a) Privacy: A comprised server or client will create imbalance in the FL system. It may cause an impairment to the availability, integrity, and confidentiality of the FL system. Attacks on the FL system can be classified as active attacks and passive attacks. The active attacks are initiated by active clients. Semi-honest servers can initiate attacks on FL systems and are known as passive attacks. In this case, the updated gradient cannot be shared among all clients.

(b) Communication: More frequently, the client and the server communicate model parameters, and gradients with each other, unlike in distributed machine learning systems. The network state of clients and communication protocols have an important bearing on the FL framework system.

(c) Heterogeneity: In distributed ML system, data are independent and identically distributed, unlike in an FL system, data are non-independent and identically distributed. The algorithm suitable for distributed ML systems, cannot be used in FL systems without modifying the same. Heterogeneity opens up new avenues or challenges for researchers.

(d) Computing cost: In the FL framework, several local models are trained simultaneously on devices with varying computing abilities and communicate the updates with a central server. An increase in the local models participating will also increase communication issues and tests for the computing. Existing federated learning models do not consider the very high cost of computing at the local devices. Asymmetric computing in Federated Learning is an open problem that needs to be addressed.

(e) Reliable server: The central server in an FL framework manages local training and admits the updates on gradients posted by clients. A reliable and trusted server may not be available. Even the differential privacy in the centralized format can enhance the protection of data. Clients are expected to trust the central server, or data leakage is possible when no trusts exist between participating entities.

(f) New FL computing frameworks consisting of redundant servers, hardware accelerators, and decentralized training models needs a rigorous study.

8. Conclusion

The articles present a brief discussion on machine learning, distributed machine learning, and federated learning thereby readers will have clear and unambiguous understanding of evolving topics in this realm. Formal definitions of machine learning and federated learning will give insight for beginners in research to focus on limitations. A gamut of machine learning algorithms and deep learning algorithms along with threats, and challenges discussed in this article will provide a glimpse into existing state-of-the-art to readers. Building a federated computing framework in compliance with data protection laws differing concerning landscape or countries will be a major challenge for researchers. The article tries to build a narrative so that readers, when experimenting with their innovations consider all challenges within the framework of Federated learning.

References

[1] Goodfellow, I., Bengio, Y. and Courville, A. April 2016. Deep Learning, MIT Press.
[2] Hatcher, W.G. and Yu, W. 2018. A survey of deep learning: Platforms, applications, and emerging research trends. IEEE Access, 6: 24411–24432. DOI: 10.1109/access.2018.2830661.
[3] Pouyanfar, S., Sadiq, S., Yan, Y., Tian, H., Tao, Y., Reyes, M.P., Shyu, M.L., Chen, S.C. and Iyengar, S.S. 2018. A survey on deep learning: Algorithms, techniques, and applications. ACM Computing Surveys (CSUR), 51(5): 1–36.
[4] Trask, A.W. February 2019. Grokking Deep Learning, Manning Publications.
[5] Hartmann, F. May 2018. Federated learning, Master thesis, Free University of Berlin. http://www.mi.fu-berlin.de/inf/groups/ag-ti/theses/download/Hartmann_F18.pdf 3.
[6] GDPR official website https://ec.europa.eu/commission/priorities/justiceand-fundamental-rights/data-protection/2018-reform-eu-data-protectionrules_en 2.
[7] DLA Piper. September 2019. Data Protection Laws of the World: Full handbook. https://www.dlapiperdataprotection.com.
[8] Yang, Q., Liu, Y., Chen, T. and Tong, Y. 2019. Federated machine learning: Concept and applications. ACM Transactions on Intelligent Systems and Technology (TIST), 10(2): 1–19.
[9] McMahan, B., Moore, E., Ramage, D., Hampson, S. and y Arcas, B.A. 2017, April. Communication-efficient learning of deep networks from decentralized data. In Artificial Intelligence and Statistics (pp. 1273–1282). PMLR.
[10] McMahan, H.B., Moore, E., Ramage, D. and y Arcas, B.A. 2016. Federated learning of deep networks using model averaging. arXiv preprint arXiv:1602.05629, 2.
[11] Konečný, J., McMahan, H.B., Yu, F.X., Richtárik, P., Suresh, A.T. and Bacon, D. 2016. Federated learning: Strategies for improving communication efficiency. arXiv preprint arXiv:1610.05492.
[12] Konečný, J., McMahan, H.B., Ramage, D. and Richtárik, P. 2016. Federated optimization: Distributed machine learning for on-device intelligence. arXiv preprint arXiv:1610.02527.
[13] Hartmann, F. August 2018. Federated Learning. https://florian.github.io/federatedlearning.

[14] Liu, Y., Yang, Q. and Chen, T. 2019. Federated learning and transfer learning for privacy, security, and confidentiality. AAAI. https://aisp-1251170195.file.myqcloud.com/fedweb/1552916850679.pdf.

[15] Yang, T., Andrew, G., Eichner, H., Sun, H., Li, W., Kong, N., Ramage, D. and Beaufays, F. 2018. Applied federated learning: Improving google keyboard query suggestions. arXiv preprint arXiv:1812.02903.

[16] Hard, A., Rao, K., Mathews, R., Ramaswamy, S., Beaufays, F., Augenstein, S., Eichner, H., Kiddon, C. and Ramage, D. 2018. Federated learning for mobile keyboard prediction. arXiv preprint arXiv:1811.03604.

[17] WeBank. 2020. Utilization of FATE in Anti Money Laundering Through Multiple Banks. https://www.fedai.org/cases/utilization-of-fate-in-anti-money-laundering-through-multiple-banks/.

[18] Intel and Consilient. 2020. Intel and Consilient Join Forces to Fight Financial Fraud with AI. https://newsroom.intel.com/news/intel-consilient-join-forces-fight-financial-fraud-ai/.

[19] MELLODDY. 2020. MELLODDY Project Meets its Year One Objective: Deployment of The World's First Secure Platform For Multi-Task Federated Learning In Drug Discovery Among 10 Pharmaceutical Companies. https://www.melloddy.eu/y1announcement.

[20] NVIDIA. 2020. Triaging COVID-19 Patients: 20 Hospitals in 20 Days Build AI Model that Predicts Oxygen Needs. https://blogs.nvidia.com/blog/2020/10/05/federated-learning-covid-oxygen-needs/.

[21] Yang, T., Andrew, G., Eichner, H., Sun, H., Li, W., Kong, N., Ramage, D. and Beaufays, F. 2018. Applied federated learning: Improving google keyboard query suggestions. arXiv preprint arXiv:1812.02903.

[22] Hartmann, F., Suh, S., Komarzewski, A., Smith, T.D. and Segall, I. 2019. Federated learning for ranking browser history suggestions. arXiv preprint arXiv:1911.11807.

[23] Ge, S., Wu, F., Wu, C., Qi, T., Huang, Y. and Xie, X. 2020. Fedner: Privacy-preserving medical named entity recognition with federated learning. arXiv preprint arXiv:2003.09288.

[24] Chai, D., Wang, L., Chen, K. and Yang, Q. 2020. Secure federated matrix factorization. IEEE Intelligent Systems, 36(5): 11–20.

[25] Liu, B., Wang, L. and Liu, M. 2019. Lifelong federated reinforcement learning: A learning architecture for navigation in cloud robotic systems. IEEE Robotics and Automation Letters, 4(4): 4555–4562.

[26] Ammad-Ud-Din, M., Ivannikova, E., Khan, S.A., Oyomno, W., Fu, Q., Tan, K.E. and Flanagan, A. 2019. Federated collaborative filtering for privacy-preserving personalized recommendation system. arXiv preprint arXiv:1901.09888.

[27] Yu, P. and Liu, Y. 2019. August. Federated object detection: Optimizing object detection model with federated learning. In Proceedings of the 3rd International Conference on Vision, Image and Signal Processing (pp. 1–6).

[28] Ramaswamy, S., Mathews, R., Rao, K. and Beaufays, F. 2019. Federated learning for emoji prediction in a mobile keyboard. arXiv preprint arXiv:1906.04329.

[29] Hard, A., Rao, K., Mathews, R., Ramaswamy, S., Beaufays, F., Augenstein, S., Eichner, H., Kiddon, C. and Ramage, D. 2018. Federated learning for mobile keyboard prediction. arXiv preprint arXiv:1811.03604.

[30] Liu, Y., Nie, J., Li, X., Ahmed, S.H., Lim, W.Y.B. and Miao, C. 2020. Federated learning in the sky: Aerial-ground air quality sensing framework with UAV swarms. IEEE Internet of Things Journal, 8(12): 9827–9837.

[31] Chen, M., Mathews, R., Ouyang, T. and Beaufays, F. 2019. Federated learning of out-of-vocabulary words. arXiv preprint arXiv:1903.10635.

[32] Qi, T., Wu, F., Wu, C., Huang, Y. and Xie, X. 2020. Fedrec: Privacy-preserving news recommendation with federated learning. arXiv preprint arXiv:2003.09592.

[33] Koren, Y., Bell, R. and Volinsky, C. 2009. Matrix factorization techniques for recommender systems. Computer, 42(8): 30–37.

[34] Gönen, M. 2012. Predicting drug–target interactions from chemical and genomic kernels using Bayesian matrix factorization. Bioinformatics, 28(18): 2304–2310.

[35] Blei, D.M. 2012. Probabilistic topic models. Communications of the ACM, 55(4): 77–84.

[36] Hofmann, T. 2013. Probabilistic latent semantic analysis. arXiv preprint arXiv:1301.6705.

[37] Weiss, S.M. and Indurkhya, N. 1995. Rule-based machine learning methods for functional prediction. Journal of Artificial Intelligence Research, 3: 383–403.

[38] Narudin, F.A., Feizollah, A., Anuar, N.B. and Gani, A. 2016. Evaluation of machine learning classifiers for mobile malware detection. Soft Computing, 20(1): 343–357.

[39] Gong, Y.J., Chen, W.N., Zhan, Z.H., Zhang, J., Li, Y., Zhang, Q. and Li, J.J. 2015. Distributed evolutionary algorithms and their models: A survey of the state-of-the-art. Applied Soft Computing, 34: 286–300.

[40] Canini, K., Chandra, T., Ie, E., McFadden, J., Goldman, K., Gunter, M., Harmsen, J., LeFevre, K., Lepikhin, D., Llinares, T.L. and Mukherjee, I. 2012. Sibyl: A system for large scale supervised machine learning. Technical Talk, 1: 113.

[41] Ji, G. and Ling, X. 2007, May. Ensemble learning based distributed clustering. In Pacific-Asia Conference on Knowledge Discovery and Data Mining (pp. 312–321). Springer, Berlin, Heidelberg.

[42] Agarwal, A., Chapelle, O., Dudík, M. and Langford, J. 2014. A reliable effective terascale linear learning system. The Journal of Machine Learning Research, 15(1): 1111–1133.

[43] Xing, E.P., Ho, Q., Xie, P. and Wei, D. 2016. Strategies and principles of distributed machine learning on big data. Engineering, 2(2): 179–195.

[44] Hsieh, K., Harlap, A., Vijaykumar, N., Konomis, D., Ganger, G.R., Gibbons, P.B. and Mutlu, O. 2017. Gaia: Geo-Distributed Machine Learning Approaching LAN Speeds. In 14th USENIX Symposium on Networked Systems Design and Implementation (NSDI 17) (pp. 629–647).

[45] Han, M. and Daudjee, K. 2015. Giraph unchained: Barrierless asynchronous parallel execution in pregel-like graph processing systems. Proceedings of the VLDB Endowment, 8(9): 950–961.

[46] Andrew, G. 2017. Bringing HPC Techniques to Deep Learning. Available: http://research.baidu.com/bringing-hpc-techniques-deep-learning/.

[47] Sergeev, A. and Del Balso, M. 2018. Horovod: Fast and easy distributed deep learning in TensorFlow. arXiv preprint arXiv:1802.05799.

[48] Jia, Y., Shelhamer, E., Donahue, J., Karayev, S., Long, J., Girshick, R., Guadarrama, S. and Darrell, T. 2014. November. Caffe: Convolutional architecture for fast feature embedding. In Proceedings of the 22nd ACM international conference on Multimedia (pp. 675–678).

[49] Seide, F. and Agarwal, A. 2016. August. CNTK: Microsoft's open-source deep-learning toolkit. In Proceedings of the 22nd ACM SIGKDD international conference on knowledge discovery and data mining (pp. 2135–2135).

[50] Dean, J., Corrado, G., Monga, R., Chen, K., Devin, M., Mao, M., Ranzato, M.A., Senior, A., Tucker, P., Yang, K. and Le, Q. 2012. Large scale distributed deep networks. Advances in Neural Information Processing Systems, 25.

[51] De Coninck, E., Bohez, S., Leroux, S., Verbelen, T., Vankeirsbilck, B., Simoens, P. and Dhoedt, B. 2018. DIANNE: A modular framework for designing, training and

deploying deep neural networks on heterogeneous distributed infrastructure. Journal of Systems and Software, 141: 52–65.

[52] Abadi, M., Barham, P., Chen, J., Chen, Z., Davis, A., Dean, J., Devin, M., Ghemawat, S., Irving, G., Isard, M. and Kudlur, M. 2016. TensorFlow: A system for large-scale machine learning. In 12th USENIX Symposium on Operating Systems Design and Implementation (OSDI 16) (pp. 265–283).

[53] Chen, T., Li, M., Li, Y., Lin, M., Wang, N., Wang, M., Xiao, T., Xu, B., Zhang, C. and Zhang, Z. 2015. Mxnet: A flexible and efficient machine learning library for heterogeneous distributed systems. arXiv preprint arXiv:1512.01274.

[54] Xing, E.P., Ho, Q., Dai, W., Kim, J.K., Wei, J., Lee, S., Zheng, X., Xie, P., Kumar, A. and Yu, Y. 2015. August. Petuum: A new platform for distributed machine learning on big data. In Proceedings of the 21th ACM SIGKDD International Conference on Knowledge Discovery and Data Mining (pp. 1335–1344).

[55] Mamidala, A.R., Kollias, G., Ward, C. and Artico, F. 2018. Mxnet-mpi: Embedding mpi parallelism in parameter server task model for scaling deep learning. arXiv preprint arXiv:1801.03855.

[56] coMind.org. Machine learning network for deep learning, https://comind.org/.

[57] coMind.org. Federated-averaging-tutorials. Available : https://github.com/coMindOrg/federatedaveraging-tutorials.

[58] Ingerman, A. and Ostrowski, K. March 2019. Introducing Tensorflow Federated. Available: https://medium.com/tensorflow/introducing-tensorflow-federated-a4147aa20041.

[59] TensorFlow/Encrypted. Available: https://github.com/tf-encrypted/tf-encrypted.

[60] Horovod. https://github.com/horovod.

[61] Ryffel, T. March 2019. Federated learning with PySyft and PyTorch, March 2019. Available: https://blog.openmined.org/upgrade-to-federated-learning-in-10-lines.

[62] OpenMined/PySyft. https://github.com/OpenMined/PySyft.

[63] OpenMined. https://www.openmined.org/.

[64] Han, B. March 2019. An overview of federated learning, March 2019. Available https://medium.com/datadriveninvestor/an-overview-of-federated-learning-8a1a62b0600d.

[65] Caldas, S., Duddu, S.M.K., Wu, P., Li, T., Konečný, J., McMahan, H.B., Smith, V. and Talwalkar, A. 2018. Leaf: A benchmark for federated settings. arXiv preprint arXiv:1812.01097.

[66] LEAF: A benchmark for federated settings. July 2019. Available: https://leaf.cmu.edu/.

[67] Baruch, G., Baruch, M. and Goldberg, Y. 2019. A little is enough: Circumventing defences for distributed learning. Advances in Neural Information Processing Systems, 32.

[68] Bhagoji, A.N., Chakraborty, S., Mittal, P. and Calo, S. 2019, May. Analysing federated learning through an adversarial lens. In International Conference on Machine Learning (pp. 634–643). PMLR.

[69] Liu, Y., Liu, Y., Liu, Z., Liang, Y., Meng, C., Zhang, J. and Zheng, Y. 2020. Federated forest. IEEE Transactions on Big Data.

[70] Cheng, K., Fan, T., Jin, Y., Liu, Y., Chen, T., Papadopoulos, D. and Yang, Q. 2021. Secure-boost: A lossless federated learning framework. IEEE Intelligent Systems, 36(6): 87–98.

[71] Hartmann, V., Modi, K., Pujol, J.M. and West, R. 2020, October. Privacy-preserving classification with secret vector machines. In Proceedings of the 29th ACM International Conference on Information & Knowledge Management (pp. 475–484).

[72] Shalev-Shwartz, S., Singer, Y. and Srebro, N. 2007, June. Pegasos: Primal estimated sub-gradient solver for svm. In Proceedings of the 24th international conference on Machine learning (pp. 807–814).

[73] Yang, K., Jiang, T., Shi, Y. and Ding, Z. 2020. Federated learning via over-the-air computation. IEEE Transactions on Wireless Communications, 19(3): 2022–2035.

[74] Jeong, W., Yoon, J., Yang, E. and Hwang, S.J. 2020. Federated semi-supervised learning with inter-client consistency & disjoint learning. arXiv preprint arXiv:2006.12097.

[75] Kingma, D.P. and Welling, M. 2013. Auto-encoding variational bayes. arXiv preprint arXiv:1312.6114.

[76] Yoon, A.S., Lee, T., Lim, Y., Jung, D., Kang, P., Kim, D., Park, K. and Choi, Y. 2017. Semi-supervised learning with deep generative models for asset failure prediction. arXiv preprint arXiv:1709.00845.

[77] Yang, S., Ren, B., Zhou, X. and Liu, L. 2019. Parallel distributed logistic regression for vertical federated learning without third-party coordinator. arXiv preprint arXiv:1911.09824.

[78] Peng, X., Huang, Z., Zhu, Y. and Saenko, K. 2019. Federated adversarial domain adaptation. arXiv preprint arXiv:1911.02054.

[79] Rong, W.A.N.G., Chunguang, M.A. and Peng, W. 2020. An intrusion detection method based on federated learning and convolutional neural network. Netinfo Security, 20(4): 47.

[80] Sattler, F., Wiedemann, S., Müller, K.R. and Samek, W. 2019. Robust and communication-efficient federated learning from non-iid data. IEEE Transactions on Neural Networks and Learning Systems, 31(9): 3400–3413.

[81] Yurochkin, M., Agarwal, M., Ghosh, S., Greenewald, K., Hoang, N. and Khazaeni, Y. 2019, May. Bayesian nonparametric federated learning of neural networks. In International Conference on Machine Learning (pp. 7252–7261). PMLR.

[82] Li, T., Sahu, A.K., Zaheer, M., Sanjabi, M., Talwalkar, A. and Smith, V. 2020. Federated optimization in heterogeneous networks. Proceedings of Machine Learning and Systems, 2: 429–450.

[83] Anwar, A. and Raychowdhury, A. 2021. Multi-task federated reinforcement learning with adversaries. arXiv preprint arXiv:2103.06473.

[84] Zeng, S., Anwar, M.A., Doan, T.T., Raychowdhury, A. and Romberg, J. 2021, December. A decentralized policy gradient approach to multi-task reinforcement learning. In Uncertainty in Artificial Intelligence (pp. 1002–1012). PMLR.

[85] Chen, F., Luo, M., Dong, Z., Li, Z. and He, X. 2018. Federated meta-learning with fast convergence and efficient communication. arXiv preprint arXiv:1802.07876.

[86] Huang, A., Chen, Y., Liu, Y., Chen, T. and Yang, Q. 2020. RPN: A residual pooling network for efficient federated learning. arXiv preprint arXiv:2001.08600.

Chapter 2

Federated Data Model - Go Local, Go Global and Go Fusion - In an Industry 4.0 Context

*Daniel Einarson** and *Charlotte Sennersten*

1. Foreword/Background

This contribution aims to discuss aspects of creating a data model in the light of federated learning and how federated learning can be achieved from an AI perspective with data fusion at the center. Federated learning in an Industry 4.0 setting with a focus on manufacturing and its sensors with robotics are the contexts we discuss, including computational and semantic aspects in this data landscape.

The primary author has a background in design of hierarchical compositions of autonomous self-aware software components, where such components cooperate within a defined environment. For such contexts, process models have been developed to represent intelligent agents, such as for robotics, but also in biologically living systems, such as for birds in certain environments. What is considered especially important here, is that the intelligent agents' awareness is represented by concepts

Dept. of Computer Science, Faculty of Natural Sciences, Kristianstad University, Sweden.
* Corresponding author: daniel.einarson@hkr.se

of AI. A key concept of AI is Machine Learning and Artificial Neural Networks. Still, in the context of intelligent agents, it has been considered both interesting and important to develop complementary models where inspiration has been received from areas of Cognitive Science, for purposes of representing reflective behavior as a response to external stimuli. Thus, such complementary modeling is regarded as interesting from both a theoretical and a technical perspective, with possible future modeling options.

The second author has founded and developed a 4D data management system at Commonwealth Scientific and Industrial Organization (CSIRO), Mineral Resources in Australia over several years with a focus on how to merge data via indexation. The data management system is called VoxelNET [1] and focuses on how to merge disparate data sets where the data includes various data types like cartography, GIS, terrain data, rock drilling data, various sensor data, with GPS locations, material movement horizontally and vertically over time (T) and ought to be scalable. The common denominator for all this data is *Volume*. In the mining industry, all is about volume calculations and what geology and minerals (Gold, Silver, Copper, Lithium, Iron, …) sits where in the rock. The mining industry is facing fewer mineral resources, meaning the mining operation has to go deeper. Due to higher temperatures at deeper levels, robots need to replace humans so we can keep safe at all times. This leads us to robotics and how federated learning has to work or we wish it to work for the ultimate benefit right over the *Manufacturing Value Chain* in any industry where many operational tasks are performed by various contractors and people/machines/robots.

This chapter sets a scene for the applicability of Federated Learning in an Industry 4.0 context, rather than addressing concrete technical solutions. Furthermore, the AI aspect of Federated Learning generally regards Machine Learning. Still, in the context of autonomous devices of Industry 4.0, the authors see a need to extend that AI concept, also to include cognitive aspects. While Federated Learning based on Machine Learning has been proven to be performed in automatic ways, this extension opens up for new challenges to solve and new opportunities in the Industry 4.0 settings.

2. Introduction

Industry with manufacturers is facing a serious dilemma; an industry with various applications knows that collaboration on data and AI is the only way forward. There are challenges around competitive concerns, intellectual property, data sovereignty, and other regulatory requirements keeping data, and ultimately innovation, siloed [2]. The common silos are the biggest challenges and bottlenecks, creating small islands of information and also not contributing to a holistic shared knowledge of a greater picture. The holistic data awareness in an application creates a competitive advantage. From a computer science lens, we usually talk about context, and this counts for both data semantics and data processing i.e., bandwidth, speed, memory, and meaning. The context needs scaling opportunities where we have ISO standards [3] and other data standards in place to be able to build the greater whole.

The term Artificial Intelligence (AI) was first coined by John MacCarthy in 1955 [4], basically, he defined AI being "The Science and Engineering of making intelligent machines, especially computer programs" [5]. AI refers to the ability of computer programs and robots to mimic the natural intelligence of humans and other animals, primarily cognitive functions such as the ability to plan a sequence of actions and to generalize [6].

Federated Learning (FL) is defined as collaborative Machine Learning (ML) without centralized training data and an FL platform is a data science system developed for dispersed noncentralized data [7]. In FL, the shared prediction model learns collaboratively and keeps all training data on the local devices, without storing all the information on a single server. FL as a concept defined by IBM [8] seen as a way to train AI models without anyone seeing or touching the data, offering a way to unlock information to feed new AI applications and as a response to the question: How is a data model trained without the need to move and store the training data centrally, i.e., server/cloud service as AWS [9] or alike. *The generic data model* representing the problem is fundamental and it is rather the changes or say the anomalies which are sent back to a central server when to update, aggregate updates to the central data model. From a local point of view, it is about cloning a generic central data model and copying this over to multiple local devices and then operating and running its raw data locally.

Industry moving from a siloed approach, an Integration-platform-as-a-Service (iPaaS) [10] can be a first step to breaking an existing silo. Depending on the complexity and what problems are approached, FL's generic data model may be combined in a distributed system running and using several microservices which are composed and integrated via several APIs.

This chapter will start with the concept of FL and how the concept connects to AI. Thereafter we set the scene in the context of Industry 4.0. To further understand FL and its interdependencies and issues we add a machine/computer aspect including a view of edge devices, and robotics. To harmonize the ML side of AI with Cognition, the authors add a short subchapter and touch quickly upon human and animal cognition. To set the main data model further in perspective and focus the authors continue with data interoperability and data pattern recognition before summarizing the chapter.

3. On Federated Learning

Federated Learning as a concept was first addressed by Google [11] as a response to the question: can a model be trained without the need to move and store the training data in a central location [12]. This approach is especially critical in cases, such as, if distributed systems contain large amounts of data existing in isolated islands [13], or in vases of local nodes with sensitive data. Many real-world scenarios, as pointed out by [14], address cases of data being distributed over a large number of devices. Applications may here require data to stay local, constrained by legal reasons, but also by limitations in network capacity or computing power. Here, FL approaches the needs pointed out by the stated question above.

Typical examples pointed out, for instance, by [15], relate to the distribution of handheld mobile devices, which is the primary computing- as well as personalized devices for many people today. Such devices often rely on trained algorithms for improving the personalized setups for high quality and intelligent user experiences, to for instance predict keyboard input. Here, as pointed out by [15], user produced keyboard data can be leveraged as training data. Still, an estimation claims that the number of SmartPhones in the world in 2023 will be 6.92 billion [16], a number that is expected to increase further in the future. Such facts clearly point out limitations in centralized algorithms for the training of user experiences, concerning communication and computation capacities. Furthermore, such user experiences are often based on personal choices where the disclosure of such personal information is governed by legislation, such as the European Union's GDPR, or through corresponding actions in the US and China [13]. Here, Federated Learning is described as a solution that allows for learning to take place at local clients, providing privacy benefits, as well as distributed computing [14].

Moreover, a question arises in how we train data from various contexts if we need contextual mapping, say in 3D, and at the same time cross correlate not only 1-dimensional data, 2-dimensional data, 3-dimensional data, and 4-dimensional data, and having an understanding of what similarities in data and what may be dissimilarities. With the collaborative approach through FL, with disparate data and its strengths and weaknesses without drowning in an API Lake of APIs we are approaching a new era of data and analytics needs and requirements.

3.1 On Techniques for Federated Learning

A key component of Federated Learning is Machine Learning (ML), where ML typically is based upon Artificial Neural Networks (ANNs). ANNs have been invented to mimic the structures and functionality of Biological Neural Networks (BNN), such as those of the human brain. Shortly, the structures of BNNs (and ANNs) consist of a number of layers of connected and communicating neurons, where such structures contribute to solving nonlinear problems. In the context of ML, *training* basically refers to building ANNs that efficiently map the input to output in complex contexts, hardly solvable by traditional programming methods (examples on this may include image processing, problems that generally are seen as quite easy to solve by BNNs of, e.g., humans). ML and ANNs are rather well-known techniques and there are several sources for further information on the subjects (please, see e.g. [17] for more on BNNs as an inspiration source for ANNs, and fairly comprehensive introductions to ML and ANNs). Training an ANN is commonly performed through a one-to-one relationship between the data and the trained model (the ANN), at one specific local device. Thus, Federated Learning can be seen as a platform for distributed ML for On-Device Intelligence coordinated by one centralized node.

Several techniques have been proposed to support Federated Learning as the response to the previously outlined question on opportunities to train a model without the data needing to be centrally located. In that context coordinating distributed training and centralized aggregation have significant roles. Coordination and distribution are, according to [18], summarized through the steps below, based on structures as outlined in Fig. 1. Here, a central node (Central Server, of Fig. 1), aggregates the outcome of the actions at the client nodes (Node 1, 2, and 3, of Fig. 1), that is performing the local training. The steps below represent one round of distributed training, and are performed iteratively to achieve reasonable accuracy:

1. *Initialization*: The nodes are activated and wait for the central server to give the calculation tasks.
2. *Client selection*: local nodes are selected to start training on local data.
3. *Configuration*: the central server orders selected nodes to undergo training on the model on their local data in a pre-specified fashion.
4. *Reporting*: each selected node sends its local model to the server for aggregation. The central server aggregates the received models and sends back the model updates to the nodes.
5. *Termination*: once a predefined termination criterion is met (e.g., a maximum number of iterations is reached, or the model accuracy is

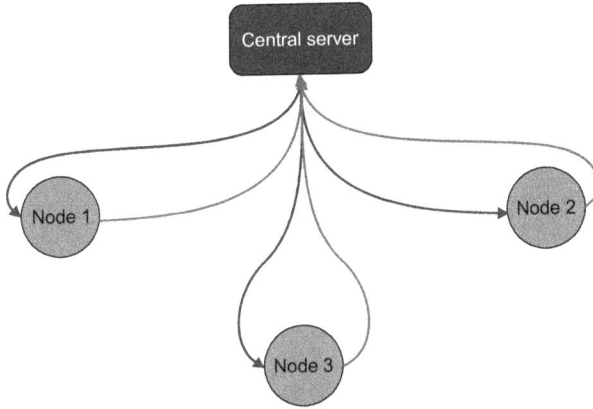

Figure 1. An example of a basic federated learning scheme [22].

greater than a threshold) the central server aggregates the updates and finalizes the global model.

Further discussions on implementation details on Federated Learning won't be elaborated on in this section. For more information on details regarding implementations of principles behind Federated Learning [19]. For developers of Federated Learning systems today, there are pre-developed platforms, where developers do not have to care about the underlying machinery, for creative customized development purposes. For instance, [20] provides information on the Python programming based TensorFlow Federated (TFF) platform, with several informative examples on how to set up a TFF environment, and how to get started using it to speed up the construction process. Moreover, similar support is provided by Oracle, but with Java as the development programming language ([21]).

3.2 *Views on Federated Learning and Artificial Intelligence*

IoT (Internet of Things) based systems are commonly addressed in contexts of the Digital Factory (more of that in a coming section), and as a concept, according to [23] this contributes to industries becoming more intelligent. Here it is pointed out that AI will have an impact on ML training on IoT devices, and with FL as a platform for distributed ML. That is, AI is seen as a higher-level concept including not only ML but also FL. The AI view of FL is also addressed in [13] which especially describes isolated islands of data, and data privacy and security, as two major AI challenges.

As a field, ML is generally seen as included in the broader field of AI. Moreover, AI is also seen in the field of Cognitive Science as a way to simulate cognitive behavior, and also as a response to external stimuli. The Cognitive side of AI is here two-folded, one sensor we often use is prolongations, and the second human cognitive functions such as vision (image, video, 3D-Lidar), hearing (sound), haptic (tactile), smell (fragrances/odors) and taste (tastebuds). To make sense of this data we often must analyze and orchestrate it as our cognitive perceptors would have experienced it in the first place firsthand.

Still, ML is a key component of the field of FL, and where further sub-concepts of AI are generally not considered an issue. However, ML has clear limitations in data imbalances [24] or even in the absence of data. The authors, therefore, extend the discussions on AI for Federated Learning, not only to focus on ML but complement the AI approach with concepts from Cognitive Science, as outlined in Fig. 2, and discussed in a further section. The authors of this chapter are aware that such approaches are outside the common frames for FL, but still want to shed light on significant principles and open up for further discussions and research themes.

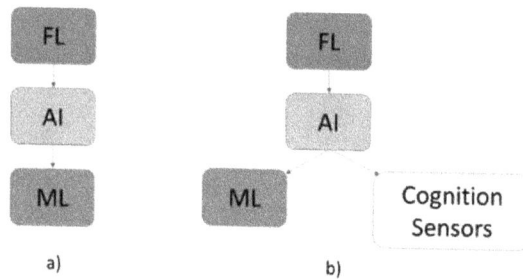

Figure 2. Scheme over federated learning, AI, ML and Cognition, a) the common view, b) the extended view.

4. Industry 4.0

The fourth industrial revolution or 4IR also called Industry 4.0 and 'The Connected Age' originates back to the 1900 century going from hand production methods to machines. A technological revolution followed where railroad and telegraph networks were extensively built out which meant faster transfer of people and ideas. The 3rd industrial revolution also known as the digital revolution advanced rapidly after the two world wars, and has completely set a new scene regarding industrialization, and has laid the foundation for the 4ir.

Industry 4.0 includes key components such as Cyber Physical Systems (CPS) where the "function of CPS has been identified as monitoring physical processes and creating a virtual copy of the physical world to support decentralized decision-making" [25]. The concept of CPS aligns with that of Digital Twins where a digital twin relates to a virtual copy of its corresponding physical system [26]. Typically, the digital twin emphasizes the connection between the physical model and the corresponding virtual model or virtual counterpart. This connection is typically established by generating real time data using sensors [25].

The fourth industrial revolution and 'always being connected' means we have a trend towards extensive automation and data exchange both privately and in manufacturing processes, it is here that the FL plays a major part. Telecommunication via 4G/5G/6G [27], Bluetooth [28], and WiFi [29] create a baseline with communication standards for our federated learning highway. Federated learning uses this network and can real time meet requirements and support in decision-making via efficient analytics in operational contexts. Industry 4.0 corresponds to the concept of a Digital Factory, that is, the digitization of the industry, where contexts of industrial robots more and more rely on technologies, such as CPS and Digital Twins. Other terminologies of use are *Smart Factory*, and *Dark Factory*, that correspond to the autonomy of industrial robots on an individual basis, as well as at a holistic level. By the holistic level, we mean the interaction between robots to fulfill a task of assembly.

4.1 Automation

Automation per definition means "the use or introduction of automatic equipment in a manufacturing or other process or facility" [30]. Automation in software development is "…applications that minimize the need for human input and can be used in a variety of ways in almost any industry. At the most basic, automation software is designed to turn repeatable, routine tasks into automated actions" [31]. This means tasks and procedures need to be correctly understood and represented when translated and digitally mapped to represent their intended operation, i.e., task and perception.

In the industry we can divide automation into '5 levels of automation', the layers usually are listed as, Sensors and Signals (Level 0), Manipulation and Control (Level 1), Supervisory Control (Level 2), Planning and Operations (Level 3) and Business Planning, Logistics, and Enterprise Level Operations [32]. In operation, most likely levels 0–3 are core focuses. This said, the whole data value chain must be able to interconnect so variations at one end can be picked up elsewhere, so the fluctuations and variations are adjusted for if it is a holistic approach, we are looking for, i.e., wanting to break up the silos.

4.1.1 Federated Automation

An automated federated data model is a "software process that collects data from diverse sources and converts it into a common model. It allows multiple databases to function as one and provides a single data source to front-end applications. Simply put, data federation allows users to access data from one place" [33, 34]. This means we have levels of automation and depending on what data layer we are in and where in the data value chain we exist we may have only parts of a full holistic view. In industry, this has arisen as a concern at times when it is mentioned as being a 'Black Box' approach [35]. The opposite of this is to shape a transparent view where the inner workings are referred to as a 'Clear Box'. To accompany federated learning, *Split Learning (SL)* is an approach that also follows a model-to-data scenario where clients train and test machine learning models without sharing raw data. The slower performance of SL has been improved by [36] via a novel approach, named split learning (SFL), computational resourcing. For the local and global approaches below are to exemplify what this means in a practical sense of this approach.

4.1.1.1 Local

Automation from a local view in a traditional sense means we use an algorithm to help us perform redundant repetitive tasks. The central data model in federated learning can be placed locally on a local server, on premise [37]. As an example, in the mining industry, a tire alone can have 400 sensors (Fig. 3) to measure pressure and 'tear and wear' for maintenance (Fig. 4), these 400 sensors are all locally 400 individual sensors and placed in a central data model for one single tire with a 'mother' data model coordinating 4 tires collectively with 1600 sensor data in total. This example means that the data model for tires does not by default include the driver of the truck and in this case, the vehicle and the person driving it, are located in two separate data models. This is an example of such a data software silo.

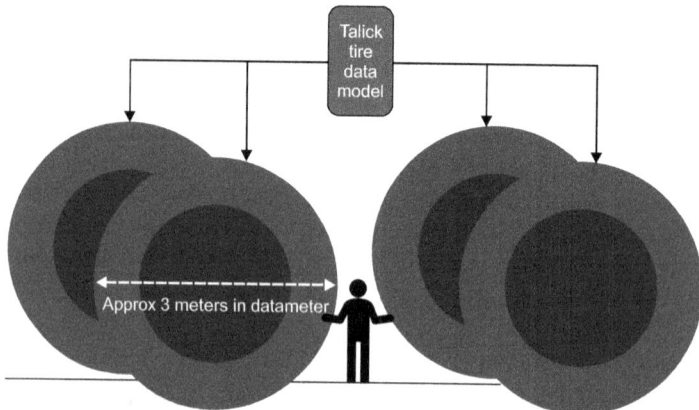

Figure 3. An exemplification of 4 tires in a single data model. Illustration by Sennersten.

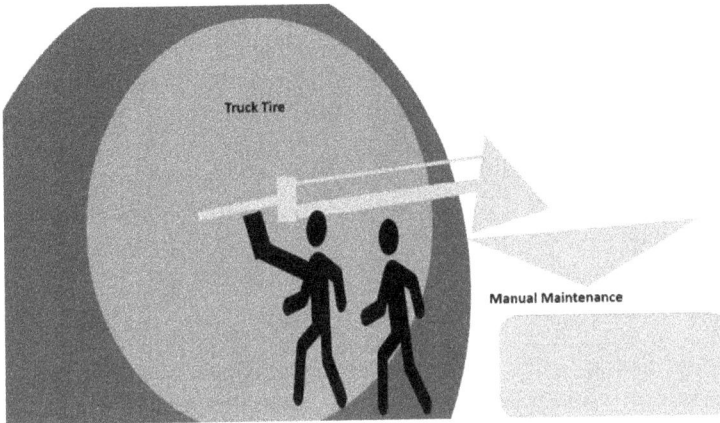

Figure 4. Example of the huge size of a tire alone. Illustration by Sennersten.

4.1.1.2 Global

To continue the previous tire example, if we now have a fleet of 30 trucks, we most likely have a data model centrally for all these trucks where each truck has a copy of this data model and can process everything locally and upload only errors, abnormal stats, driven time, fuel consumption, wear and tear due terrain conditions, etc. Centrally the main 'mother' data model aggregates all client data being sent to it and this happens on a central server, i.e., cloud service. The data sent can still be siloed and is just updating one siloed data model. A data model may sit with x hundreds of other sub data models which in their turn are aggregated to a higher-level data model. To have all these data models linked and talking to one another we want to achieve an interoperable environment achieving *interoperability* across the data value chain across operations and being able to cross correlate data and benefit from it. Interoperability is "the ability of computer systems or software to exchange and make use of information" [38].

5. Machine View

A machine or rather a computer generally needs input to be able to give an output. In between input and output, a data model/algorithm needs to perform calculations based on mathematical formulas, rules, and/or logic so it does represent an awareness of environmental context and domain knowledge. These models vary depending on context and environment, meaning an algorithm or model can be applied to any data but may miss out on specifics if not included in the general model. We can say all environments and situations have commonalities but also vary due to context. IBM defines federated learning as "Federated learning is a way to train AI models without anyone seeing or touching your data, offering a way to unlock information to feed new AI applications" [8]. This means we take humans out of the loop and have to mimic human capabilities if needed and if we can. Human thinking and human expertise have to be understood and mapped rigorously and is not a matter of a simple yes/no mapping only.

5.1 Edge Devices

Edge devices can play various roles depending on the purpose, but they essentially serve as network entr – or exit – points. These devices can function and perform transmission, routing, processing, monitoring, filtering, translation, and storage of data passing between networks. In terms of data, and referring back to IBM's definition, understanding that a network of devices situated in various locations with specific contextual particulars, various time zones, sensors capturing and sending data, computational network resourcing we have a network of data and operational demands in real time with software execution requirements. Often this network of devices is mentioned as being part of the Internet of Things (IoT) where these are not just capturing and sending data but also processing the data and this way contributing and resourcing computational demands over the network as such. ML can classify already known repetitive information, i.e., data and this way operates very efficiently.

5.1.1 Federated Capability of a Device

The analytics performed on the device using a particular data model has challenges depending on its purpose and context. If 'thinking' data as being 1-dimensional we have a quite uniform understanding of the problem, and it is quite straightforward to comprehend and analyze this type of dataset. In terms of federated learning and having a central data model copied over to local devices, the central data model needs to understand the local particulars if it is not just constrained to repetitive redundant tasks. If we extend the complexity of the data model, it also should be able to understand its environmental context. This can contain terrain in a 360-degree manner incorporating the vertical aspect, adding 2-dimensions, the 3-dimensional aspect, and time to the data model. All of a sudden, we have quite many data challenges in terms of knowing where we are and what sits where at what time.

5.1.1.1 Global Point-of-view

In a Digital Twin Industry 4.0 setting with various types of sensors, the challenge is how to build a comprehensive top-down/bottom-up data model representing not just isolated data per see but an artificial intelligence without human presence meaning we prolong the human via vision, auditive capabilities, (haptic, taste and smell). The discussion is in the light of a Digital Twin setting where we map the physical world as a 3-dimensional mapped reality with sensor inputs.

5.1.1.2 Local Point-of-view

The IoT devices have GPS coordinates. Localizing via a GPS point does not tell us anything about an object's direction in a 360-degree space, rotation of where an object/device sits where in context, how it behaves in this Digital Twin mapped world beyond basic actions as to move from A to B, how the sensors are directed, i.e., pointed into space in relation to its position.

5.2 *The Robotics View*

A robot is "a machine that resembles a living creature in being capable of moving independently (as by walking or rolling on wheels) and performing complex actions (such as grasping and moving objects)" [39]. To be autonomous, a robot is gained by having an internal 'mental model' of itself and the environment where it acts, to be able to fulfill its main task. Here, a CPS of a robot may not only carry a digital model of the robot's main tasks in correspondence to its environment, but also those of its robot counterparts. This view is illustrated in Fig. 5, by an example of two robots acting in an industrial park, where one robot internally simulates the actions of another robot to avoid collisions [40]. It shall be pointed out that this example is based on simulations only, and is used as a proof of concept, where internal digital twins are used to anticipate future actions.

Industrial robotics often generate huge amounts of diverse data to be understood both at local levels of individual robots and equipment, and at global, fused levels for compound and holistic interpretations.

The five primary areas of robotics in terms of 'understanding' are, Operator interface, Mobility or Locomotion, Manipulators and Effectors, Programming, and Sensing and Perception [41]. Controlled industrial robotics is one field in robotics while 'Field Robotics' is another. Controlled industrial robotics deal with the most controllable so called *'Knowns'* while Field Robotics deals with quite a few *'Unknowns'*. The unknowns can consist

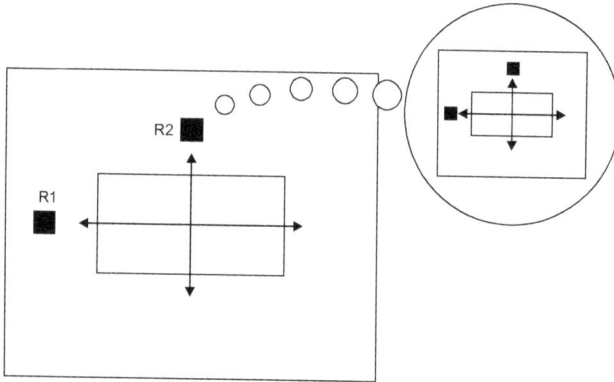

Figure 5. Robot R2 plans its actions through internally simulating the actions of R1 and itself [40].

of terrain challenges, environmental issues, pressure, obstacles, weather, … There are many different challenges depending on the purpose, goal, and problem area but a couple of challenges can be in swarm robotics, improvements in navigation and exploration, and developing AI that can "learn how to learn" [42].

Robotic Operative System (ROS) has evolved as an open-source library framework developed and progressed by its users, i.e., industry and academia [43]. ROS is a set of libraries and tools that help you build robot applications [44]. Much development is shared but some are also kept confidential depending on Intellectual Property (IP), patents and commercialization reasons.

5.2.1 Federated Capability in Navigation and Beyond

A robot with an internal 'mental model' as mentioned above contains redundant navigation tasks such as moving forward, left, right and reverse using onboard instrumentation to be able to move around. The robot/edge device in this case is a vehicle with an origin at a geolocation and moving along a path, i.e., a void from A to B. The vehicle is equipped with a navigation system so it can navigate with help of a Global Positioning System (GPS), an Inertial Navigation System (INS) [45]/Inertial Measurement Unit (IMU)[46] Visual Odometry (VO) [47, 48] used in both GPS denied environments and with GPS, Simultaneous Localisation and Mapping (SLAM) which builds a map (a mental model) and localize the vehicle/robot in that map at the same time. SLAM algorithms allow the vehicle to map out unknown environments [49]. Underground and indoor environments do not always have access to GPS coverage. Before setting a robot or a vehicle into its actual environment simulations can be conducted to mimic the conditions as close and realistically as possible. Federated learning is here a coordination of several processes at once to steer the robot in a safe route ahead. The battery or energy resourcing is also part of this where the robot needs to have enough energy to be able to return to the start destination.

5.2.1.1 Global

Robots and autonomous vehicles like autonomous cars are helped by cartography [50] and Geographic Information Systems (GIS) [51, 52] and ArcGIS [53]. For a robot in a field, a topographic map [54] with terrain elevation needs to be added in terms of navigation to understand the vertical aspect because the world is not flat. A fleet of robots may not all have the same purpose, and they may need to be coordinated stepwise or in synchronization. Several attempts and papers are published where various FL architectures and focus are tested. The authors in [55] propose a novel Federated Learning (FL) based Deep Reinforcement Learning (DRL) training strategy (FLDDPG) for use in swarm robotic applications under a limited communication bandwidth scenario. Often in remote areas and extreme locations, baseline strategies suffer from low bandwidth.

For robots not to bump into one another or with obstacles they need to be equipped with collision avoidance [56, 57] to be able to navigate in an environment. Sensors and collated data in these contexts have various data models and the semantics of this data can stay siloed with many software and systems in parallel but there is also a benefit of federated learning here so we can truly design for intelligence in the system. With all disparate data and when creating a data model for data fusion, geo-volumetric indexation becomes truly important both for computational resourcing and semantics.

5.2.1.2 Local

As an analogy in the data landscape and following from the above data layout, federated learning is assisting us in not, literally, forcing an elephant through a keyhole. We have to understand how we quantize and discretize data from both a computational stand and from semantic meaning.

The robots and the vehicles in navigation mode are carriers of a lot of sensors and these shall map, represent, and show a digital reality that metaphorically shows how a human may walk/move through an environment seeing its surroundings and shaping the mental model of their presence in real time. This means we glue/build together our mental data map via the 3D 360-degree space with its navigation executed with the help of maps, GIS, GPS, INU/IMU, VO, SLAM, data placed and executed in this 3D space, invisible data we may not see but can measure and can visualize, sounds, fragrances, smells, tastes, psychology, biology, ... An enormous design landscape we try to understand and intelligently architect. The purpose is to share data efforts so end-users remotely understand and can make use of captured data.

Federated learning with ML on one hand, processing and cloning data models on the other where the data community still figures out how to map informal human and animal knowledge into a complete data model including human reasoning for algorithmic comprehension. The next subchapter continues on the federated learning track where discussing how to approach human and animal cognition going toward intelligent agents and softbots.

6. Data Interoperability

Interoperability is very important when many people are to work in a context where everyone is highly specialized and where data contributions need to penetrate through to benefit a system, in the best cases real time. Interoperability is the ability to access and process data from multiple sources without losing meaning and then integrate that data for mapping, visualization, and other forms of representation and analysis. Interoperability enables people to find, explore, and understand the structure and content of datasets [58]. In Australia and the mining industry with companies like Volvo [59], SAAB [60], ABB [61], Atlas Copco [62], SANDVIK [63], Komatsu [64], Caterpillar [65], and others where all these data models need to contribute to geolocation with its operation and economy. Federated learning is even more essential when operating in remote areas and online connection is not even possible at times. This creates an operation where data models are forced to train on edge devices and not lose benefits from not being connected to the cloud at all times.

6.1 Computation and Semantics

Computational semantics is the study of how to automate the process of constructing and reasoning with meaning representations of natural language expressions. It consequently plays an important role in natural language processing and computational linguistics [66] also explaining what computation means without semantics [67] and the interoperability perspective on federated learning [68].

6.1.1 Federated Computation

A federation is a group of computing or network providers collectively agreeing upon standards of operation. The term may be used when describing the interoperation of two distinct, formally disconnected, telecommunications networks that may have different internal structures [69]. The question arises of *how* we compute versus *why* [70].

6.1.2 Federated Semantics

If we share data, the purpose is to understand one another: semantic interoperability. This is a major issue within the Federated project. Since 2019 Federated has done a lot of work on semantics, i.e., the Federated Semantic model. The importance is to enable any logistics stakeholder to access various (standard based) data silos through a Federated bridge [71, 72].

6.1.2.1 The Data Fusion

Data fusion is the process of getting data from multiple sources to build more sophisticated models and understand more about a project. It often means getting combined data on a single subject and combining it for central analysis [73]. Data fusion also means the joint analysis of multiple interrelated datasets that provide complementary views of the same phenomenon. The process of correlating and fusing information from multiple sources generally allows more accurate inferences than those that the analysis of a single dataset can yield. Data fusion is a multifaceted concept with clear advantages but at the same time with numerous challenges that need to be carefully addressed [74]. The joint analytics can include everything from maps, cartography, terrain data, text files, CAD files, imagery, satellite imagery, 3D Lidar scans, descriptions, timed data, navigation data, … This creates a huge demand on staff and personnel to understand how to fuse data to not lose precision and foremost not lose knowledge. Limitations in knowledge and knowing how to fuse and not yet create another silo is a challenge.

6.1.2.2 Federated Fusion

We can ask why federated fusion is interesting at all and there is a reason for why. Training the learning model centrally requires a central cloud with immensely powerful computing capabilities and storage capacities. The standard ML model is not readily applicable to large-scale IoT networks and cannot exploit the availability of distributed computing. This calls for a new learning model that leaves the training data distributed across individual IoT devices instead of being centralized. Google invented the concept of FL for on-device learning and data privacy preservation using the FL approach, each IoT device can train its model based on locally collected data [75].

6.1.2.3 Data Indexation

To be able to fuse data and be able to cross correlate the same, data standards are crucial to aligning the data landscape so we all can benefit from all and everyone's efforts. We are most likely envisioning a database and its indexation when we look at indexation. Indexing is the way to get an unordered table into an order that will maximize the query's efficiency while searching. When a table is unindexed, the order of the rows will likely not be discernible by the query as optimized in any way, and your query will therefore have to search through the rows linearly. In other words, the queries will have to search through every row to find the rows matching the conditions, a process that certainly is very time consuming [76].

6.1.2.4 Federated Indexation

When conducting a query and accessing several data sources at once by making a single query is what we aim for. The federator gathers results from one or more search engines and then presents all of the results in a single user interface [77].

7. Data Pattern Recognition

With all data at hand and analyzed, the aim is to discover patterns and trends in the data so we can understand what happens beyond what we instantly can see. Pattern recognition is the use of computer algorithms to recognize data regularities and patterns. This type of recognition can be done on various input types, such as biometric recognition, colors, image recognition, and facial recognition. It has been applied in various fields such as image analysis, computer vision, healthcare, and seismic analysis [78].

7.1 Data Logging

Without data we cannot read or reveal any patterns, so we must have access to valuable data to be able to draw conclusions from it. Data is an important source and can sometimes be very hard to get access due to confidentiality, integrity, etc. Data logging is the process of collecting and storing data over a while to analyze specific trends or record the data-based events/actions of a system, network, or IT environment. It enables the tracking of all interactions through which data, files or applications are stored, accessed, or modified on a storage device or application [79].

7.1.1 Algorithmic Understanding

Algorithmic understanding is kind of twofold. When we analyze data, we can see how we algorithmically need to build an algorithm and so can a computer. When a human at first instance is going to translate a data finding s/he needs to convert it to code via programming so the computer can understand how it is going to execute the algorithm on incoming data or maybe predict. An algorithm is a set of commands that must be followed for a computer to perform calculations or other problem-solving operations. According to its formal definition, an algorithm is a finite set of instructions carried out in a specific order to perform a particular task [80].

7.2 Steering

Steering a robot or a vehicle to access certain data the algorithm may look at executing an action so a gripper on the robot can grip and pick a typical object or mineral we ask it for. Steering data can of course be used for pure navigation [81], data thresholds, and frequency of certain events or actions.

7.2.1 Algorithmic Execution

Algorithmic trading is a process for executing orders utilizing automated and pre-programmed trading instructions to account for variables such as price, timing, and volume. An algorithm is a set of directions for solving a problem. Computer algorithms send small portions of the full order to the market over time [82].

8. Cognition in System and Agent System Opportunities for Agents with Cognition

8.1 Human Cognition

In light of federated learning and data models used for ML, we will introduce here the cognitive side of the AI model from Fig. 2. Cognitive functions refer to multiple mental abilities, including learning, thinking, reasoning, remembering, problem solving, decision making, and attention

[83, 84]. Neuroscience studies the structure or function of the nervous system and brain [85]. 'Symbolic AI' is an approach mimicking how the brain learns [86] as a top-down approach whereas the 'Connectionist AI' is an approach from more of an attempt to understand how the brain works at the neural level, i.e., how learning and memory work from a bottom-up approach [87]. These two approaches 'symbolic vs. connectionist approaches' [88] do somewhat compete with one another in terms of how knowledge is perceived. The symbolic approach as rule-based AI looks at how to insert human knowledge and behavioral rules into a computer via code [89] via if-then statements as 'replicating intelligence by analyzing cognition independently of the biological structure of the brain' rather than 'creating artificial neural networks in imitation of the brain's structure' [88].

Gärdenfors introduces the approach of conceptual spaces [90] to meet problems of the symbolic and connectionist views on the modeling of cognitive tasks. He addresses two dominating perspectives in modeling representations of cognitive tasks, *symbolic* and *associationism*. Gärdenfors' *symbolic* view corresponds to the manipulation of symbols while *associationism* corresponds to associations between different elements of information. Here, *connectionism* is seen as a specific case of *associationism*, typically represented by neural networks. Gärdenfors argues that both perspectives have limitations and sees complementary modeling as more promising than those two perspectives being competing. Marcus [91], on the other hand, rather takes the approach of integrating Connectionism and Cognitive Science as a core way to model such tasks, merely through neural networks. The authors of this contribution have a rather pragmatic standpoint, still relying on the principles of [90]. That is, the connectionist view, through neural networks (here ANN), are used in typical situations of reasonable amounts of training data for well-defined purposes (such as image recognition).

8.1.1 Agent

An intelligent agent is anything that perceives its environment, takes actions autonomously to achieve goals, and may improve its performance with learning or may use knowledge [92]. A softbot is a computer program that acts on behalf of a user or another program [93]. In terms of perceiving or perception means the ability to see, hear, or become aware of something through the senses. Our five senses are vision, hearing, haptic, smell, and taste. Visual attention in this means an agent creates an understanding from what it sees and in a digital setting the perception is perceived via optical sensors such as cameras, video, 3D Lidar, CAD, etc. Visual attention is the ability to prepare for, select, and maintain awareness of specific locations, objects, or attributes of the visual scene [94]. Eye tracking, usually used for 2-dimensional stimuli, can also be used for 3D scenes where objects in the scene are logged via ray casting to be able to log gazed objects especially attended [95].

8.2 *Animal Cognition*

In robotics, it is very common to look at animal swarm behavior and especially ants to find out how they move around and coordinate themselves in and around tasks [96]. Animals' cognitive capacities and problem-solving abilities are used when to find food, avoid predators, and seek shelter. Cognition involves processing information, from sensing the environment to making decisions based on available information. Such cognitive capacities include, among others, the ability to navigate through space, account for the passage of time, determine the quantity, and remember events and locations [97–99].

The authors here choose to present an example of AI-based simulations of animal behavior. Even though the main theme in this chapter relies on robotics in industrial processes, it is interesting to reflect upon such examples, for purposes of abstracting the scene of behavior of intelligent agents.

8.2.1 *Reactive/Response vs. Reflective*

Reaction is largely driven by external stimuli. Reflection, on the other hand, is a metacognitive function, which is a higher-order executive thinking skill that requires one's awareness and regulation of one's thinking process [100]. Reactive behavior often refers to an immediate response to feelings about an uncontrollable situation, a problem, or other issues [101]. In robotics, we rather talk about response behaviors than reactions and especially in a stimuli-response framing [102]. An emphasis here is to be aware of how federated learning with ML and its data model may exclude reflective capacity where we may use cognitive software models such as ACT-R, BDI, etc., to cover these behaviors. Adaptive Control of Thought—Rational (ACT-R) is a cognitive architecture: a theory for simulating and understanding human cognition [103, 104]. The belief–desire–intention software model (BDI) is a software model developed for programming intelligent agents [105].

8.2.1.1 Test Case Following the Symbolic vs. the Connectionist Views

As a test case and to look at a combined scenario, the authors used an Artificial Neural Network (ANN) applied to data collected for birds. Individual software representations of birds should carry independently trained ANN as a basis for simulated movements, representing their individual biological neural networks. One problem that clearly could be seen, was that the data points for 'flying' were a lot less than those for 'swimming', with the implication that the ANNs were imbalanced and incapable of representing flights, e.g., between lakes. Imbalances in datasets are a problem that shows a need to use ANNs with care [106, 107]. Thus, a solution to the problem considered an individual software representation to include not only an ANN part, for the common *reactive behavior*, but also a part including reasoning about the state (such as, 'time to fly, as a response to the choice to fly by a neighboring bird') that is considered a *reflective behavior*. While the reactive behavior originates from Machine Learning, approaching the reflective behavior, to compensate for the lack of information from the ANN, is inspired by Cognitive Science. Figure 6 illustrates a Multi-Agent System, with reactive and reflective parts. Here, two intelligent agents (represented by green frames) reside within an environment, and where the system state is updated through continuous mutual communication.

The representation of the intelligent agents contains an ANN (upper left part), that provides information on possible next moves (the reactive behavior). Furthermore, the upper right symbol represents logic for reflective behavior, that is, reasoning on alternative moves. The included circle represents the life-code of the intelligent agent, under which the current internal state is observed, and the next, internal, state produced.

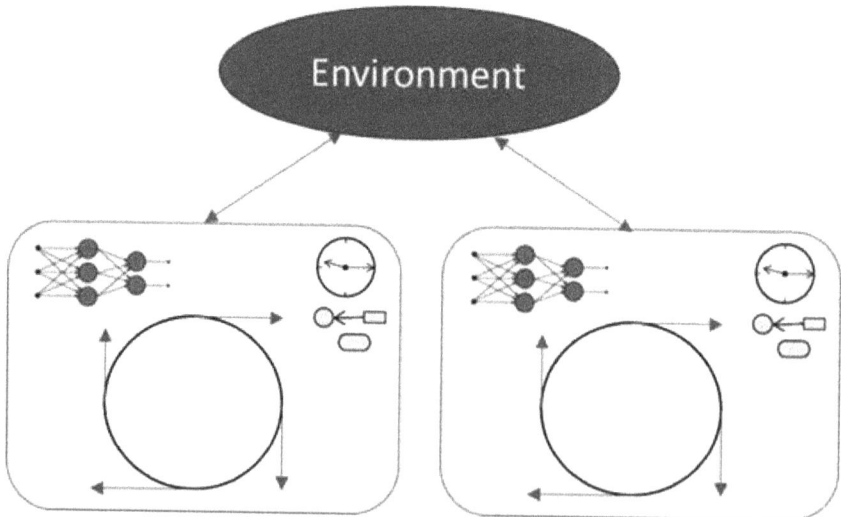

Figure 6. A communicating multi-agent system representing reactive and reflective behavior.

The approach of representing reflective behavior mainly inspired from ACT-R. ACT-R, is a conceptual architecture for cognitive studies, especially on human behavior ([108–111]), but is generalized here to be used for studies of birds, and even more general, on concepts of intelligent agents.

8.2.1.2 Local

On one hand, at a local level, in the context of FL, the scene should be quite clear. That is, each bird is performing an individual ANN training phase, for instance, according to Fig. 1. Still, the behavior of the birds is not merely defined through that step, but also through that of, what we propose to point out as the *reflective behavior*. Such actions may be harder to generalize since they are more hard-coded on individual bases, rather than through ANNs that are integrated into the actions of FL.

8.2.1.3 Global

As has been addressed in this chapter, FL has reached a point of maturity where the actions of FL can be automated via pre-developed APIs. The step of training the ANNs of all the individual birds should therefore be seen as fully doable, in accordance with Fig. 1. Still, the aggregation of the training processes can only be based on the ML part, the *reactive* part, of the birds' behavioral patterns, not the reflective part. We the authors, see this problem as a challenge for future studies, and opportunities for further research. Again, the purpose behind this example is to point out concepts of FL and intelligent agents at an abstract level. The ongoing ecological study will be scaled up to regard further aspects of the birds behind swimming and flying, and the current view upon this is that the complementary modeling techniques will be even more significant, with further interesting inspiration to FL and an intelligent agent context.

9. Summary

This chapter has approached the concept of Federated Learning from two perspectives, that is, on the one hand, from an experience-based perspective, and on the other hand, from an elaborative perspective. Experience-based approaches are here based on previous work in contexts of Industry 4.0, and more precisely in the mining industry and its robotics context where the fusion of data and needs for what we can call 'fused' analytics. Federated Learning, with data standards and data models can clearly contribute to advancement in that industry. Remote environments with various data analytical challenges benefit from a Federated Learning paradigm when bandwidth and data processing is limited. Artificial Intelligence (AI), supported by Machine Learning (ML), on one hand, and AI supported by Cognitive Science approaches to sensor-based equipment on the other, can assist in a duality prolonging human ability and assist humans in robotics.

A second perspective of this contribution has had an emphasis to further guide and show how the cognitive aspect being part of the AI besides ML can build a holistic data view, and contribute to inherently autonomous, self-aware behavior of intelligent agents. Here, it has been discussed that the current Federated Learning processes are primarily based on ML and Artificial Neural Networks and that more studies must be approached to integrate the cognitive aspects of digital devices with such Federated Learning processes. The authors have discussed how Federated Learning and a local data model on an edge device needs to be understood at a central aggregated level to not lose local or global awareness on when to fuse disparate data to build a holistic data model.

References

[1]   VoxelNET 4D data integration platform – CSIRO.
[2]   Apheris - Collaborative Data Ecosystems in manufacturing. [Online] 2022 [cited 2022 December 29]. Available from: https://www.apheris.com/resources/learning-hub/article/guide-to-collaborative-data-ecosystems.
[3]   ISO - Standards. [Online] 2022 [cited 2022 December 29]. Available from https://www.iso.org/standards.html.
[4]   Gregersen, E. 2022. John MacCarthy – American Mathematician and Computer Scientist. [Online] 2022 [cited 2022 December 13]. Available from: https://www.britannica.com/biography/John-McCarthy Encyclopedia Britannica.
[5]   MacCarthy, J. 2022. What is AI? Basic Questions, [Online] [cited 2022 December 13]. Available from: http://jmc.stanford.edu/artificial-intelligence/what-is-ai/index.html .
[6]   WIKI (in Swedish): https://sv.wikipedia.org/wiki/Artificiell_intelligens.
[7]   Jamil, S. and Rahman, M. 2022. Fawad. A comprehensive survey of Digital Twins and Federated Learning for Industrial Internet of Things (IIoT), Internet of Vehicles (IoV) and Internet of Drones (IoD). Appl. Syst. Innov., 5: 56.
[8]   What is federated learning? | IBM Research Blog [Online] [cited 2022 November 23]. Available from:https://research.ibm.com/blog/what-is-federated-learning.
[9]   Free Cloud Computing Services - AWS Free Tier (amazn.com) [Online] [cited 2022 November 23]. Available from:https://aws.amazon.com/free.
[10] Frends iPaaS | iPaaS [Online] [cited 2022 November 23]. https://frends.com/landing/ipaas.
[11] McMahan, H.B., Moore, E., Ramage, D., Hampson, S. and y Arcas, B.A. 2017. Communication-efficient learning of deep networks from decentralized data. Proceedings of the 20th International Conference on Artificial Intelligence and Statistics (AISTATS). JMLR: W&CP volume 54. https://doi.org/10.48550/arXiv.1602.05629.

[12] Han, B. March 31, 2019. An Overview of Federated Learning, Published in Data Driven Investor. https://medium.datadriveninvestor.com/an-overview-of-federated-learning-8a1a62b0600d.

[13] Yang, Q., Liu, Y., Chen, T. and Tong, Y. March 2019. Federated Machine Learning: concept and applications. Published in Association for Computing Machinery, 10(2). ISSN:2157-6904. https://dl-acm-org.ezproxy.hkr.se/doi/pdf/10.1145/3298981.

[14] Zec, E.L., Mogren, O., Martinsson, J., Sütfeld, L.R. and Gillblad, D. 2020. Federated Learning using a Mixture of Experts. https://arxiv.org/abs/2010.02056.

[15] Konečný, J., McMahan, H.B., Ramage, D. and Richtárik, P. October 2016. Federated optimization: Distributed machine learning for on-device intelligence. https://doi.org/10.48550/arXiv.1610.02527.

[16] HOW MANY SMARTPHONES ARE IN THE WORLD? (Source: https://www.bankmycell.com/blog/how-many-phones-are-in-the-world).

[17] Géron, A. 2019. Hands-On Machine Learning with Scikit-Learn, Keras & TensorFlow — Concepts, Tools. And Techniques to build Intelligent Systems, O'Reilly, ISBN 978-1-492-03264-9, 2019.

[18] Bonawitz, K., Eichner, H., Grieskamp, W., Huba, D., Ingerman, A., Ivanov, V., Kiddon, C., Konečný, J., Mazzocchi, S., McMahan, B., Overveldt, T.V., Petrou, D., Ramage, D. and Roselander, J. 2019. Towards Federated Learning at scale: System design. Proceedings of Machine Learning and Systems, 1 (MLSys 2019). Available at https://arxiv.org/abs/1902.01046.

[19] Yang, Q., Liu, Y., Cheng, Y., Kang, Y., Chen, T. and Yu, H. 2020. Federated Learning. In Springer Book Series: Synthesis Lectures on Artificial Intelligence and Machine Learning, ISBN: 978-3-031-01585-4. Available at https://link.springer.com/book/10.1007/978-3-031-01585-4#about-this-book.

[20] TensorFlow. April 2021. *TensorFlow Federated*. Available at TensorFlow Federated.

[21] Oracle Developer Resource Center. May 2022. Federated Deep Learning using Java on the Client and in the Cloud. Available at https://developer.oracle.com/learn/technical-articles/federated-deep-learning-using-java.

[22] Federated Learning: Collaborative Machine Learning with a Tutorial on How to Get Started (exxactcorp.com) Available at: https://www.exxactcorp.com/blog/Deep-Learning/federated-learning-training-models.

[23] Hosseinzadeh, M., Hemmati, A. and Rahmani, A.M. April 2022. Federated learning-based IoT: A systematic literature review. International Journal of Communication Systems. Available through https://doi.org/10.1002/dac.5185.

[24] Kulkarni, A., Chong, D. and Batarseh, F.A. 2021. Foundations of data imbalance and solutions for a data democracy. https://arxiv.org/abs/2108.00071.

[25] Kenett, R.S. and Bortman, J. 2021. The Digital Twin in Industry 4.0: A Wide-angle Perspective. Wiley Online Library. DOI: https://doi.org/10.1002/qre.2948, 2021.

[26] Pires, F., Cachada, A., Barbosa, J., Moreira, A.P. and Leitão, P. 2019. Digital Twin in Industry 4.0: Technologies, applications and challenges. Published In: 2019 IEEE 17th International Conference on Industrial Informatics (INDIN), DOI: 10.1109/INDIN41052.2019.8972134, 2019.

[27] What is 5G? How will it transform our world? - Ericsson [Online].; [cited 2022 November 23]. https://www.exxactcorp.com/blog/Deep-Learning/federated-learning-training-models

[28] Innovations with impact - Ericsson [Online] [cited 2022 October 20]. https://www.ericsson.com/en/about-us/company-facts/innovation-history.

[29] Bringing WiFi to the world – CSIRO [Online] [cited 2022 April 13]. https://www.csiro.au/en/research/technology-space/it/wireless-lan.

[30] Automation - Wikipedia [Online] [cited 2022 June 22]. https://www.csiro.au/en/research/technology-space/it/wireless-lan.

[31] Franco, N., Van, H.M., Dreiser, M. and Weiss, G. 2021. Towards a self-adaptive architecture for federated learning of industrial automation systems. 2021 International Symposium on Software Engineering for Adaptive and Self-Managing Systems (SEAMS), pp. 210–216, doi: 10.1109/SEAMS51251.2021.00035.

[32] Five Levels in Industrial Automation - Inst Tools (instrumentationtools.com) [Online] [cited 2022 November 23]. https://instrumentationtools.com/five-levels-in-industrial-automation/.

[33] Vogel-Heuser, B., Legat, C., Folmer, J. and Schütz, D. 2015. An assessment of the potentials and challenges in future approaches for automation software. Science Direct, Industrial Agents, Emerging, pp. 137–152, Applications of Software Agents in Industry, https://doi.org/10.1016/B978-0-12-800341-1.00008-5.

[34] Agarwal, A. 2022. RPA Federated Model: Automation Capabilities and Growth, SS&C blueprism, RPA Federated Model: Automation Capabilities and Growth | SS&C Blue Prism [Online] [cited 2022 November 23].

[35] Black box [Online] [cited 2022 November 23].https://en.wikipedia.org/wiki/Black_box.

[36] Thapa, C., Mahawaga Arachchige, P.C., Camtepe, S. and Sun, L. 2022. SplitFed: When Federated Learning meets split learning. Proceedings of the AAAI Conference on Artificial Intelligence, 36(8): 8485–8493. https://doi.org/10.1609/aaai.v36i8.20825.

[37] Cloud-Based Vs. On-Premise Servers (forbes.com) [Online] [cited 2022 November 23]. https://www.forbes.com/sites/forbestechcouncil/2019/03/22/cloud-based-vs-on-premise-servers/.

[38] Lewis, S. 2019. DEFINITION interoperability, TechTarget Network. https://www.techtarget.com/searchapparchitecture/definition/interoperability [Online].; [cited 2022 November 23].

[39] Robot Definition & Meaning - Merriam-Webster [Online] [cited 2022 November 23]. https://www.merriam-webster.com/dictionary/robot.

[40] Einarson, D. 2001. Hierarchical Models of Anticipation, CASYS'01. This contribution received a best paper award, with a crystal Belgium, and was chosen to be published in AIP (American Institute of Physics) Conference Proceeding, 627.

[41] Das, S. 2022. What are the 5 Major Fields of Robotics? thecoderworld, 2022, What are the 5 Major Fields of Robotics? - thecoderworld.

[42] Hart, A. 2022. Field robotics solves problems we can't even imagine yet – but to capitalise, it needs a more diverse data set, Cosmos. Field robotics solves vital future problems, but is it fair? (cosmosmagazine.com).

[43] What is ROS? | Ubuntu [Online] [cited 2022 October 23]. https://ubuntu.com/robotics/what-is-ros.

[44] ROS: Home [Online] [cited 2022 November 23]. https://ubuntu.com/robotics/what-is-ros.

[45] VN-200 GNSS/INS - world's first single packaged SMD GNSS/INS · VectorNav [Online].; [cited 2022 November 23]. https://www.vectornav.com/products/detail/vn-200?gclid=EAIaIQobChMI-MGY3q21_AIVj7myCh0U0ARZEAAYASAAEgIMCvD_BwE.

[46] Mahdi, A.E., Azouz, A., Abdalla, A.E. and Abosekeen, A. 2022. A machine learning approach for an improved inertial navigation system solution. Sensors, 22: 1687. https://doi.org/10.3390/s22041687.

[47] Aqel, M.O.A., Marhaban, M.H., Saripan, M.I. and Ismail, N.B. 2016. Review of visual odometry: Types, approaches, challenges, and applications. SpringerPlus volume 5, Article number: 1897, https://doi.org/10.1186/s40064-016-3573-7.

[48] Zuo, Y., Yang, J., Chen, J., Wang, X., Wang, Y. and Kneip, L. 2022. DEVO: Depth-Event Camera Visual Odometry in Challenging Conditions. 2022 International

Conference on Robotics and Automation (ICRA), pp. 2179–2185, doi: 10.1109/ICRA46639.2022.9811805.

[49] What Is SLAM (Simultaneous Localization and Mapping) – MATLAB & Simulink – MATLAB & Simulink (mathworks.com) [Online] [cited 2022 November 23] https://se.mathworks.com/discovery/slam.html.

[50] Unearth Labs, Modern Cartography – History, Tools & Applications, Unearth, 2022, https://www.unearthlabs.com/blogs/modern-cartography.

[51] What is a geographic information system (GIS)? | U.S. Geological Survey (usgs.gov) [Online] [cited 2022 November 23]. https://www.usgs.gov/faqs/what-geographic-information-system-gis.

[52] GIS tooling for your success – CYBERTEC | Datavetenskap & PostgreSQL [Online] [cited 2022 November 23] (cybertec-postgresql.com) https://www.cybertec-postgresql.com/sv/tjaenster/spatial-services-2/gis-tooling-2/?gclid=EAIaIQobChMIptzgrLK1_AIV17LVCh1xiw4SEAAYASAAEgKJAvD_BwE.

[53] Web GIS Mapping Software [Online] [cited 2022 November 23]. https://www.esri.com/en-us/arcgis/products/arcgis-online/overview.

[54] Get Maps [Online] [cited 2022 November 23]. https://ngmdb.usgs.gov/topoview/viewer/#4/40.01/-99.93.

[55] Na, S., Roucek, T., Ulrich, J., Pikman, J., Krajnik, T., Lennox, B. and Arvin, F. 2022. Federated reinforcement learning for collective navigation of robotic swarms, computer science and robotics, https://arxiv.org/abs/2202.01141.

[56] Collision avoidance system - Wikipedia [Online] [cited 2022 November 23]. https://en.wikipedia.org/wiki/Collision_avoidance_system.

[57] Continental Automotive - Collision Avoidance (continental-automotive.com) https://www.continental-automotive.com/en-gl/Passenger-Cars/Autonomous-Mobility/Functions/Low-Speed-Maneuvering/Collision-Avoidance.

[58] Team Intertrust, What is data interoperability? Intertrust, 2023. https://www.intertrust.com/blog/what-is-data-interoperability/.

[59] Volvo Lastvagnar Sverige (volvotrucks.se) [Online] [cited 2022 November 23]. https://www.volvotrucks.se/sv-se/.

[60] Saab Archives - International Mining (im-mining.com) [Online] [cited 2022 November 23]. https://im-mining.com/tag/saab/.

[61] ABB i Gruvor [Online] [cited 2022 November 23]. https://new.abb.com/mining/sv.

[62] Atlas Copco: Home of industrial ideas - Atlas Copco Sweden [Online] [cited 2022 November 23].https://www.atlascopco.com/sv-se.

[63] Utrustning och verktyg för gruvor och infrastruktur — Sandvik Group (home.sandvik) [Online] [cited 2022 November 23]. https://www.home.sandvik/se/produkter-tj%C3%A4nster/utrustning-och-verktyg-f%C3%B6r-gruvor-och-infrastruktur/.

[64] Home | Komatsu [Online] [cited 2022 June 12]. https://www.home.sandvik/se/produkter-tj%C3%A4nster/utrustning-och-verktyg-f%C3%B6r-gruvor-och-infrastruktur/.

[65] Caterpillar | Caterpillar [Online] [cited 2022 DEcember 24]. https://www.home.sandvik/se/produkter-tj%C3%A4nster/utrustning-och-verktyg-f%C3%B6r-gruvor-och-infrastruktur/.

[66] Erk, K. 2018. Computational Semantics, Oxford Research Encyclopedias. https://doi.org/10.1093/acrefore/9780199384655.013.331.

[67] Fresco, N. 2010. Explaining computation without semantics: Keeping it simple. Minds & Machines, 20: 165–181. https://doi.org/10.1007/s11023-010-9199-6.

[68] Roschewitz, D. et al. 2021. IFedAvg: Interpretable data-interoperability for federated learning. ArXiv abs/2107.06580 (2021): n. pag.

[69] Florissi Patricia, Federated Computing Will Shape the Future of Computing, InformationWeek, 2021 [Online] [cited 2022 November 23]. https://www.informationweek.com/big-data/federated-computing-will-shape-the-future-of-computing.

[70] Federated Computing Will Shape the Future [Online] [cited 2022 November 23]. https://www.informationweek.com/big-data/federated-computing-will-shape-the-future-of-computing.

[71] FEDeRATED semantic interoperability (federatedplatforms.eu) [Online] [cited 2022 November 23]. http://www.federatedplatforms.eu/index.php/federated-semantic-interoperability.

[72] Hofman, W. et al. 2022. Towards a mdobility Data Space: Data sharing via linked semantic data, an example for eFTI. Transportation Research Procedia, 00(2019): 000–000, Science Direct.

[73] What is Data Fusion? [Online] [cited 2022 November 23]. https://www.techopedia.com/definition/32735/data-fusion.

[74] Chatzichristos, C., Eyndhoven, S., Kofidis, E. and Van Huffel, S. 2022. Chapter 10 – Coupled tensor decompositions for data fusion. pp. 341–370. *In*: Yipeng Liu (ed.). Tensors for Data Processing, Academic Press. ISBN 9780128244470, https://doi.org/10.1016/B978-0-12-824447-0.00016-9.

[75] Boobalan, P. et al. 2022. Fusion of Federated Learning and Industrial Internet of Things: A survey. Computer Networks, 212: 109048. ISSN 1389–1286, https://doi.org/10.1016/j.comnet.2022.109048.

[76] How Does Indexing Work | Tutorial by Chartio [Online] [cited 2022 November 23]. https://chartio.com/learn/databases/how-does-indexing-work/.

[77] What is Federated Search and why is it important? | Sinequa [Online] [cited 2022 November 23]. https://www.sinequa.com/blog/intelligent-enterprise-search/what-is-federated-search/.

[78] Understanding Pattern Recognition in Machine Learning | Engineering Education (EngEd) Program | Section [Online] [cited 2022 November 23]. https://www.section.io/engineering-education/understanding-pattern-recognition-in-machine-learning/.

[79] What is Data Logging? – Definition from Techopedia[Online] [cited 2022 November 23]. https://www.techopedia.com/definition/596/data-logging.

[80] What is Algorithm | Introduction to Algorithms - GeeksforGeeks [Online] [cited 2022 November 23]. https://www.geeksforgeeks.org/introduction-to-algorithms/.

[81] 4.2 Create Steering Data (opendtect.org) [Online] [cited 2022 November 23]. https://doc.opendtect.org/6.6.0/doc/dgb_userdoc/Default.htm#dip-steering/create_steering_data.htm.

[82] Denisov, M., Anikin, A., Sychev, O. and Katyshev, A. 2021. Program execution comprehension modelling for algorithmic languages learning using ontology-based techniques. *In*: Yang, X.S., Sherratt, S., Dey, N. and Joshi, A. (eds.). Proceedings of Fifth International Congress on Information and Communication Technology. Advances in Intelligent Systems and Computing, vol 1184. Springer, Singapore. https://doi.org/10.1007/978-981-15-5859-7_25.

[83] Kiely, K.M. 2014. Cognitive function. *In*: Michalos, A.C. (ed.). Encyclopedia of Quality of Life and Well-Being Research. Springer, Dordrecht. https://doi.org/10.1007/978-94-007-0753-5_426.

[84] What are Cognitive Functions (neuronup.us) [Online] [cited 2022 November 23]. https://neuronup.us/areas-of-intervention/cognitive-functions/.

[85] What is neuroscience? | School of Neuroscience | King's College London (kcl.ac.uk) [Online] [cited 2022 November 23]. https://www.kcl.ac.uk/neuroscience/about/what-is-neuroscience

[86] What is symbolic artificial intelligence? – TechTalks (bdtechtalks.com) [Online] [cited 2022 November 23]. https://bdtechtalks.com/2019/11/18/what-is-symbolic-artificial-intelligence/

[87] Artificial intelligence - Connectionism | Britannica [Online] [cited 2022 November 23]. https://www.britannica.com/technology/artificial-intelligence/Connectionism.

[88] Copeland, B. 2022. Artificial Intelligence. Encyclopedia Britannica [Online] 2022 [cited 2022 November 11]. Available from: https://www.britannica.com/technology/artificial-intelligence.

[89] AI for Beginners - The Difference Between Symbolic & Connectionist AI (re-work.co) [Online] [cited 2022 November 23]. https://blog.re-work.co/the-difference-between-symbolic-ai-and-connectionist-ai/.

[90] Gärdenfors, P. January 2004. Conceptual spaces – The geometry of thought. The MIT Press, ISBN-13-978-0262572194.

[91] Marcus, G.F. January 2003. The algebraic mind: Integrating connectionism and cognitive science (Learning, Development, and Conceptual Change), Bradford Books, ISBN 13-978-0262632683.

[92] Intelligent agent [Online] [cited 2022 November 23]. https://en.wikipedia.org/wiki/Intelligent_agent.

[93] softbot [Online] [cited 2022 November 23]. https://en.wiktionary.org/wiki/softbot.

[94] McMains, S.A. and Kastner, S. 2009. Visual attention. *In*: Binder, M.D., Hirokawa, N., Windhorst, U. (eds.). Encyclopedia of Neuroscience. Springer, Berlin, Heidelberg. https://doi.org/10.1007/978-3-540-29678-2_6344.

[95] Vad är eyetracking? [Online] [cited 2022 November 23]. https://corporate.tobii.com/sv/om-oss/vad-ar-eyetracking.

[96] Cheraghi, A.R., Shahzad, S. and Graffi, K. 2022. Past, present, and future of swarm robotics. *In*: Arai, K. (ed.). Intelligent Systems and Applications. IntelliSys 2021. Lecture Notes in Networks and Systems, vol 296. Springer, Cham. https://doi.org/10.1007/978-3-030-82199-9_13.

[97] Animal Cognition [Online] [cited 2021 November 13]. https://www.nature.com/scitable/knowledge/library/animal-cognition-96639212/.

[98] Stevens, A.N.P. and Stevens, J.R. 2012. Animal cognition. Nature Education Knowledge, 3(11): 1.

[99] Animal Cognition [Online] [cited 2021 April 3]. http://www.animalcognition.org/.

[100] Huang, S. 2021. Oranisational Psychologist, MBPsS, Reaction vs Reflection: What's the difference? (5) Reaction vs Reflection: What's the difference? | LinkedIn [Online] [cited 2021 May 2].

[101] Indeed Editorial Team, Reactive vs. Proactive Behavior: What's the Difference? indeed, 2023, Reactive vs. Proactive Behavior: What's the Difference? | Indeed.com.

[102] Gianni, M., Kruijff, GJ.M. and Pirri, F. January 2015. A stimulus-response framework for robot control. ACM Transactions on Interactive Intelligent Systems, Volume 4, Issue 4, January 2015 Article No.: 21: 1–41. https://doi.org/10.1145/2677198.

[103] ACT-R (John Anderson) [Online] [cited 2022 June 23]. https://www.instructionaldesign.org/theories/act/.

[104] ACT-R [Online] [cited 2022 November 23]. http://act-r.psy.cmu.edu/.

[105] Belief–desire–intention software model [Online] [cited 2022 November 23]. https://en.wikipedia.org/wiki/Belief%E2%80%93desire%E2%80%93intention_software_model.

[106] Einarson, D. and Mengistu, D. 2022. Deep learning approaches for crack detection in bridge concrete structures. Proceedings of the 2022 International Conference on Electronic Systems and Intelligent Computing, ICESIC 2022. Institute of Electrical and Electronics Engineers (IEEE).

[107] Kulkarni, A., Batarseh, F.A. and Chong, D. July 2021. Foundations of Data Imbalance and Solutions for a Data Democracy, Cornell University Arxiv. https://doi. org/10.48550/arXiv.2108.00071.

[108] ACT R [Online] [cited 2022 November 23]. http://act-r.psy.cmu.edu/about/.

[109] Ritter, F., Tehranchi, F. and Oury, J. 2018. ACT-R: A cognitive architecture for modeling cognition. Wiley Interdisciplinary Reviews: Cognitive Science, 10: e1488. 10.1002/wcs.1488.

[110] Pentecost, D., Sennersten, C., Ollington, R., Lindley, C.A. and Kang, B. 2016. Predictive ACT-R (PACT-R): Using A physics engine and simulation for physical prediction in a cognitive architecture. Eighth International Conference on Advanced Cognitive Technologies and Applications (COGNITIVE 2016), March 20–24, Rome, Italy.

[111] Pentecost, D., Sennersten, C., Ollington, R., Lindley, C.A. and Kang, B. 2016. Using a physics engine in ACT-R to aid decision making-using a physics engine and simulation for physical prediction in a cognitive architecture. International Journal on Advances in Intelligent Systems, Academy, Research and Industry Association (IARIA). December 2016 (An invited journal paper followed from a "Best Paper Award").

Chapter 3

Federated Learning Architectures, Opportunities, and Applications

Pradipta kumar Mishra,[1] *Rabinarayan Satapathy*[1]
and *Debashreet Das*[2,*]

1. Introduction

To improve privacy and security, federated machine learning, an emerging field in machine learning, allows many parties to train a machine learning model jointly without disclosing their data. The writing on federated machine learning can be broadly divided into three categories: Applications, Systems, and Algorithms.

Algorithms: A sizable body of study has been devoted to developing algorithms for federated machine learning. The algorithms tackle numerous issues, such as diverse data, privacy and security, communication effectiveness, and convergence. Some algorithms most used in federated machine learning include federated averaging, federated transfer learning, federated optimization, and federated deep learning.

---
[1] Faculty of Emerging Technology, Sri Sri University, Cuttack, Odisha, India.
[2] School of Computer Science and Engineering, Vellore Institute of Technology, Vellore.
Emails: pradipta.m@gmail.com; rabinarayan.s@srisriuniversity.edu.in
* Corresponding author: debashreet.das@vit.ac.in

Systems: In addition to methods, researchers have focused on developing federated machine learning systems. The solutions are made to provide a strong and expandable framework for distributed machine learning. Some of the most well-known federated machine learning systems include Tensor Flow Federated, PySyft, and OpenMined. These systems provide numerous frameworks and tools for developing and using federated machine learning models.

Applications: Just a few of the many areas that scholars have examined using federated machine learning include the Internet of Things, banking, and healthcare. Healthcare professionals have used federated machine learning to identify novel drugs, predict diseases, and assess patient risk. For customer segmentation, fraud detection, and credit scoring in the financial industry. It has been applied to the Internet of Things for energy consumption forecasting, monitoring smart homes, and traffic forecasting.

Data never leaves its initial source—this is federated learning's central tenet. The inputs are used to train the model there. The model is modified or trained without modifying or sharing any of the user's data. A trustworthy server located centrally is crucial to this learning process. To begin the training process, the participating machines obtain a model from the main trustworthy server. Every machine that participates trains the model, updates it, and sends it to the server, which aggregates the data and creates a global updated model.

Then, according to the requirements of the participating machines, this model is decentralized and distributed [1]. Most heterogeneous data generated by IoT devices, mobile phones, and other devices can be processed using this technique [2]. Without gathering raw data, federated learning uses machine learning. The trained data is uploaded to the server (cloud) after the model has been autonomously trained [5]. It follows the principles of edge computing and connects the device to the cloud, also referred to as a centralized computer. Consider services like Netflix and others that suggest shows based solely on our viewing habits. Our heart rate and walking distance can be measured by applications that are like fitness trackers simply by getting training from the data that the tracker programme produces. Federated Learning is also used in G board's predictive handwriting. It can only predict the entire word by typing the first two characters. Another intriguing use case for an edge gadget is facial unlock. Fundamentally, Federated Learning keeps the raw data private and secure while automatically training the model. If Federated Learning trains a single model from thousands of data sources, it is regarded as being standard [3]–[4]. A machine learning technique called federated learning educates associate devices or servers that keep native information samples without transferring them. It makes it possible for numerous players to collaborate on building a reliable, common machine-learning model. Permit to manage important issues such as information security, privacy, and access rights, as well as access to diverse information [8]. Its applications span a wide range of sectors, including defense, telecommunications, the Internet of Things, and medicine [9]. Almost every industry and sphere of life is demonstrating the benefits of AI technology. As we look back on the history of AI, it's obvious that there have been many ups and downs [8].

Permission to print or electronically copy the entirety of this work, or just a part of it, for private or educational purposes. if copies contain this notice and the entire citation on the first page, and if it is not obvious that copies were made or disseminated to profit commercially or for financial gain. Any components of this work covered by copyrights must be credited to the creator. Would it not be feasible to send the data across organizations and combine it into a single, standardized website? It is frequently very challenging, if not impossible, to remove the barriers between various information sources. Any AI endeavor, in general, needs a variety of data. For instance, in an AI-driven product suggestion service, the product vendor has information about the product and is aware of the user's purchase but does not have information about the user's capacity to obtain and ability to pay. The idea behind federated learning is to share parameters and new data, aggregate them, find the right rules to combine parameters and ensure that we are following the right procedures so that at the end of the process, all the iterative aggregation and re-estimation results in something that is meaningful and makes sense. So, the concept is powerful and solves the problem of data protection [5].

One application is in healthcare because data is so delicate. Federated learning provides a way to safeguard user privacy by decentralizing data and bringing the advantages of AI to areas with sensitive data and heterogeneity. Machine learning is a technique for spotting patterns in medical imaging. In the field of healthcare, significant developments in machine learning have made it feasible to obtain medical data. Through federated learning, a trained neural network made up of all the heterogeneous data from various hospitals has been formed, and the entire training procedure is controlled by a centralized server. The institutions inform the network about their raw data while updating the model and communicating it with the server. The updated global model is then sent back to the hospital by the computer after it has assembled all the modified models. This ensures that the model will include data from many clients without changing the initial data [6]. With the help of evolving technology, comprehensive medical imaging is created. The trained model made possible by Federated Learning will enable this medical imaging to deliver better healthcare. Deep learning algorithms have significantly assisted in the early detection of disorders by improving the images from MRI or CT scans. This has helped numerous medical facilities [7]. Mobile devices have entered a new intelligent age thanks to the advancement of internet technology. Devices are now more vulnerable to data hacking due to the fast pace of development. Data security is aided by federated learning [3]–[4].

2. Types of Federated Learning

A distributed machine learning method called federated learning allows multiple parties to train a machine learning model jointly without sharing their data. There are various varieties of cooperative learning, such as

2.1. Horizontal Federated Learning: In this kind of federated learning, a model is trained across a variety of clients or platforms with comparable data distributions. By combining the data from various sources, the model's accuracy is intended to be increased.

2.2. Vertical Federated Learning: In this form of federated learning, a model is trained using a variety of clients or devices with various data distributions but the same general set of features. By combining the data from various sources, the model's accuracy is intended to be increased.

2.3. Federated Transfer Learning: In this kind of federated learning, a model that has already been taught is transferred to a new setting where data is dispersed among numerous devices. The model will be adjusted for the new job and environment.

2.4. Federated Multi-Task Learning: In this kind of federated learning, a model is trained on various tasks using various clients or devices. By combining the data from various jobs, the model's performance is intended to be improved.

2.5. Federated Meta-Learning: In this kind of federated learning, a model is trained to discover how to discover various activities and environments. The objective is to increase the model's capacity for future job and environment adaptation.

3. Review of Literature

Federated Machine Learning (FML) is a subset of machine learning (ML) in which several participants—typically clients or nodes—can work together to build a single ML model without having to share any private or sensitive information. FML aims to allow decentralized, privacy-preserving, and scalable ML for use in a variety of industries, including healthcare, finance, and IoT. Maintaining the global model's accuracy while protecting the security and privacy of the data of individual participants is one of the primary challenges of FML. Numerous techniques, including federated averaging, federated transfer learning, and federated deep learning, have been suggested in the literature as ways to accomplish this. The article "Communication-Efficient Learning of Deep Networks from Decentralized Data" by McMahan et al., first introduced the federated averaging algorithm, also known as FedAvg, which is a widely used algorithm in FML. In this approach, the update is transmitted to each participant after a central server computes the average of the locally trained models. The updated model is then used by the players to refine their local models.

The discipline of FML is expanding quickly, and it provides a promising answer for decentralized and privacy-preserving ML. A substantial corpus of the literature concentrates on different FML algorithms, techniques, and applications. However, there are still many issues that need to be resolved, including the need to protect anonymity, improve communication, and deal with the heterogeneity of data.

In the field of machine learning, federated machine learning (FML) is a new area that concentrates on creating algorithms for training models across a decentralized network of nodes, each of which has its own local data. The purpose of FML is to avoid the need to store and process massive amounts of sensitive data in a centralized location while still training a global model while preserving the privacy and security of the local data. The existence of non-IID (Independent and Identically Distributed) data across the nodes, which makes it challenging to combine the local models into a global one, is one of the major challenges in FML. Several methods, including distributed averaging, transmission learning, and multi-task learning, have been suggested to deal with this problem.

Federated Averaging is a straightforward and popular method that updates the global model by averaging the local models' parameters. This strategy has been used in a variety of applications, including healthcare, mobile, and IoT, and has been proven to be successful in several situations. Federated Transfer Learning is a more sophisticated method that makes use of knowledge transfer across nodes from the source job to the target task. It has been demonstrated that this strategy works well in situations where the local data is scant or unrepresentative. A method called federated multi-task learning trains a shared representation for all tasks while considering various tasks across the nodes. This strategy has been proven to work well in situations where the nodes are tasked with related but dissimilar kinds of data.

Stragglers, or nodes with sluggish response times, present in FML present another difficulty by delaying the entire training process.

Numerous methods, such as federated distillation, federated momentum, and federated adaptive gradient descent, have been suggested to handle this problem. A method called federated distillation uses the expertise of the fast nodes to train the slow nodes, minimizing the effect of stragglers. Federated Momentum is a method that updates the model based on the momentum term while maintaining a running average of the model's parameters, lessening the effect of stragglers. A method called Federated Adaptive Gradient Descent reduces the effect of stragglers by adjusting each node's learning rate based on its communication speed. In conclusion, FML is an active and expanding field of study with the ability to fundamentally alter how machine learning models are developed and applied in distributed systems. Despite these difficulties, new developments in FML have produced encouraging results and shown that it is possible to train machine learning models across a decentralized network of nodes while maintaining security and privacy. In the field of machine learning known as "federated machine learning," or "FML," models are trained using decentralized data, such as that which is dispersed across various platforms or organizations. FML aims to train a machine learning model without requiring the data to be centralized, preserving anonymity and security. FML has been the subject of an extensive recent study, and numerous papers have been written about it. A few pieces that are connected to FML are listed below:

A machine learning model can be trained on data from various devices using Google's Federated Learning of Cohorts (FLoC) technology without the need to centralize the data. Federated Averaging Algorithm is a simple and effective method for federated optimization, where the goal is to find the parameters that minimize a given objective function.

Federated learning protects privacy, with a particular emphasis on making sure that the training process protects the privacy of the data used in the training process. The goal of Secure Multi-Party Machine Learning is to address the challenge of training machine learning models in situations where numerous parties have data and wish to work together to do so while maintaining the privacy of the data. Federated Transfer Learning is concerned with transferring knowledge from a centralized machine learning model to a decentralized environment in which the data is dispersed across various platforms. These are only a few examples of the numerous works that are closely linked to FML. There has been a significant advancement in this area, and the industry is still developing and expanding.

The field of federated machine learning has seen the publication of numerous surveys and studies to give a thorough summary of the state of the field and highlighting the most significant advancements and difficulties. Federalized Machine Learning: Concept and Applications, is one of the most well-known studies in this field. This study offers a thorough overview of the idea and its potential uses. It is one of the first surveys in federated machine learning. It discusses federated learning's technological difficulties as well as its privacy and security concerns. In a survey on Federated Learning for Mobile, and the Internet of Things, the context of mobile devices and the Internet of Things, this survey concentrates on the difficulties and applications of federated learning (IoT). It offers a thorough analysis of the state of the field at the present and covers the different algorithms and techniques used in federated learning. In the context of edge computing, this study focuses on the unique opportunities and challenges presented by federated learning. It gives insights into the path this field will take in the future and covers the various federated learning algorithms, systems, and applications. A study on federated learning classification offers a thorough analysis of the subject of federated learning, covering its various issues, solutions, and future prospects. Additionally, it offers a thorough explanation of the privacy and security issues connected to federated learning and emphasizes possible uses for this technology across a range of industries. The numerous studies that have been published in federated machine learning are numerous; these are just a few examples. They offer insightful information about the present state of the field, the difficulties and opportunities, and the future course of this fascinating field of study. In the machine learning community, federated machine learning is a burgeoning field that seeks to train machine learning models on data that is decentralized and kept on various devices or network nodes. The following is a survey of related works in federated machine learning.

Federated Averaging: One of the oldest and best-known methods of federated learning is federated averaging. According to this method, each network component averages its locally computed gradients before sending the average to the central server, which updates the overall model.

Secure Aggregation: Secure Aggregation is a distributed averaging extension that takes privacy issues into account. Prior to transmitting their locally computed gradients to the central server, each node in secure aggregation encrypts them. The worldwide model is then updated by the server aggregating the encrypted gradients.

Federated Transfer Learning: In federated settings, the issue of non-IID (independent and identically distributed) data must be addressed. This method makes use of transfer learning. In this method, efficiency is enhanced by fine-tuning a trained model using local data.

Federated Distillation: This strategy makes use of knowledge distillation to transmit information from the world model to local models. Using the data that is readily accessible to them, the local models then make predictions, which are then combined at the central server.

Federated Stochastic Gradient Descent: This method of federated averaging updates the overall model using stochastic gradient descent. This method computes gradients from a mini-batch of samples taken from each node's data and sends them to the central computer.

Federated Learning with Non-IID Data: This is a relatively new area of study that seeks to solve the problems associated with federated learning with non-IID data. In this method, data is re-sampled, re-weighted, or modified before training to make it more comparable across nodes.

Privacy-Preserving Federated Learning: This active research field concentrates on creating algorithms that maintain privacy while carrying out federated learning. In this method, sensitive data is protected by using privacy-preserving methods like differential privacy or secure multi-party computation.

A new area of machine learning called federated machine learning seeks to train models on data sources that are not centralized but rather are stored and processed locally on individual devices. Recent years have seen an increase in its relevance because of growing privacy worries and the volume of data produced by edge devices.

Federated optimization, federated transfer learning, and federated learning systems are three general categories into which the related works in federated machine learning can be divided.

Federated Optimization: Federated Optimization is a subcategory of optimization that concentrates on developing algorithms that can improve machine learning models on distributed data sources. Designing algorithms that can successfully handle non-IID (Independent and Identically Distributed) data, which is a frequent problem in federated settings, is the main challenge in this category. Popular techniques in this area include FedProx, FedAvg, and Federated Averaging.

Federated Transfer Learning: Federated Transfer Learning focuses on moving knowledge gained from one data source to another, usually from a centralized data source to a federated setting. Designing techniques that can efficiently transfer information while protecting privacy is the main challenge in this category. This group includes some well-liked techniques like federated transfer learning and federated fine-tuning.

Federated Learning Systems: Federated Learning Systems: The development and deployment of federated machine learning systems are the main topics of this category. Designing systems that can handle communication overhead and privacy issues in federated settings effectively is the main challenge in this category. Popular programmes in this area include OpenMined, PySyft, and TensorFlow Federated.

Federated machine learning is a rapidly expanding area with a wide range of opportunities and challenges. The connected works in federated machine learning cover a broad spectrum of subjects, such as system implementation, transfer learning, and optimization. These are a few of the major contributions to shared machine learning. New strategies and methods are being created to handle the difficulties of federated learning, including scalability, communication effectiveness, privacy, and robustness.

Federated machine learning is a subset of machine learning that allows multiple parties to jointly train a model without directly exchanging data. Instead, each party builds a local model on its data, updates it, and transmits it to a central server, which aggregates the updates and returns a global model. This strategy can support data privacy preservation while still enabling cooperative model training.

The performance of federated machine learning approaches as opposed to conventional centralized machine learning approaches has recently been the subject of several comparison studies. Here are a few illustrations:

1. Researchers compared the effectiveness of federated learning approaches to centralized learning approaches for training convolutional neural networks (CNNs) on image classification tasks in a paper published in the Journal of Machine Learning Research in 2021. They discovered that, with fewer communication rounds and lower communication costs, federated learning methods could achieve comparable performance to centralized approaches.

2. For training recommendation models on sizable datasets, a study published in IEEE Transactions on Services Computing in 2020 compared the efficacy of federated learning approaches to conventional machine learning approaches. The researchers discovered that while maintaining the privacy of the user data, federated learning approaches could reach performance levels that were comparable to or even superior to those of centralized approaches.

3. Another study compared the effectiveness of federated learning methods to conventional machine learning approaches for training natural language processing (NLP) models on text classification tasks. It was published in the IEEE Proceedings in 2021. In particular, when dealing with highly unbalanced datasets, the researchers discovered that federated learning methods could achieve comparable or even better performance than centralized approaches.

In conclusion, these studies show that federated machine-learning techniques can be efficient for a variety of machine-learning tasks while also protecting the privacy of sensitive data. However, several variables can affect how well-federated learning methods perform, including communication costs, the number of participants, and the complexity of the model being trained.

Federated machine learning (FML) is a method of machine learning that allows multiple parties to work together on a model without having to share their data. There have been several case studies in recent years that show how effective FML is in different applications. Such a case would be:

Case Study: Federated Learning for Predictive Maintenance in the Manufacturing Industry

Siemens and Google Cloud worked together in 2021 to put into practice a federated learning solution for preventive maintenance in the industrial sector. The objective was to create a machine-learning model that could forecast equipment failures in a manufacturing facility without requiring participants to exchange confidential data.

The shared learning system was divided into two parts. A global model was trained in the first stage using a tiny subset of data from each manufacturing facility. The second part began with each plant training a local model with its data using the global model as a starting point. The final model, which was more precise than the global model, was made by combining these local models. The federated learning solution provided several benefits, including:

- Improved privacy and security: Since the information was stored locally, there was no need to transfer confidential information to other manufacturing facilities or outside service providers.
  - Better predictions of equipment failures were made thanks to the local models' training on data that was more representative of each facility.
- Lower communication expenses: Because only local model changes were sent between the plants and the cloud, communication expenses were drastically cut.

Overall, this case study demonstrates the potential of federated learning to enable collaboration and improve machine learning models while protecting sensitive data.

4. Architecture of Federated Machine Learning

A distributed machine learning paradigm called federated machine learning (FML) allows multiple parties to work together and train a machine learning model while maintaining local control over their data. The following elements usually make up the FML architecture:

Clients: These are the objects or people who have access to the info. They could be edge computers, smartphones, IoT devices, or any other data-generating equipment. The clients securely store their info locally.

Server: The server serves as a coordinator, gathering the model inputs from the clients and applying them to the world model. The server also communicates with the clients to coordinate the training process and disseminate the most recent model parameters.

Communication channels: The server and clients share information and coordinate the training process using the communication channels. The channels of contact ought to be private and secure.

Model: The model is the federated machine learning model that is being taught. On the server, the model is set up, and after that, it is taught using an amalgamation of client update data. A prediction or other move is then made using the final trained model.

Privacy-Preserving Techniques: Since some clients might not want to share their raw data with others, privacy is a major issue in FML. To combat this, FML usually makes use of a variety of privacy-preserving methods, including differential privacy, homomorphic encryption, secure multi-party computation, and others, to guarantee that the data is kept private and secure throughout the training process.

4.1 Horizontal Federated Learning

The term "horizontal federated learning" refers to a form of federated learning in which the data samples used for model training are dispersed across several devices with comparable qualities, such as the same kind of device or comparable usage patterns. In horizontal federated learning, the updates from each device are combined to create a global model, and the data samples from each device are used to train the model in a decentralized way. This method protects the privacy of individual data samples while enabling the training of models on sizable datasets that can't be kept on a single device. Figures 1 and 2 show the federated learning design.

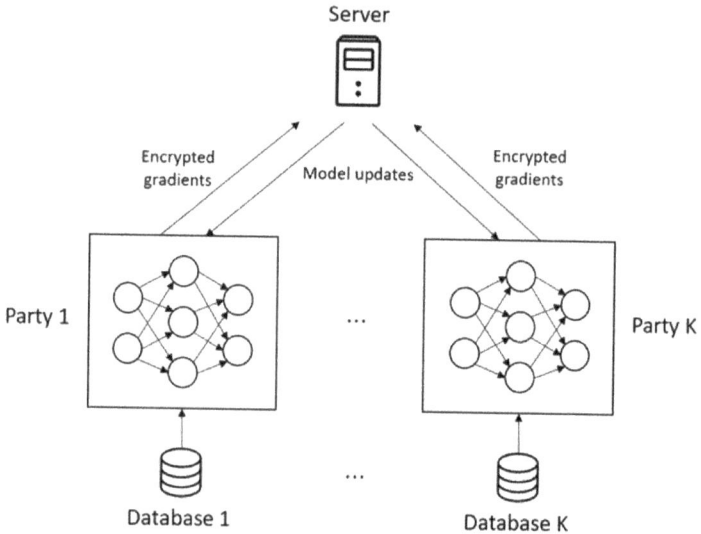

Figure 1. Architectural model for Federated Learning.

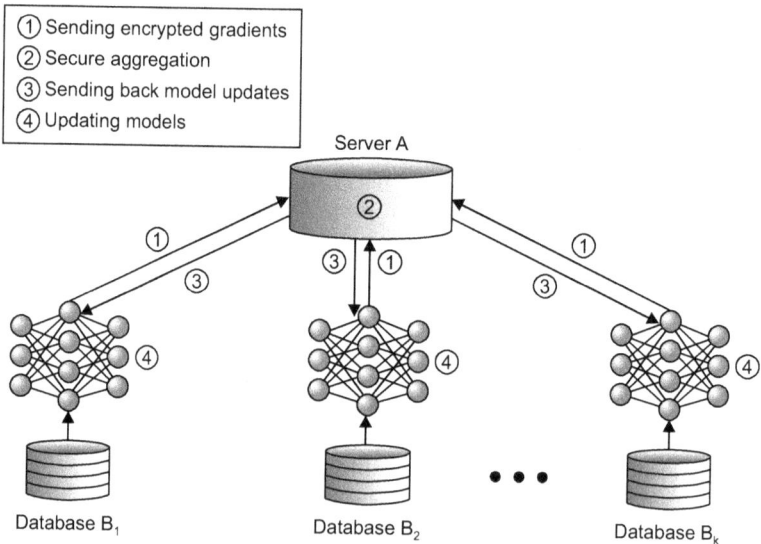

Figure 2. Architectural model for vertical Federated Learning.

For instance, in a smartphone application, various devices can contribute their camera images to the training of a deep learning model for image classification. Personalized suggestions can then be given to each user based on their usage trends using the trained model without jeopardizing the privacy of their data.

By allowing model training on enormous amounts of decentralized data while protecting user privacy, horizontal federated learning has the potential to revolutionize machine learning.

This algorithm jointly trains the data from datasets of samples with the same features. It primarily utilizes uniform data.

4.2 *Vertical Federated Learning*

Multiple organizations work together to train a common machine learning model on their own local data, which is vertically partitioned by a particular feature, in a process known as vertical federated learning(s). Each organization in this setup has its data that includes details about a specific subset of features, and these features vary between the organizations. The objective is to train a model that can efficiently use data from all involved organizations while adhering to their data sovereignty and privacy standards.

Think about a situation where several hospitals want to create a shared model to forecast patient results, for instance. Every hospital has its local data, which may contain details on various elements like patient demographics, medical histories, lab findings, and therapeutic outcomes. Each hospital can maintain its data locally and add to a shared model in a vertical federated learning setup. This shared model can then be used for prediction tasks by all hospitals.

Verifying that the model can utilize vertically partitioned data from various organisations while adhering to their privacy requirements is the primary challenge in vertical federated learning. For this, advanced privacy-preserving methods like homomorphic encryption, safe multiparty computation, or differential privacy must be used.

The process of combining these various features, calculating the training loss, and keeping everything private to create a centralized model is known as vertical federated learning [10]–[18].

5. Challenges of Federated Learning

In federated learning, numerous edge devices, such as smartphones or Internet of Things (IoT) devices, are coordinated by a central server to jointly build a machine learning model without exchanging raw data. While federated learning has several advantages, including the ability for edge devices to learn from one another and the preservation of user anonymity, it also has several drawbacks:

1. Heterogeneity of devices: It can be difficult to guarantee uniform model updates across all edge devices in federated learning due to

variations in network conditions, programme versions, and hardware configurations.

2. Data quality and distribution: The quality, size, and distribution of the data on the edge devices may differ, which could cause bias and fairness problems with the model's performance.

3. Security and privacy: Federated learning depends on sharing aggregated data and model updates between devices and the central server, which could be risky in terms of security and privacy.

4. Communication and latency: Low-bandwidth networks or high-latency connections may make it difficult for edge devices and the central server to communicate frequently, resulting in communication overhead and delays in model changes.

5. Model complexity and scalability: To handle the complexity of the model, distribute the model to edge devices, and guarantee scalability to support many devices, federated learning needs a strong infrastructure.

6. State of Federated Machine Learning based on Different Criteria

Keep in mind that this is only a high-level comparison and that various federated machine-learning algorithm variations may have different advantages and disadvantages. Furthermore, the best-federated learning algorithm relies on the particular use case and system requirements.

7. Applications of Federated Learning in Edge Computing

A machine learning strategy called federated learning allows for cooperative learning on dispersed data while protecting user privacy [10]. It could be used for a variety of things, such as:

1. Healthcare: Machine learning models can be trained on dispersed healthcare data using federated learning while protecting patient anonymity. For instance, medical facilities can work together to build a machine-learning model to identify diseases, forecast patient outcomes, or customize care.

2. Financial services: Machine learning models can be trained on dispersed financial data using federated learning while maintaining individual privacy. For instance, banks can work together to identify fraudulent transactions, forecast credit risk, or provide individualized financial advice.

Table 1. Federated Learning in edge computing.

Criteria	Federated averaging	Federated stochastic gradient descent	Federated distillation
Data Privacy [19]	By training models locally and only transmitting model updates to the central server, it protects user anonymity.	The privacy of users is protected by locally training models and sending gradient changes to the main server.	Uses a small, compressed model to extract information from a larger model trained on centralized data, protecting user privacy.
Communication Overhead [20, 21]	High communication costs as a result of frequently sending model changes to the main server.	Lower communication costs because gradient changes are sent to the main server.	Reduced transmission costs as a result of distillation using a compact, small model.
Convergence Speed [7, 16]	Faster convergence because local model changes are combined at the central server.	Slower convergence speed as a result of the central server's central device's connection delay.	The increased pace of convergence as a result of knowledge extraction from a bigger, more centralized model.
Scalability [14]	Able to scale to big datasets and a lot of edge devices.	Flexible enough for a reasonable amount of edge devices.	Flexible enough for a reasonable amount of edge devices.
Robustness [6]	Due to the use of weighted averaging, the method is robust to non-IID data and outliers.	Less resilient to stragglers but more robust to non-IID data.	Less resistant to outliers and non-IID data.
Algorithmic Complexity [17, 14]	Simple and low-complexity algorithmic structure.	Moderate complexity, more complicated algorithmic framework.	A more complicated algorithmic framework with intricacy ranging from moderate to high.

3. Smart cities: While protecting user privacy, federated learning can be used to build machine learning models on distributed data from sensors, cameras, and other IoT devices. This strategy can enhance public safety, traffic control, urban planning, and environmental tracking.

4. Federated learning for natural language processing: Federated learning can be used to train natural language processing models on distributed input from various languages and dialects. This method can increase the precision and effectiveness of sentiment analysis, machine translation, and other language-related activities.

5. Federated learning for image recognition: Machine learning models can be trained on distributed image data, such as photos or videos, using federated learning while maintaining individual anonymity. This method can increase the precision of visual activities such as object detection, image recognition, and others.

6. Federated learning for recommendation systems: With the protection of user privacy, federated learning can be used to develop machine learning models on distributed user behavior data. This method can enhance suggestion systems' precision and personalization in e-commerce, social media, and other areas.

These are only a few of the numerous possible uses for federated learning. We can anticipate more creative use cases in a range of sectors and domains as this technology develops.

8. Security in Federated Learning

Edge devices, such as smartphones, sensors, and Internet of Things (IoT) devices, can cooperatively train a machine learning model without sharing raw data thanks to the distributed machine learning method known as edge federated learning. Edge federated learning is advantageous in that it protects user anonymity, lowers communication costs, and allows edge devices to learn from one another [10], but it also presents several security issues that must be resolved, such as:

1. Malicious attacks: It's possible for malicious assaults like data poisoning, model poisoning, and Byzantine attacks to target edge devices. The integrity, accessibility, and confidentiality of the model and the data may be jeopardised by these assaults.

2. Data privacy: Edge devices may keep private information like health records, location data, and personal information. Federated learning necessitates sharing aggregated data and model updates between devices, which could compromise user privacy if not properly secured.

3. Model privacy: Sensitive information, including trade secrets, intellectual property, and proprietary algorithms, may be present in the model parameters and updates sent between edge devices and the central computer. To avoid theft, tampering, or unauthorised entry, these must be secured.

4. Authentication and access control: To prohibit unauthorised access to edge devices, the main server, and the model, federated learning needs a strong authentication and access control mechanism.

5. Communication security: To prevent data eavesdropping, data interception, and data tampering, federated learning needs secure communication channels between edge devices and the central server.

Diverse methodologies and frameworks, such as secure multi-party computation, homomorphic encryption, differential privacy, and blockchain-based solutions, are being created to address these security issues. These methods and tools seek to provide edge federated learning that is safe, private, and maintains the advantages of collaborative learning on distributed data.

9. Future Direction of Federated Machine Learning

A potential method for enabling collaborative machine learning on distributed data without sacrificing user privacy is federated learning. Future directions for federated learning could go in the following ways:

1. Privacy-enhancing technologies: To increase the security and privacy of federated learning, research is being done to create privacy-enhancing tools. For instance, federated learning on sensitive data can be made possible while maintaining privacy using methods like secure multi-party computation and homomorphic encryption.

2. Federated learning for edge computing: To allow collaborative machine learning on edge devices, federated learning can be expanded to edge computing. This strategy can enhance the scalability of federated learning, decrease latency, and increase productivity.

3. Federated learning for the Internet of Things (IoT): IoT devices may be able to learn from one another cooperatively through federated learning without sending their raw data to the cloud. This strategy can decrease network bandwidth usage, cut down on communication overhead, and enhance anonymity.

4. Federated learning for personalized machine learning: By enabling edge devices to learn from their data while utilising the knowledge of other devices in the network, federated learning can facilitate personalised machine learning. This strategy can increase the privacy-protected accuracy of machine learning algorithms for specific users.

5. Federated learning for autonomous systems: Federated learning can be used to allow cooperative machine learning on distributed data in autonomous systems, such as self-driving vehicles, drones, and robots. This strategy can increase the autonomous systems' dependability, efficiency, and safety.

References

[1] Zhang, C., Xie, Y., Bai, H., Yu, B., Li, W. and Gao, Y. 2021. A survey on federated learning. Knowledge-Based Systems, 216: 106775.

[2] Li, T., Sahu, A.K., Talwalkar, A. and Smith, V. 2020. Federated learning: Challenges, methods, and future directions. IEEE Signal Processing Magazine, 37(3): 50–60.

[3] Li, L., Fan, Y., Tse, M. and Lin, K.Y. 2020. A review of applications in federated learning. Computers & Industrial Engineering, 149: 106854.

[4] Yang, Q., Liu, Y., Chen, T. and Tong, Y. 2019. Federated machine learning: Concept and applications. ACM Transactions on Intelligent Systems and Technology (TIST), 10(2): 1–19.

[5] Yang, Q., Liu, Y., Cheng, Y., Kang, Y., Chen, T. and Yu, H. 2019. Federated learning. Synthesis Lectures on Artificial Intelligence and Machine Learning, 13(3): 1–207.

[6] Sánchez, P.M.S., Celdrán, A.H., Beltrán, E.T.M., Demeter, D., Bovet, G., Pérez, G.M. and Stiller, B. 2022. Analyzing the robustness of decentralized horizontal and vertical federated learning architectures in a non-IID scenario. arXiv preprint arXiv:2210.11061.

[7] Hegedűs, I., Danner, G. and Jelasity, M. 2021. Decentralized learning works: An empirical comparison of gossip learning and federated learning. Journal of Parallel and Distributed Computing, 148: 109–124. 20.

[8] Ye, Y., Li, S., Liu, F., Tang, Y. and Hu, W. 2020. EdgeFed: Optimized federated learning based on edge computing. IEEE Access, 8: 209191–209198.

[9] Xia, Q., Ye, W., Tao, Z., Wu, J. and Li, Q. 2021. A survey of federated learning for edge computing: Research problems and solutions. High-Confidence Computing, 1(1): 100008.

[10] Yang, J., Zheng, J., Zhang, Z., Chen, Q.I., Wong, D.S. and Li, Y. 2022. Security of federated learning for cloud-edge intelligence collaborative computing. International Journal of Intelligent Systems, 37(11): 9290–9308.

[11] Abreha, H.G., Hayajneh, M. and Serhani, M.A. 2022. Federated learning in edge computing: A systematic survey. Sensors, 22(2): 450.

[12] Zeng, R., Zeng, C., Wang, X., Li, B. and Chu, X. 2021. A comprehensive survey of an incentive mechanism for federated learning. arXiv preprint arXiv:2106.15406.

[13] Liu, J., Huang, J., Zhou, Y., Li, X., Ji, S., Xiong, H. and Dou, D. 2022. From distributed machine learning to federated learning: A survey. Knowledge and Information Systems, 64(4): 885–917.

[14] Bonawitz, K., Eichner, H., Grieskamp, W., Huba, D., Ingerman, A., Ivanov, V., Kiddon, C., Konený, J., Mazzocchi, S., McMahan, B., Van Overveldt, T. and Roselander, J. 2019. Towards federated learning at scale: System design. Proceedings of Machine Learning and Systems, 1: 374–388.

[15] Zhang, C., Xie, Y., Bai, H., Yu, B., Li, W. and Gao, Y. 2021. A survey on federated learning. Knowledge-Based Systems, 216: 106775.

[16] Niknam, S., Dhillon, H.S. and Reed, J.H. 2020. Federated learning for wireless communications: Motivation, opportunities, and challenges. IEEE Communications Magazine, 58(6): 46–51.

[17] Połap, D. and Woźniak, M. 2021. Meta-heuristic as manager in federated learning approaches for image processing purposes. Applied Soft Computing, 113: 107872.

[18] Aledhari, M., Razzak, R., Parizi, R.M. and Saeed, F. 2020. Federated learning: A survey on enabling technologies, protocols, and applications. IEEE Access, 8: 140699–140725.

[19] Alazab, M., Swarna Priya, R.M., Parimala, M., Maddikunta, P.K.R., Gadekallu, T.R. and Pham, Q.V. 2021. Federated learning for cybersecurity: Concepts, challenges, and future directions. IEEE Transactions on Industrial Informatics, 18(5): 3501–3509.

[20] Zhang, T. and Mao, S. 2022. An introduction to the federated learning standard. GetMobile: Mobile Computing and Communications, 25(3): 18–22.

[21] Posner, J., Tseng, L., Aloqaily, M. and Jararweh, Y. 2021. Federated learning in vehicular networks: Opportunities and solutions. IEEE Network, 35(2): 152–159.

Chapter 4

Secure and Private Federated Learning through Encrypted Parameter Aggregation

K Vijayalakshmi,[1,*] *PM Sitharselvam,*[2] *I Thamarai,*[3]
J Ashok,[4] *Goski Sathish*[5] *and S Mayakannan*[6]

1. Introduction

ML/DL has demonstrated promising results in a variety of application domains, especially when vast volumes of data are collected in one location, such as a data center or a cloud service [1]. But if data is collected

---

[1] Assistant Professor, Department of Electrical and Communication Engineering, College of Engineering, National University of Science and Technology, Sultanate of Oman, Muscat.
[2] Dean, RVS Educational Trust's Group of Institutions, Dindigul, Tamilnadu.
[3] Associate Professor, Department of Computer Science and Engineering, Panimalar Engineering College, Chennai City Campus, Chennai, Tamil Nadu, 600029.
[4] Professor, School of Business and Management, CHRIST (Deemed to be University) Bengaluru, Karnataka.
[5] Assistant Professor, Department of Information Technology, St. Martin's Engineering College Secunderabad Telangana India.
[6] Assistant Professor, Department of Mechanical Engineering, Vidyaa Vikas College of Engineering and Technology, Tiruchengode, Namakkal, India.
Emails: sitharselvam@gmail.com; thamarai.panimalar@gmail.com; ashok.j@christuniversity.in; gsathishit@smec.ac.in; kannanarchieves@gmail.com
* Corresponding author: vijayalakshmi@nu.edu.om

and stored centrally by a (third-party) cloud service, companies (cloud service clients) may be exposed to significant legal liability in the event of a data breach [2, 3]. This is especially important to remember when dealing with information about people's health, conversations, footage from private homes, money transfers, and other such matters [4]. After being uploaded, data "loss of control" is a common problem with centralized data collection. Is the data being utilized by the cloud service in the manner in which it was advertised? a frequent question with a frustrating response. Do my files really get erased when it says it does? Because of government regulations, even businesses that were not persuaded by the hazards of privacy breaches and loss of control have limited the amount of data they share with outside providers (such as HIPAA and GDPR).

The goal of FL is to improve the quality of ML/DL models while minimizing their drawbacks. Participating devices in an FL task could range in size from a single smartphone or watch to a global corporation housing multiple data centers [5, 6]. FL is necessary when training a single ML/DL model on data from multiple "local" entities. This procedure develops a set of "local models," which are then integrated with minimal data loss by simply swapping over a few parameters (e.g., models' weights in neural networks) [7]. There are two main types of FL algorithms: centralized and decentralized. Similarly, a coordinator is needed to compile information from all the many regional models into a unified global one (overlay multicast, broadcast, etc.) [8, 9].

It was originally believed that just a little amount of information about the original training data would be carried over into subsequent model updates as FL interactions occurred [10]. In this light, it was decided that the dissemination of model updates was "privacy preserving." In any case, the updates to the models still contain traces of the training data, even if this is not immediately apparent. The privacy assurances of FL have been put into doubt by recent studies that show how simple it is to infer private features and reconstruct huge chunks of training data by exploiting model changes. This holds even when dealing with trustworthy but inquisitive aggregation servers.

2. Assumptions and Focus

Two common federated learning deployments are device-agnostic and silo-agnostic [11]. While there are many participants in the cross-device scenario (over a thousand), each one has just a few data points, limited computing power, and a finite supply of available energy (e.g., smartphones and Internet-of-Things devices) [12, 13]. They are really unpredictable, always wandering off and then back to the group. For example, a large firm may benefit from the knowledge saved on employees' personal devices, and a device manufacturer could use the data kept privately on millions of its own devices to train a model [14]. When training across several devices, it's helpful to have a central point of contact who can collect data and oversee logistics [15]. In contrast, in the cross-silo scenario, only a few are involved, but they all have access to vast quantities of data and powerful computational resources (such as constant power or hardware ML accelerators) [16, 17]. Each participant is present and active for the duration of the FL training process, yet sensitive information is more likely to be compromised [18]. Some applications of this principle include the training of tumor detection models on radiographs by multiple hospitals, the training of credit card fraud detection models by different banks, and so on. There is no obvious go-to reliable source in situations involving multiple silos [19]. There is no hierarchy between the trainers and trainees. Public cloud hosting or infrastructure provisioning by one of the parties is a common component of these installations. This chapter is dedicated to cross-silo private parameter aggregation [20].

Authors presume that everyone involved is on board with the idea that federated learning is a good thing because it increases accuracy and robustness [21, 22]. The authors point out that it is an unsolved research problem to persuade people to work together by showing them hypothetical improvements in performance that could result from federated learning. At the centre of our attention is the "honest but curious" model of trust [23]. To gain insight into the data of others, however, the participant may be motivated to reproduce the model parameters. Regarding participants' personal information, authors presume that the coordinator is not only trustworthy, but also really interested. There should be as little reliance on the coordinator as possible, and the participants want to make that happen. Authors also suppose that no one is intentionally trying to introduce spurious weights that would distort the overall model's results. This trust model is simplified when its members are prevented from sharing information or adjusting each other's model parameters due to legal or regulatory constraints (for example EU data protection guidelines, FedRAMP).

In other words, the world's largest financial institution BankA has multiple locations across the globe (BankA India, BankA USA, BankA UK, etc.), but it is unable to use data from any of its subsidiaries to train a fraud detection model [24–26]. One way in which this paradigm of trust is used frequently is in federated learning in the workplace. Regional data centers cater to the needs of each subsidiary, while global data centers may house coordinators. The hospitals don't trust each other and don't want to provide data to a centralised service, yet they still want to work together to train a tumour detection model [27]. It's also possible that several (competing) corporate clients of a cloud-hosted ML service (such as Azure ML) may not trust one another, but have faith in the cloud service's capacity to safely and securely allow FL.

3. Private Federated Learning

The differential privacy framework is concerned with restricting the release of private information while sharing the outcomes of computations or queries performed on a dataset [28–30]. To be considered differentially private, a computation must ensure that it is impossible for an outside observer to tell which individual's data was used in the calculation just by looking at the results it produced on the dataset [31]. Differential privacy is commonly understood to be the mathematical characterization of the privacy loss incurred due to the disclosure of any dataset derivative.

Recently, many researchers have begun to employ differential privacy while training models in a federated setting [32–35]. In an FL environment, it is normal practice to add noise to every model/data derivative before making it public. The participant's need for anonymity will determine how much random noise should be included. The trained model's accuracy is harmed by the introduction of this noise. Adding a little bit of noise and carefully selecting the training hyperparameters has been shown to have a negligible effect on accuracy, but there has been no systematic research on how exactly the noise level affects model convergence [36]. Determining the feasible level of privacy without compromising too much precision may need a lengthy empirical approach. In this subsection, the authors provide a more in-depth description of this issue and supporting data [37].

3.1 Background of Differential Privacy (DP)

Recently, there has been a lot of attention paid to the differential privacy literature in the field of computer science. Due to space limitations, the authors will only briefly discuss the differential privacy (DP) structure in this chapter. DP is described in greater depth in [38–40]. According to the literature on differential privacy, a function f (·) is ε DP if and only if it returns the same value for all t's when applied to a pair of data X and X' that vary by exactly one.

$$\left| \ln \frac{P(f(X)=t)}{P(f(X')=t)} \right| \leq \varepsilon \tag{1}$$

Lower values for the parameter ϵ indicate less potential for a breach of privacy. In this study, the authors employ a practical generalization of the DP concept introduced in [41], which the authors denote for the function $f(\cdot)$ as (ϵ,δ) DP:

$$P(f(X) = t) \le e^\epsilon \, P(f(X') = t) + \delta \qquad (2)$$

By inspection, it appears that $f(\cdot)$ has differential privacy equal to $1-\delta$, as suggested by this definition. To achieve DP, the authors adjust the result of $f(\cdot)$ by adding a noise term whose variance depends on the parameters ϵ and δ. In addition, it has been shown that adding a zero-mean, zero-variance Gaussian noise term

$$\sigma^2 = \Delta f \cdot \frac{2\ln \dfrac{1.25}{\delta}}{\epsilon} \qquad (3)$$

guarantees (ϵ, δ)-DP [42], with f denoting the level of confidentiality required for $f(\cdot)$. The sensitivity of a computation is the amount by which it can shift when a single variable in the underlying dataset is altered. A formal expression for the sensitivity Δf is

$$\Delta f = \sup_{(X, X')}(\|f(X) - f(X')\|_\ell) \qquad (4)$$

Although the above definitions are limited to the DP of a single query, several findings of the data are actually performed throughout the training of a neural network. Therefore, authors must take into account the cumulative privacy loss throughout the training process, rather than relying on a single value of ε. When many (ε,δ)-DP are used, the total amount of privacy lost is equal to the product of those used, as proven by the Theorem of the Privacy Budget (β). To train for T iterations where ε_0 remains constant, authors require a privacy budget of $\beta = \Sigma_{i=1}^{T} \epsilon_0 = T\epsilon_0$.

One need not be so naive as to spend all their privacy money in this way. Several authors have pondered numerous ways to spend privacy budgets to make a variety of advancements. Using a generalization of differential privacy, the work [43] achieves tighter restrictions on privacy loss per query, whereas it reduces the number of searches, and therefore the overall privacy loss.

3.2 *Insertion of Differential Privacy and Stochastic Gradient Descent*

To train machine learning algorithms, particularly deep learning models, SGD is a common optimization approach. For distributed SGD to work, the mini-data batch must be divided into smaller pieces and transmitted to each participant, who is then tasked with calculating the gradient of the model. The combined gradients are then sent to a centralized parameter server. The parameters model is adjusted once the parameter server finds a global average of the learners' gradients.

In an FL setting, users often do gradient calculations of the model parameters on their own local dataset before submitting the results to a centralized server. To stop data from leaking out through the shared gradients, the SGD algorithm incorporates methods from the DP field. For more information on the DP, SGD that arises from this tweak, is as given below.

Our next topic is the application of (ϵ, δ)-DP to federated learning based on Stochastic gradient descent. Using $f(\cdot)$ on a local dataset which is equivalent to computing the gradient models.

Mini-batch SGD's update rule, given a batch size of S and a learning rate of η, is

$$\theta_{k+1} = \theta_k - \eta \frac{1}{S} \sum_{i=1}^{S} g(x_i) \tag{5}$$

X is a set of data points, and g(xi) represents the gradient of the loss function at the ith point. During gradient exchange, the following quantity is communicated: $f(x) = \eta \frac{1}{S} \sum_{i=1}^{S} g(x_i)$. In the above formulation, the sensitivity would be determined by the gradient of the single data point that is unique between the two datasets X and X'.

$$\Delta f = \left\| f(X) - f(X') \right\|_\ell$$

$$= \frac{\eta}{S} \left\| \sum_{i=1}^{S} (g(x_i) - g(x_i')) \right\|_\ell$$

$$= \frac{\eta}{S} \left\| g(x_j) - g(x_j') \right\|_\ell$$

$$\leq \frac{\eta}{S} \cdot 2C$$

where C is a constant that sets a maximum value for the gradient. Therefore, adding a Gaussian noise factor with variance $2C \frac{\eta}{S} \frac{\ln \frac{1.25}{\epsilon}}{\epsilon}$ to the gradients makes them (ϵ, δ) DP. Herein, authors detail the abridged version of the method for differentially private SGD that the authors use in practice.

Algorithm 1 (ϵ, δ)-DP SGD

Inputs: η = learning rate, C = clipping length, S = batch size, $T\epsilon_0$ = privacy budget, θ_0 = primary weights

for t = 0, 1, 2, 3,…, T do

calculate $f(X) = \Sigma_{i=1}^{S} g(x_i)$

gradient clip: $f(X) \leftarrow \dfrac{f(X)}{\max\left(1, \dfrac{\|f(X)\|}{C}\right)}$

sample Z_k distribution $N\left(0, 2C\dfrac{\eta}{S}\dfrac{\ln\frac{1.25}{\delta}}{\epsilon}\right)$

$$\theta_{t+1} \leftarrow \theta_t - \eta f(X) + Z_k$$

end for

3.3 Discussions

Let's dive into the nitty-gritty of the difficulties of implementing DP in federated education by discussing several relevant trials. The CIFAR-10 dataset may be used to train a Resnet-18 model whereas the SVHN dataset could be used to build a Resnet-50 model. All models are trained for 200 iterations using both the original SGD and the DP variation. It was intended that the first 80 epochs would be trained at a learning rate of 0.1, the next 40 at a rate of 0.01, and the last 80 at a rate of 0.001.

3.3.1 Accuracy vs ϵ

The resulting convergence charts for varying batch sizes and ϵ are displayed in Figs. 1 and 2, respectively. The first thing to notice is that when authors reduce the value of ϵ, the models converge to less precision. Training methods with less noise generated approach closer to the accuracy of non-private training, as illustrated in Fig. 1 for a certain batch size, say 1024. As more noise is introduced, the accuracy quickly decreases for $\epsilon = 0.01$ and below (equal ϵ to 0.05). This pattern holds across all models for each given batch size, as displayed in both Figs. 1 and 2.

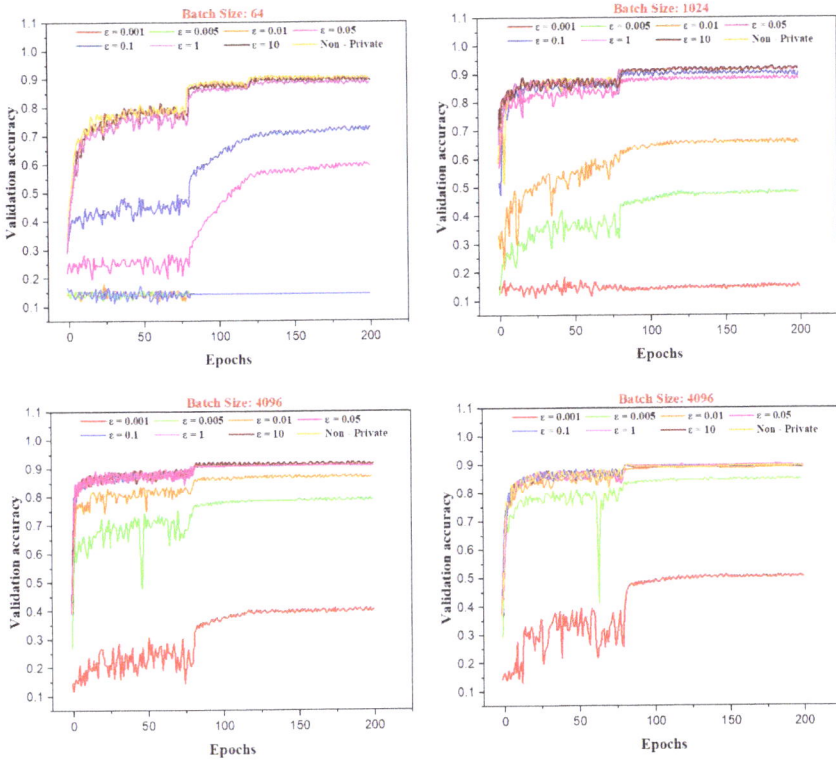

Figure 1. Comparison of Resnet18's validation accuracy on CIFAR10 across batch sizes (64, 1024, 4096, and 8192) and ε (from 10 down to 0.001). Smaller ϵ indicates more noise privacy.

3.3.2 Batch Size Versus Accuracy

In Figs. 1 and 2, authors can see how the model's accuracy varies widely depending on the batch size S used for private training, while it remains reasonably consistent throughout a wide range of batch sizes for the public versions ($\varepsilon = 0$). To make private models perform as well as public ones, even at small values of ϵ, we've found that raising the batch size is sufficient. Specifically, at $\epsilon = 0.005$ and 0.01 (orange and green plots in Fig. 1, seen left to right), CIFAR10+Resnet18 shows this behavior, although this holds true for all values of ϵ.

When the batch size is increased from 1024 to 8192, the final accuracy of CIFAR10+Resnet18 improves by around 37% and 78.5% for more private settings corresponding to $\varepsilon = 0.01$ and 0.005, respectively. Figure 2 displays analogous tendencies. Although differential privacy shows promise, it appears to call very precise hyperparameter optimization

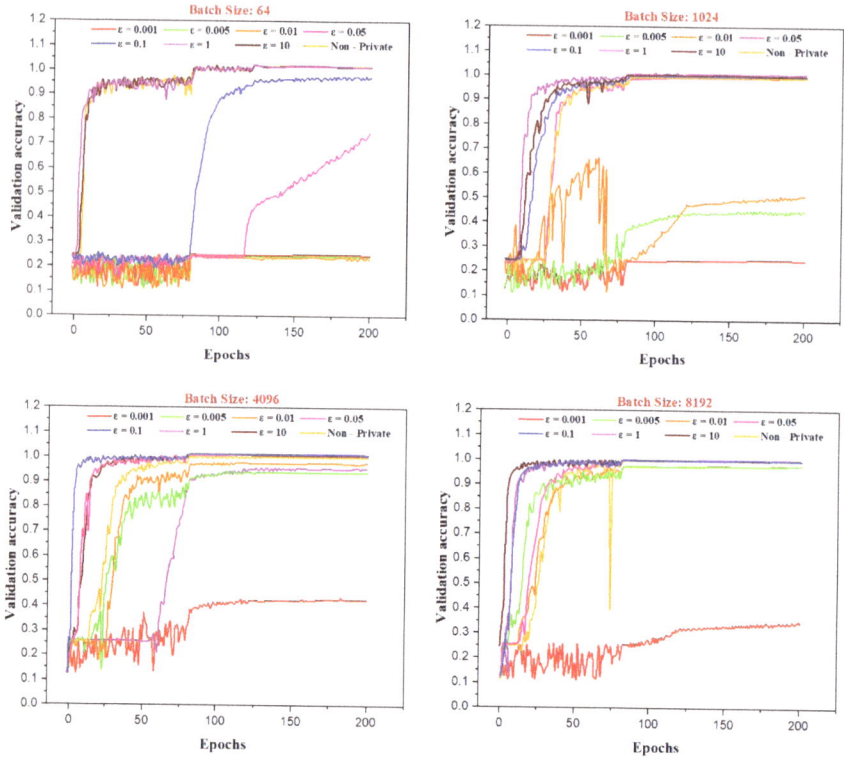

Figure 2. Resnet18 per-epoch validation accuracy on CIFAR10 at 64, 1024, 4096, and 8192-batch sizes and ε. Smaller ϵ indicates more noise added and privacy.

to ensure optimal performance. To do this, it may be necessary to launch numerous FL jobs, each with its own set of hyperparameters.

4. Additive Homomorphic Encryption

Researchers have either (1) employed novel applications of cryptography (safe multi-party computing with homomorphic encryption) or (2) altered model parameters or gradients by introducing statistical noise. Both can be combined in certain procedures.

Homomorphic encryption enables computation to be conducted on ciphertexts, with the encrypted output matching the plaintext's value after decryption. Fully homomorphic encryption is effective, but it is time-consuming and resource-intensive due to the length of the ciphertext. On the other hand, with federated gradient descent, averaging gradient vectors is as easy as summarizing their components (Before or after the encrypted aggregation method, the total number of participants might be subtracted.). Federated training gradients can be encrypted with additive homomorphic encryption, such as the Paillier cryptosystem. When it comes to public-key cryptography, the Paillier cryptosystem provides an additional asymmetric option. By combining the encrypted versions of m_1 and m_2, along with the public key, authors may arrive at the encrypted version of the sum. Homomorphic encryption's definition provides a beautiful illustration of the difficulty of using it for private gradient aggregation. All participants must use the same public key to encrypt their gradients and parameters. A reliable source of keys and their distribution is required. However, it may be possible to decrypt the data if all players communicate with this key generator/distributor and send their Paillier-encrypted gradients to it. Therefore, aggregation should take place independently of the key generator/distributor wherever possible. In addition, the key generator/distributor must be trusted implicitly not to disclose the private key to any third parties. There are a variety of designs that could work within these parameters. Finally, evaluate our method against other approaches to differential privacy and secure multi-party (SMC) computation.

It's worth noting that Mystiko is not the only system that uses Paillier cryptography to safeguard collected data. The collaborators encrypt the gradient vectors using a Paillier key pair and send them to a coordinator they don't trust, until when it's time to combine the weights. The collected gradient vectors can then be analysed by the participants. To produce the Paillier keys collectively, participants need to have high levels of trust in both themselves and the untrusted coordinator so that they may decode their individual gradient vectors independently. The leak of the Paillier keys by a single unreliable actor can compromise the entire system and compromise users' privacy.

The goal of the area of cryptography known as secure multi-party computing (SMC) is to find ways for several parties to safely and independently compute the same function over their respective inputs. In this paradigm, cryptography is used to safeguard the privacy of the participants from each other, as opposed to the conventional cryptographic professions where it is employed to preserve the security and integrity of communication and storage and the adversary is external to the system of participants (an eavesdropper on the sender and receiver). Traditional SMC methods can be improved with SPDZ and its derivatives. Promising features of such protocols include their invulnerability to attacks from small groups of colluding peers, their ability to scale to a high number of peers, and their insensitivity to changes in the ultimate correctness of the trained model. One potential issue is that SMC methods are computationally expensive.

4.1 Learners and Administrative Domains, Participants

A natural place to begin is to define FL algorithms in terms of administrative domains. A group of computers can be thought of as an administrative domain (Servers, virtual machines, computers, laptops, etc.). Within an administrative domain, all parties can put their trust in one another, malicious intent is not a factor, and there are no legal or regulatory roadblocks to the free flow of data and information produced from data. It's important to keep in mind that a company or non-profit does not have to be an administrative domain. It could be an internal company project dealing with sensitive information, or it could be an externally hosted service linked to a public cloud account. Multiple spheres of control might exist within a single organization. Each node in a federated learning algorithm represents a separate regulatory sphere. During the process of training a neural network, the computational learning process itself (often running on a GPU) is referred to as a Learner. Each learner in the participant processes (a subset of) the data to compute the gradient vector.

4.2 Architecture

In a federated learning system that prioritizes participant privacy, the aggregation of model parameters or gradients is conditional on both the encrypting technique participants employ and the communication channel players take to reach the system's coordinator. Using a shared Paillier public encryption key, all MYSTIKO users encrypt their data before combining their encrypted gradient vectors to generate a single encrypted result. As a result, confidentiality is maintained because only the participants can access their information. The issue today is (1) keeping everyone's private key secure while sharing a shared Paillier public key. secondly, how to safeguard against the decryption of specific weights. What follows is an explanation of each of these.

For the sake of brevity, let's assume that each participant is a single student. Usually, MYSTIKO is used as a cloud service that acts as a go-between for several different people. For this purpose, authors need a Job Manager, a Membership Manager, a Key Generator, and a Decryptor. Additionally, to register users for MYSTIKO, the Membership Manager also keeps tabs on who is taking part in each FL activity. The FL job's Job Manager is responsible for fixing bugs, updating membership, finding missing people, and deciding on optimal values for the job's hyperparameters. This chapter focuses on assaults on data privacy within a federation, but standard communication security is also necessary to thwart attacks on a federation from the outside. To do this, MYSTIKO and the students (participants) will share a PKI. Protecting the privacy of data sent between the MYSTIKO components and the students is one of the many ways in which the PKI helps the Paillier infrastructure get off the ground. To establish a chain of trust between the students and MYSTIKO, the PKI supplies CAs, as well as intermediate and Root CAs. Using the public key infrastructure (PKI), MYSTIKO establishes an encrypted TLS channel for sending and receiving control data. Standard but potent (non-homomorphic) cryptographic methods (such as Key agreement/ exchange/authentication using RSA), message confidentiality using AES, and message authentication using SHA are utilized to build the TLS channel. After a learner is registered, ranked, given a Paillier public key, and the decrypted aggregated gradient vector is communicated during topology formation in MYSTIKO, all of these operations take occur via a TLS channel.

4.3 MYSTIKO Algorithms

Authors begin with the simplest ring-based method and work our way up to the most complex algorithms available today, including the MYSTIKO family of algorithms, which combines parallelism and resilience via broadcast.

4.3.1 Algorithm Based on Ring

Figures 3 and 4 depict an example of the fundamental aggregation method based on a ring. Each of the P people involved in this algorithm is a learner with their administrative domain (L). At the outset of the procedure, each player must sign up for a MYSTIKO account. MYSTIKO coordinates everything that happens. Since MYSTIKO generates secure encryption pairs of keys, safeguards secret keys, and adheres to protocol, it can be relied on by students.

Once all expected students have joined MYSTIKO, the Membership Manager will initiate the FL procedure. Beginning with ring topology, the authors organize the classes (Fig. 3). There are a few straightforward approaches to this, such as (1) by location, to minimize players' actual physical distance from one another, (2) by employing a name-based hierarchy (either ascending or descending), and (3) by consistently hashing [24] the names and identities of the participants. The students are then ranked (1–P) based on where they sit in the ring they have been assigned to.

Every FL task in MYSTIKO has its own private and public key to the Paillier Key Generator. A Paillier key pair is unique to each FL task and is

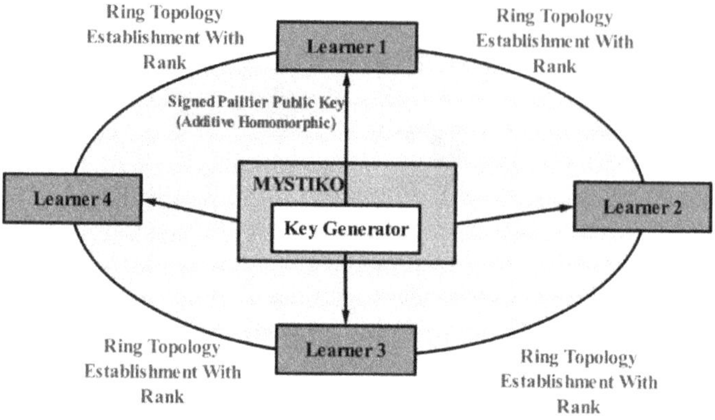

Figure 3. Establishing a topology and key distribution.

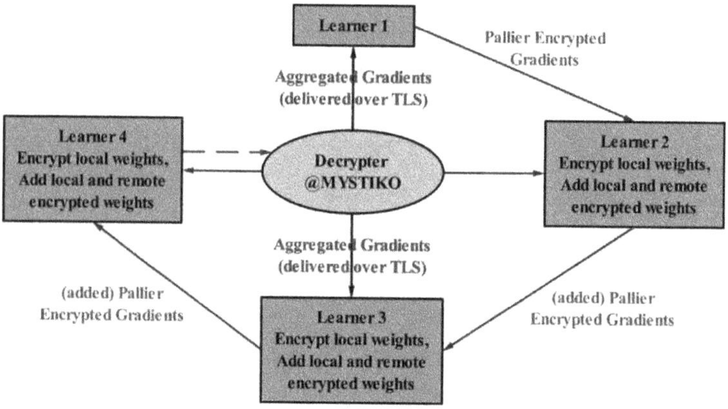

Figure 4. Standard aggregation method using a ring network topology.

securely disseminated (through TLS) to all participants. If the operation is expected to run for a while, a new key pair can be generated at regular intervals, such as every h minutes or every epoch (this is configurable). The aggregated gradient vector encrypted with Paillier is received by MYSTIKO's Decryptor and decrypted before being sent on to the learners. Each node in the ring of learners utilizes the Paillier public key to decode a shared gradient vector, encrypt its gradient vector, and then combine the two encrypted vectors. This last ring learner receives the aggregated, Paillier-encrypted gradient vector. Next, the final ring learner sends MYSTIKO the aggregated encrypted gradient vector. Before transmitting information via TLS to the students, the MYSTIKO Decryptor encrypts the combined vector.

Security Analysis

Information on a student, and by extension, any educational institution, is never given out to outside parties. Data privacy is ensured provided that robust anti-intrusion mechanisms are in place on every server within the administrative domain. In this system, gradient vectors are never transmitted outside of the learner in an unencrypted form. The learner only releases the aggregated Paillier encrypted gradient vectors if P is less than 1. MYSTIKO's private key is required for decryption. Through Paillier-encrypted transmission, the second ring student receives the unaggregated gradient vector from the first ring student, which the first ring student is unable to decrypt. Once aggregated, the complete gradient vector is hidden from view. To ensure that the aggregated gradient vector reaches all of the participants safely, it is encrypted before being sent over TLS. When multiple gradient vectors are averaged from large datasets, it becomes extremely difficult to achieve the goals of reverse engineering attacks such as those described, which are to determine whether or not a given participant has access to a particular data record, or to identify data items that cause a given change in gradient vectors. To protect the secrecy of the gradients before they are averaged, they are encrypted first. Due to this, MYSTIKO can only see the overall results and not the data or gradients of individual pupils.

Colluding Participants

The fundamental ring-based algorithm can't be cheated on by more than P–2 people. That is, if MYSTIKO is used in an uncorrupted and honest manner, it can be cracked only if P–1 pupils collaborate. Learners L1, L2..., Li–1, Li+1..., and LP must collaborate to decode gradients from learner Li by passing along encrypted gradient vectors they get from the preceding learner without appending their gradients.

Fault Tolerance

A ring-based aggregation approach has the potential drawback of shattered rings; for optimal performance, it is necessary to maintain robust connectivity between each learner and MYSTIKO. There is the option of using tried-and-true methods of failure detection such as monitoring for irregular heart rates or calculating average round-trip timings. Every iteration in distributed synchronous gradient descent averages the gradients over the previous iteration. If a learner's failure is discovered, the MYSTIKO's Membership Manager will either remove the learner from the ring or wait for the student to reconnect before continuing with the averaging of gradient vectors. When the MYSTIKO connection is briefly interrupted, it is possible to pause the gradient averaging process.

4.3.2 Broadcast Algorithm

The difficulty in setting up and maintaining the ring topology is a major downside of the ring-based approach. The se of group membership and broadcast is one solution to this problem. Except for the fact that the topology is now established, nothing else has changed in the situation. The MYSTIKO Membership Manager is where students sign up, agree on a shared PKI, and learn the names and numbers of all the other people who will be using the system. The public key for each Paillier public-private key pair connected with a federated task in MYSTIKO is securely communicated to all participating students.

The Paillier learning algorithm encrypts each learner's gradient vector before sending it to other learners. To decrypt the encrypted vectors sent to them by P–1 other learners, a learner submits the Paillier encrypted sum to the MYSTIKO. Decrypted copies of the combined gradient vector are sent to each student via TLS safely and securely manner. Since each student calculates the sum, the broadcast approach is inefficient and unnecessary. With redundancy, though, the authors may tolerate more setbacks before having to start over. Unlike broadcast, where all aggregated gradients would still be available if a single node failed, the ring would suffer a partial loss in the event of a node failure.

Colluding Participants

A given LA's plaintext gradient vector represents the algorithm's intended weak spot. It takes P–1 cooperation to break this algorithm, making it highly collusion-resistant. Similarities in the broadcasted Paillier ciphertexts would indicate cooperation if P–1 students were secretly Paillier-encrypting 0 vectors in place of their true gradient vectors. In fact, it is suspicious if even two learners produce identical Paillier encrypted gradient vectors, as their data would likely be different.

The ring-based aggregation method is described and implemented in parallel with Ring-based All-Reduce. The diagrammatic representation of this is shown in Fig. 5. Authors introduce the ring protocol, but it has a major flaw in that each student must wait for the preceding one to finish the cycle before they can continue. All-Reduce gets around the issue by slicing the gradient vector into P pieces, where P is the total number of subjects. The Paillier-encrypted pieces are then given to the students, who reassemble them independently. In Fig. 5, for instance, there are three learners, so each gradient vector consists of three subcomponents. Without having to wait for Learner-1 to finish providing its vector, Learner-2 can proceed. Instead, Learner-1 sends the first chunk to Learner-3 and then sends its second chunk to Learner-3 at the same time. Following Step 1,

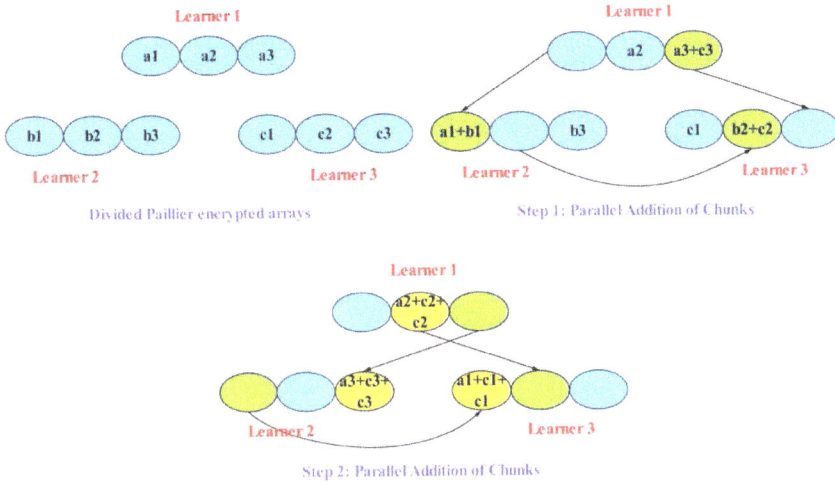

Figure 5. Encrypted MYSTIKO rings using all-reduce over paillier.

Learner-2 will transmit the first partially aggregated chunk to Learner-3, who will then transmit it to Learner-1. When Step 2 is complete, each student will have Paillier-encrypted, aggregated chunks to send to MYSTIKO's Decryptor, where they will be concatenated and decoded.

Security Analysis

It is important to highlight that All-Reduce is the fastest MYSTIKO procedure. When given P learners, All-Reduce can be thought of as a parallel implementation of the P1 rings. The first ring in Fig. 5 begins with Learner-1's first piece, the second with Learner-2's second chunk, and the third with Learner-3's third chunk. This means that All-Reduce offers the same security assurances as the simplest ring protocol.

4.4 Multi Learners Per Administrative Domain

For ease of analysis, researchers have assumed that each individual is a unique learning process. Multiple servers and training processes (learners) can share the same dataset, with the help of a gradient vector aggregator running in the same administrative domain. This is required due to several considerations, including the magnitude of the datasets and the relative simplicity of the analysis cost of computer resources, and the importance placed on shortening the training time. The protocols developed for MYSTIKO can be easily adapted to this scenario, with the only change being that the communication takes place between local aggregators (LAs) rather than students. Local aggregation is not safe

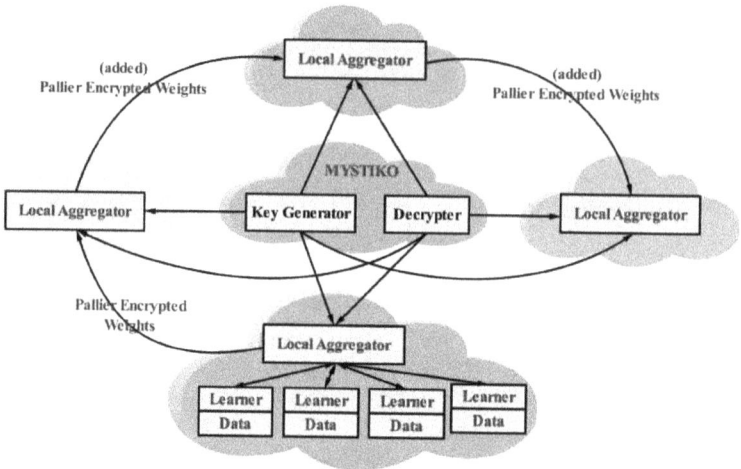

Figure 6. Federating gradient descent.

since trusted computing resources inside an administrative domain can transport even raw data, but remote aggregation is Paillier-encrypted and thus private. However, MYSTIKO protocols are used for aggregating data between LAs. Figure 6 depicts this idea visually.

5. Trusted Execution Environments

A primary processor's safe space is known as a trusted execution environment (TEE). Isolated execution, application integrity, and data asset confidentiality are just some of the many security characteristics that TEEs (trusted execution environments) offer. Through the creation of a separate execution environment, TEEs shield vital code and data against privileged software attacks on the native OS. Common examples of native OSes are Linux and Microsoft Windows. Systems that use TEE integrate hardware and software safeguards for confidential information, and examples include IBM Hyper protect, ARM Trust Zone, and Intel SGX. In a TEE, the server's CPU, I/O, and RAM are all reserved for programs that have been verified to be safe before they are loaded. The TEE's hardware isolation shields its contents from threats posed by other programs running on the host OS, such as viruses and malware. When there are numerous programs running within a TEE, they are often shielded from each other by means of software and cryptographic isolation.

The TEE uses something called a "hardware root of trust" to shield itself from being simulated by malicious or rogue server-side software. The trusted execution environment can have a one-time use set of private keys written into it during the manufacturing process using programmable memory. Even after a restart or a complete factory reset, authors will be unable to modify these settings. For the firmware and cryptography/access-control circuits to be signed by the trusted party, the manufacturer stores the public key equivalents of the trusted party and a non-secret hash (often the chip vendor). The TEE hardware and its special capabilities are only accessible to programs that have been digitally signed using the trusted party's key. The chip already contains the vendor's public key, which is hashed and compared at runtime. Using the public key, trusted vendor-controlled firmware (such as an Android device's chain of bootloaders or SGX's "architectural enclaves") can be verified digitally. The remote attestation is then implemented using the trusted firmware.

Using Trusted Execution Environments (TEEs), the authors explain how one system, TRUDA, provides security for the fusion of models. Other examples with several optimizations. Here, the authors focus solely on the fundamentals of TEE aggregation using TRUDA as an illustration. TRUDA employs AMD's Secure Encrypted Virtualization (SEV) to run each aggregator as an encrypted virtual machine (EVM). During model aggregation, all memory-resident data is kept encrypted at runtime. Using a two-step process and a suite of tools for combining and systematizing secret computing in Federated learning, expedites the process of building trust between parties and aggregators. Aggregator trustworthiness can be checked by any FL training participant at any time. After attestation, parties and EVMs are connected via end-to-end secure channels that encrypt all data between them, including any revisions to the models.

5.1 Trustworthy Aggregation

For FL aggregation, TRUDA mandates cryptographic isolation via SEV. Aggregators are run on EVMs. The memory of each EVM is encrypted with a unique, temporary VM Encryption Key (VEK). In this way, TRUDA can keep model aggregation secret from prying eyes like server administrators and privileged apps. The validity of SEV hardware and firmware can be confirmed with the use of AMD's provided attestation primitives. To quickly build confidence between parties and aggregators in a decentralized FL environment, TRUDA makes use of a novel attestation protocol built on top of the primitives. There are two parts to this FL attestation procedure:

Phase 1: Launching Trustworthy Aggregators

Launching SEV EVMs linked to an aggregator securely is the initial stage towards TRUDA. To ensure the safety of the EVM, it is vital to (1) check that the hardware platform is real AMD SEV-enabled hardware with the required security features, and (2) check that the Open Virtual Machine Firmware (OVMF) image used to launch the EVM is genuine. TRUDA will send the aggregator's secret along with the EVM after the remote attestation is complete. Until the Phase 2 aggregator authentication is finalized, the secret remains hidden in the EVM's encrypted physical memory. A valid AMD root certificate can be used to verify the authenticity of the entire chain of certificates. The attestation report contains a digest of the original OVMF image in addition to the full certificate chain.

After the report has been attested, it is transmitted to the attestation server, which has been set up with the necessary AMD root certificates. By validating the certificates in a chain, an attestation server can determine whether or not OVMF firmware has been tampered with and whether or not a certain hardware platform is genuine. Attestation servers create both the launch blob and the Guest Owner Diffie-Hellman Public Key (GODH) certification, which verifies the identity of the host machine. Soon after an aggregator's machine submits a DHKE request to an SP for a Transport Encryption Key (TEK) and a Transport Integrity Key (TIK), the EVMs begin processing data.

After halting the EVM upon launch, TRUDA can obtain the OVMF runtime measurement from the SP. For verification, this metric is transmitted to the host system alongside the SEV API version and the EVM deployment method. The attestation server produces an ECDSA prime251v1 key as part of the packaged secret to assure the precision of the measurements. In order to reveal the identity of a trusted aggregator, an EVM must be launched, at which point the hypervisor must inject the secret into the physical memory area of the EVM.

Phase 2: Aggregator Authentication

It is imperative for FL parties to only work with recognized aggregators that use safe runtime memory encryption. During the first stage of an EVM deployment, an ECDSA key is provided in secret by the attestation server to the aggregators so that they can authenticate. This signature on the challenge request verifies that the request is coming from a trusted aggregator. A challenge response protocol aggregator authentication is required before a participant may take part in FL. The sender of a communication will sometimes append a randomly generated nonce to the message. After signing the nonce with its own ECDSA key, the aggregator sends it back to the asking user. To guarantee the nonce was signed with the correct ECDSA key, authors verify its signature. After you've been verified, the next step in joining FL is to sign up with the aggregator. After each aggregator has had its turn, the process is carried out.

After signing up, the aggregators and the parties involved in the exchange of model updates can communicate via encrypted channels from beginning to end. TRUDA enables TLS to work with aggregators that require two-way authentication. As a result, all model changes are secure while in EVM and in transit.

In this chapter, the authors examine the three Mystiko algorithms in relation to SPDZ, a leading protocol for dependable multi-party computation, and strategies for differential privacy (DP) via the introduction of statistical noise. Researchers use many different neural network models and datasets of varying sizes for image processing: Authors start with a 1MB 5-layer CNN that was trained on the MNIST dataset. Here is a list of the four models available: Four models were used: (2) Resnet-18 (medium, 50MB) trained on the SVHN dataset (600K images of street digits), (3) Resnet-50 (medium, 110MB) based on the CIFAR-100 dataset (60K images of color, 100 classes), (4) VGG-16 (large, 600MB) trained on the Imagenet-1K dataset, and (5) AlexNet (small, 10MB) trained on the MNIST dataset (14.2 million images of 1000 classes).

Using a cluster of 40 computers, the authors did trials with groups of people from 2 to 40 to test and compare all the algorithms. No more than one worker was ever scheduled to operate on any one computer, and each computer included 8 Intel Xeon E5-4110 (2.10 GHz) processor cores, 64 GB of RAM, a single NVIDIA V100 graphics processing unit, and a 10 Gigabit Ethernet connection. All the machines were located in four different data centers, and in each and every test, the participants were spread out evenly throughout all four. The dataset was evenly and randomly distributed among the participants in each experiment. One data center provided a single machine for the execution of Mystiko. Moving forward, the authors will average together results from 10 separate trials to arrive at each new data point.

Comparing MYSTIKO and SPDZ

Each learner (or local aggregator) in federated learning undergoes individual training for a fixed number of iterations before sharing their training data and updating the model together. Systems like MYSTIKO and SPDZ become involved in privacy-threatening ways during the gradient aggregation stage. To accomplish, the authors will utilize the two criteria listed below to evaluate MYSTIKO and SPDZ: There are two types of time synchronization: Both seconds and milliseconds can be used to determine how well two devices are in sync with one another Two parts of the synchronization process contribute to the total: (1) the time needed to make confidential adjustments to the gradients (MYSTIKO's Paillier encryption, SPDZ's share generation, etc.) and (2) the time needed to broadcast the updated gradients to the members of the federation.

In Fig. 7, the authors compare the overall time required for federation-wide synchronization and communication versus the total number of federation members using a scatter plot of all of our model/dataset combinations. All-Reduce is the most scalable option as the user count grows (as seen in Fig. 7). Due to the protocol's inherent parallelism, each learner/LA need only continuously broadcast a subset of the gradient array, drastically reducing the total amount of data being delivered. Due to its sequential nature, the basic ring protocol is the least scalable network topology.

Broadcast's performance and scalability far outstrip those of the standard ring protocol because each node transmits independently of the others. Since with SPDZ, each participant's gradient vectors are partitioned into secret shares before being broadcast to the other participants, SPDZ is less efficient and scalable than broadcast. With MYSTIKO, the dual broadcast is unnecessary because data is encrypted using Paillier theory and then decrypted centrally using aggregated gradients. The effects of switching to alternative protocols on two different types of performance are shown in Table 1.

While All-Reduce promises "enormous" speedups, when the entire synchronization time is included, those speedups disappear. There has been no change in the general tendencies of the four scalability protocols; the reductions in overall synchronization time are still sizable (with two examples shown in Table 4.1). In comparison to the benefits made available by enhanced communication, however, the improvements are rather modest. This illustrates that the time spent changing gradients before sharing them is the primary cost of private gradient descent compared to non-private gradient descent in MYSTIKO and SPDZ. Figure 7 and Table 1 illustrate how communication time becomes increasingly crucial for less complex models (5-Layer CNN and Resnet18). However, large models benefit more from gradient modification (like Resnet-50 and VGG-16).

When comparing private and non-private gradient descent, authors find that synchronization time for the former is significantly longer when epoch time is used as a proxy for training time. One reason for this is that while gradient modification is performed on CPUs, training is done on V100 GPUs (which have hundreds of cores). Researchers are unaware of a GPU-accelerated version of the Paillier technique, despite the fact that totally homomorphic encryption does exist (with poorer performance than Paillier on CPUs).

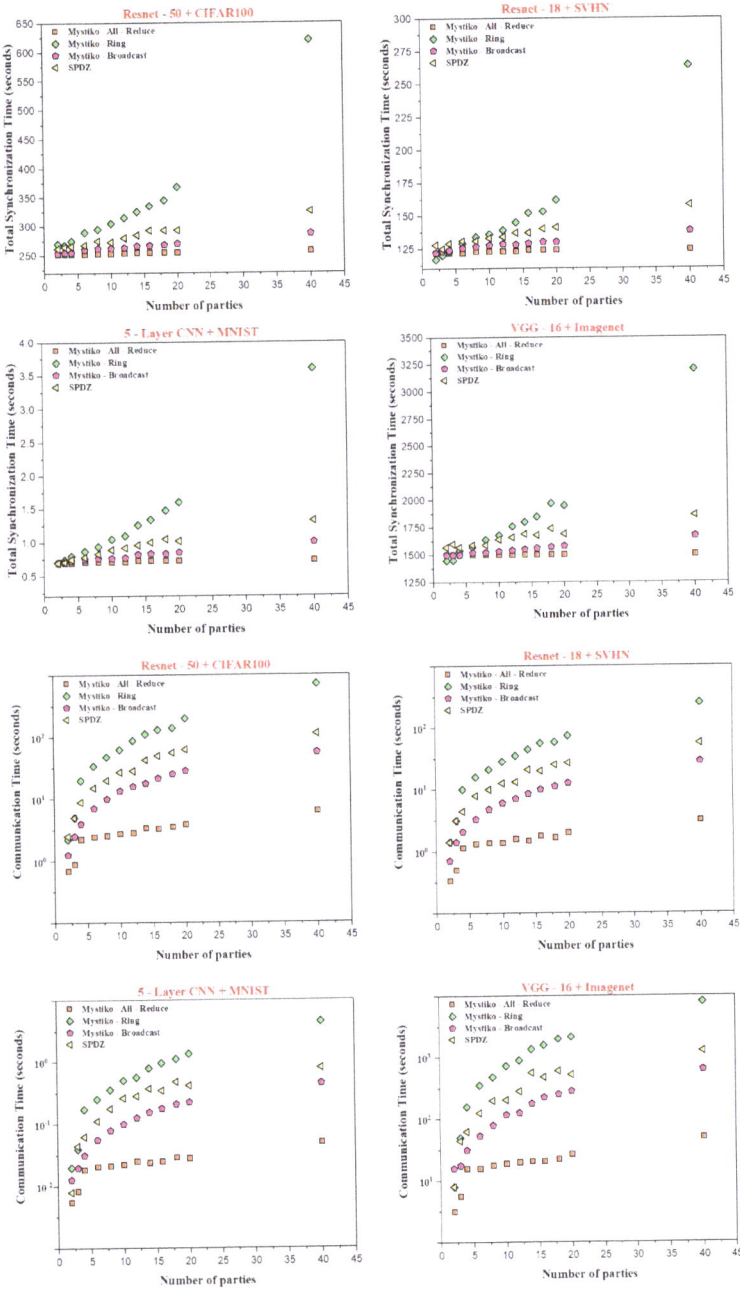

Figure 7. MYSTIKO: total synchronization time (seconds) versus number of parties (top plots) and total communication time (seconds) versus number of parties (bottom). Time spent communicating and transforming gradients are both components of the overall synchronization time.

Table 1. MYSTIKO: Decreased performance of broadcast, ring, and spdz compared to all-reduction.

Communication time				
		MYSTIKO		
# Parties	All-reduce	Broadcast	Ring	SPDZ
20	1	6.4–7.2×	38.2–39.9×	13.1–14.2×
40	1	7.9–8.6×	70.5–80.8×	13.8–14.7×
Total synchronization time				
		MYSTIKO		
20	1	12–51%	27–100%	2.2–6.1×
40	1	6–25%	15–59%	1.3–2.6×

5.2 *Trusted Execution Environments: AMD SEV*

The minimal overheads are the primary benefit when compared to MYSTIKO and SPDZ. Using TEEs causes more work because the EVMs must conduct aggregation. Two to four percent of extra end-to-end latency was incurred for each federated synchronization round. The convergence rate and accuracy were both identical.

6. Conclusion

Options for private parameter aggregation in federated learning have been explored in this chapter. It is evident that every approach has its benefits and drawbacks, and the hunt for the optimal answer remains an ongoing field of study. Although differential privacy shows promise for preventing data leaks, it does so at the cost of model accuracy and necessitates non-trivial hyperparameter tuning (learning rate schedule, batch size) for optimal results (or even near optimal results). Given that hyperparameter tweaking relies on the availability of all participants over extended periods and because doing many trials increases total delay, it may not be practical in federated situations. Neither the accuracy nor the convergence rate needs to be tuned for when utilizing homomorphic encryption or a secure multiparty computation scheme. However, they have significant overhead in terms of runtime; as shown by MYSTIKO, additive homomorphic encryption can greatly reduce this overhead if the aggregation protocol is well-designed. Finally, if anyone is willing to invest in some specialized hardware, TEEs can reduce aggregation overhead to almost nothing, with no negative effects on accuracy or convergence.

References

[1] Abdel-Basset, M., Hawash, H. and Moustafa, N. 2022. Toward privacy preserving Federated Learning in internet of vehicular things: Challenges and future directions. IEEE Consum. Electron. Mag., 11(6): 56–66, doi: 10.1109/MCE.2021.3117232.

[2] Ibitoye, O., Shafiq, M.O. and Matrawy, A. 2022. Differentially private self-normalizing neural networks for adversarial robustness in Federated Learning. Comput. Secur., 116, doi: 10.1016/j.cose.2022.102631.

[3] Moudoud, H. and Cherkaoui, S. 2022. Toward Secure and private Federated Learning for IoT using blockchain. In 2022 IEEE Global Communications Conference, GLOBECOM 2022−Proceedings, pp. 4316–4321. doi: 10.1109/GLOBECOM48099.2022.10000623.

[4] Guo, K., Chen, Z., Yang, H.H. and Quek, T.Q.S. 2022. Dynamic scheduling for heterogeneous Federated Learning in private 5G edge networks. IEEE J. Sel. Top. Signal Process, 16(1): 26–40, doi: 10.1109/JSTSP.2021.3126174.

[5] Cholakoska, A., Pfitzner, B., Gjoreski, H., Rakovic, V., Arnrich, B. and Kalendar, M. 2021. Differentially private federated learningfor anomaly detection in eHealth networks. In UbiComp/ISWC 2021 - Adjunct Proceedings of the 2021 ACM International Joint Conference on Pervasive and Ubiquitous Computing and Proceedings of the 2021 ACM International Symposium on Wearable Computers, pp. 514–518. doi: 10.1145/3460418.3479365.

[6] Ziller, A. et al. 2021. PySyft: A library for easy Federated Learning. Studies in Computational Intelligence, 965: 111–139. doi: 10.1007/978-3-030-70604-3_5.

[7] Augustine, C., Anand, J., Kaleeswaran, D., Muneeswari, G. and Santhoshkumar, S.P. 2022. 6G-based mobile IPTV using adaptive hybrid transmission. In Handbook of Research on Design, Deployment, Automation, and Testing Strategies for 6G Mobile Core Network, IGI Global, pp. 72–91.

[8] Liang, Y., Guo, Y., Gong, Y., Luo, C., Zhan, J. and Huang, Y. 2021. FLBench: A benchmark suite for Federated Learning. Communications in Computer and Information Science, 1385 CCIS. pp. 166–176. doi: 10.1007/978-981-16-1160-5_14.

[9] Kong, L., Tao, H., Wang, J., Huang, Z. and Xiao, J. 2020. Network coding for Federated Learning Systems. Lecture Notes in Computer Science (including subseries Lecture Notes in Artificial Intelligence and Lecture Notes in Bioinformatics), 12533 LNCS. pp. 546–557, doi: 10.1007/978-3-030-63833-7_46.

[10] Dhanalakshmi, R., Anand, J., Sivaraman, A.K. and Rani, S. 2022. IoT-based water quality monitoring system using cloud for agriculture use. In Cloud and Fog Computing Platforms for Internet of Things, Chapman and Hall/CRC, pp. 183–196.

[11] Wang, H., Li, A., Shen, B., Sun, Y. and Wang, H. 2020. Federated multi-view spectral clustering. IEEE Access, 8: 202249–202259, doi: 10.1109/ACCESS.2020.3036747.

[12] Yang, Z., Zhou, M., Yu, H., Sinnott, R.O. and Liu, H. 2023. Efficient and secure Federated Learning with verifiable weighted average aggregation. IEEE Trans. Netw. Sci. Eng., 10(1): 205–222, doi: 10.1109/TNSE.2022.3206243.

[13] Zhao, P., Cao, Z., Jiang, J. and Gao, F. 2003. Practical private aggregation in Federated Learning against inference attack. IEEE Internet Things J., 10(1): 318–329, doi: 10.1109/JIOT.2022.3201231.

[14] Mansouri, M., Önen, M. and Jaballah, W.B. 2022. Learning from failures: Secure and fault-tolerant aggregation for Federated Learning. In ACM International Conference Proceeding Series, pp. 146–158. doi: 10.1145/3564625.3568135.

[15] Devi, K.N., Anand, J., Kothai, R., Krishna, J.M.A. and Muthurampandian, R. 2002. Sensor based posture detection system. Mater. Today Proc., 55: 359–364.

[16] Kumar, R. et al. 2022. Blockchain and homomorphic encryption based privacy-preserving model aggregation for medical images. Comput. Med. Imaging Graph, 102, doi: 10.1016/j.compmedimag.2022.102139.

[17] Jin, G., Wei, X., Wei, S. and Wang, H. 2022. FPCBC: Federated Learning Privacy Preserving Classification system based on crowdsourcing aggregation. Jisuanji Yanjiu yu Fazhan/Computer Res. Dev., 59(11): 2377–2394, doi: 10.7544/issn1000-1239.20220528.

[18] Jiang, Z.L., Guo, H., Pan, Y., Liu, Y., Wang, X. and Zhang, J. 2021. Secure neural network in Federated Learning with model aggregation under multiple keys. In Proceedings—2021 8th IEEE International Conference on Cyber Security and Cloud Computing and 2021 7th IEEE International Conference on Edge Computing and Scalable Cloud, CSCloud-EdgeCom, pp. 47–52. doi: 10.1109/CSCloud-EdgeCom52276.2021.00019.

[19] Kalapaaking, A.P., Stephanie, V., Khalil, I., Atiquzzaman, M., Yi, X. and Almashor, M. 2022. SMPC-based Federated Learning for 6G-enabled internet of medical things. IEEE Netw., 36(4): 182–189, doi: 10.1109/MNET.007.2100717.

[20] Zhu, S., Li, R., Cai, Z., Kim, D., Seo, D. and Li, W. 2002. Secure verifiable aggregation for blockchain-based federated averaging. High-Confidence Comput., 2(1), doi: 10.1016/j.hcc.2021.100046.

[21] Ku, H., Susilo, W., Zhang, Y., Liu, W. and Zhang, M. 2022. Privacy-preserving federated learning in medical diagnosis with homomorphic re-Encryption. Comput. Stand Interfaces, 80, doi: 10.1016/j.csi.2021.103583.

[22] Du, W., Li, M., Han, Y., Wang, X.A. and Wei, Z. 2022. A homomorphic signcryption-based privacy preserving federated learning framework for IoTs. Secur. Commun. Networks, 2022, doi: 10.1155/2022/8380239.

[23] Bhansali, P.K., Hiran, D. and Gulati, K. 2022. Secure data collection and transmission for IoMT architecture integrated with federated learning. Int. J. Pervasive Comput. Commun. doi: 10.1108/IJPCC-02-2022-0042.

[24] Chen, L., Fu, S., Lin, L., Luo, Y. and Zhao, W. 2022. Privacy-preserving swarm learning based on homomorphic encryption. Lecture Notes in Computer Science (including subseries Lecture Notes in Artificial Intelligence and Lecture Notes in Bioinformatics), 13157 LNCS. pp. 509–523. doi: 10.1007/978-3-030-95391-1_32.

[25] Sotthiwat, E., Zhen, L., Li, Z. and Zhang, C. 2021. Partially encrypted multi-party computation for federated learning. In Proceedings - 21st IEEE/ACM International Symposium on Cluster, Cloud and Internet Computing, CCGrid 2021, pp. 828–835. doi: 10.1109/CCGrid51090.2021.00101.

[26] Chen, F., Li, P. and Miyazaki, T. 2021. In-network aggregation for privacy-preserving Federated Learning. In 2021 International Conference on Information and Communication Technologies for Disaster Management, ICT-DM 2021, pp. 49–56. doi: 10.1109/ICT-DM52643.2021.9664035.

[27] Li, Q. et al. 2020. InvisibleFL: Federated Learning over non-informative intermediate updates against multimedia privacy leakages. In MM 2020—Proceedings of the 28th ACM International Conference on Multimedia, pp. 753–762. doi: 10.1145/3394171.3413923.

[28] Zhang, C., Li, S., Xia, J., Wang, W., Yan, F. and Liu, Y. 2020. BatchCrypt: Efficient homomorphic encryption for cross-silo federated learning. In Proceedings of the 2020 USENIX Annual Technical Conference, ATC 2020, pp. 493–506.

[29] Hahn, C., Kim, H., Kim, M. and Hur, J. 2023. VerSA: Verifiable Secure Aggregation for cross-device Federated Learning. IEEE Trans. Dependable Secur. Comput., 20(1): 36–52. doi: 10.1109/TDSC.2021.3126323.

[30] Sun, Z., Wan, J., Yin, L., Cao, Z., Luo, T. and Wang, B. 2022. A blockchain-based audit approach for encrypted data in federated learning. Digit. Commun. Networks, 8(5): 614–624. doi: 10.1016/j.dcan.2022.05.006.

[31] Ma, J., Naas, S.A., Sigg, S. and Lyu, X. 2022. Privacy-preserving Federated Learning based on multi-key homomorphic encryption. Int. J. Intell. Syst., 37(9): 5880–5901. doi: 10.1002/int.22818.

[32] Shen, S., Zhu, T., Wu, D., Wang, W. and Zhou, W. 2022. From distributed machine learning to Federated Learning: In the view of data privacy and security. Concurr. Comput. Pract. Exp., 34(16): 2022, doi: 10.1002/cpe.6002.

[33] Fu, A., Zhang, X., Xiong, N., Gao, Y., Wang, H. and Zhang, J. 2022. VFL: A Verifiable Federated Learning with privacy-preserving for big data in industrial IoT. IEEE Trans. Ind. Informatics, 18(5): 3316–3326. doi: 10.1109/TII.2020.3036166.

[34] Ibrahem, M.I., Mahmoud, M., Fouda, M.M., Elhalawany, B.M. and Alasmary, W. 2022. Privacy-preserving and efficient decentralized Federated Learning-based energy theft detector. In 2022 IEEE Global Communications Conference, GLOBECOM 2022 - Proceedings, pp. 287–292. doi: 10.1109/GLOBECOM48099.2022.10000881.

[35] Wang, Z., Wang, P. and Sun, Z. 2022. SDN traffic anomaly detection method based on convolutional autoencoder and Federated Learning. In 2022 IEEE Global Communications Conference, GLOBECOM 2022 - Proceedings, pp. 4154–4160. doi: 10.1109/GLOBECOM48099.2022.10001438.

[36] Hsu, R.H. and Huang, T.Y. 2022. Private data preprocessing for privacy-preserving Federated Learning. In Proceedings of the 2022 5th IEEE International Conference on Knowledge Innovation and Invention, ICKII 2022, pp. 173–178. doi: 10.1109/ICKII55100.2022.9983518.

[37] Zhou, D., Chen, Z., Wu, D., Yu, Y., Gan, Q. and Xu, B. 2022. HEND-FL: Accurate Federated Learning using homomorphic encryption and a new distributed protocol. doi: 10.1109/DASC/PiCom/CBDCom/Cy55231.2022.9928012.

[38] Jebreel, N.M., Domingo-Ferrer, J., Blanco-Justicia, A. and Sanchez, D. 2022. Enhanced security and privacy via fragmented Federated Learning. IEEE Trans. Neural Networks Learn Syst., pp. 1–15. doi: 10.1109/TNNLS.2022.3212627.

[39] Zhang, C., Ekanut, S., Zhen, L. and Li, Z. 2022. Augmented multi-party computation against gradient leakage in Federated Learning. IEEE Trans. Big Data, pp. 1–10. doi: 10.1109/TBDATA.2022.3208736.

[40] Ma, Z., Ma, J., Miao, Y., Li, Y. and Deng, R.H. 2022. ShieldFL: Mitigating model poisoning attacks in privacy-preserving Federated Learning. IEEE Trans. Inf. Forensics Secur., 17: 1639–1654. doi: 10.1109/TIFS.2022.3169918.

[41] Huang, W., Liu, J., Li, T., Ji, S., Wang, D. and Huang, T. 2022. FedCKE: Cross-domain knowledge graph embedding in Federated Learning. IEEE Trans. Big Data, pp. 1–12. doi: 10.1109/TBDATA.2022.3205705.

[42] Zhao, B., Liu, X., Chen, W. and Deng, R. 2022. CrowdFL: Privacy-preserving mobile crowdsensing system via Federated Learning. IEEE Trans. Mob. Comput. doi: 10.1109/TMC.2022.3157603.

[43] Zhou, X., Liang, W., Ma, J., Yan, Z. and Wang, K.I.K. 2022. 2D Federated Learning for personalized human activity recognition in cyber-physical-social systems. IEEE Trans. Netw. Sci. Eng., 9(6): 3934–3944. doi: 10.1109/TNSE.2022.3144699.

Chapter 5

Navigating Privacy Concerns in Federated Learning

A GDPR-Focused Analysis

G Anitha[1], and *A Jegatheesan*[2]

1. Introduction

The vast majority of today's software and services, especially those used in the medical, transportation, and monetary sectors, rely on data collected by sophisticated machine learning (ML) algorithms powered by artificial intelligence (AI) [1]. Artificial intelligence has been making progress in every sphere of life, and its impact on the globe is predicted to be greater than any other event in human history. More than only the power grid [2, 3]. The entire potential of AI has not yet been realised, and there are still significant obstacles to the development of AI/ML-based applications. Centralized storage and processing is a major factor in these problems [4]. Data, especially personal data, is typically produced and stored in isolated locations, such as individual users' devices or central data centres, in

[1] Research Scholar, Institute of Computer Science and Engineering, Saveetha School of Engineering, Saveetha Institute of Medical and Technical Sciences, Chennai, India.
[2] Professor, Institute of Computer Science and Engineering, Saveetha School of Engineering, Saveetha Institute of Medical and Technical Sciences, Chennai, India.
Email: jegatheesanphdcse@gmail.com
* Corresponding author: gani3086@gmail.com

the real world. Training data is often consolidated in a data server for most conventional ML algorithms [5]. To put it succinctly, collecting, consolidating, and integrating disparate data from many sources is not an easy task, much alone managing and securely analysing the resulting mountain of information [6]. Transferring massive volumes of data at high speeds across organisations with variable degrees of accuracy is already challenging without adding the complexity of data protection rules and constraints like the GDPR [7]. When processing massive amounts of data using traditional ML algorithms on a single, highly-capable cloud server, security breaches and single points of failure are both possible [8].

The lack of transparency and provenance that comes with centralized data processing and administration may make it difficult to win consumers' confidence and fulfil the GDPR [9]. Professionals and academics alike have taken notice of Google's creation of Federated Learning (FL) in 2016 to overcome such obstacles. Executing the ML algorithm on many local datasets kept at isolated data sources like smartphones, tablets, etc. FL eliminates the requirement for a centralized data server [10, 11]. With FL, local nodes store the training dataset and compute together to train a centralized ML model [12].

With FL, service providers can implement certain types of machine learning algorithms into their applications and services without collecting any personally identifiable information from their customers, which may make it easier for them to fulfill data protection regulations (DPR) like the GDPR [13, 14]. Inappropriately, private data can be secretly retrieved from local training despite FL's privacy-protecting distributed collaborative learning methodology [15]. Therefore, Florida-based service providers are obligated to take precautions in line with the GDPR while dealing with citizens of India [16]. In this article, the authors take a look at previous FL research, focusing on privacy-protecting methods from the standpoint of GDPR compliance [17]. To begin, we present FL as a feasible way to overcome the difficulties of data privacy protection in traditional centralized ML systems [18, 19]. Second, cutting-edge privacy safeguards for centralized FL are outlined, along with an examination of how these methods reduce vulnerabilities in terms of data security and privacy. Third, we offer a thoughtful discussion, with possible solutions for how an FL system might be designed to meet GDPR requirements. We also present potential directions for future study and highlight the remaining challenges that hinder a Federation learning system from meeting the necessities of the GDPR.

2. ML-based Systems of GDPR-Compliance and Privacy Preservation

2.1 Contextual Properties

The innovative field of machine learning (ML) allows machines to do tasks without being explicitly programmed, allowing them to learn and improve through experience. To do this, a machine learning system constructs a mathematical model using a dataset (the training data) with the intention of fine-tuning the model's parameters over time [20, 21]. This allows the algorithm to make more accurate forecasts or judgments when presented with novel data. Finding the maximum value of an objective function is often the end aim of a mathematical optimization problem, which may be used to describe an ML work. For this reason, any ML-based system must have an optimization technique.

2.1.1 Algorithm of Gradient Descent

Gradient descent lies at the base of both ML and FL, making it a popular optimization approach for ML [22]. It is an iterative optimization algorithm of the first order for locating the minimal of an objective function $f(\theta)$ with user-specified parameters. $D = (x_1, y_1), (x_2, y_2), \ldots, (x_m, y_m)$, and the objective function $f(\theta)$; Each parameter in a model is updated in the contradictory direction of the gradient of the objective function $f(\theta)$ using the gradient descent approach, as shown in the following equation.

$$w_j \leftarrow w_j - \eta \nabla \frac{1}{m} \Sigma_{i=1}^m L\left(f(x_i) - y_i\right) \tag{1}$$

where

w_j j th parameter of θ

η learning rate hyper-factor.

MSE and cross-entropy losses are examples of loss functions that may be denoted by the letter L. Parameters are updated repeatedly using Equation 1 until a local minimal is reached or the change in loss between stages is small enough to be ignored.

2.1.2 Gradient Descent Variants

Based on the quantity of data needed to calculate the objective function $f(\theta)$, gradient descent algorithms are often divided into one of three categories. Batch gradient descent is the first type and it includes computing gradients throughout the whole training dataset D with just one update. In disparity to batch gradient degradation, stochastic gradient descent (SGD) updates parameters depending on the gradient of a subset of data selected at random from D [23]. The third type of mini-batch gradient descent involves recalculating the parameters after processing every n batch of training samples. Each iteration of gradient descent requires a compromise between the accuracy of the calculation and the accuracy of the parameter update. Mini-batch gradient descent corrects SGD's potential for gradient oscillation and the inefficiencies of batch gradient descent [24, 25]. The additional hyper-parameter batch-size n introduced by this method, however, sometimes necessitates professional knowledge, significant experimentation, and even manual adjustment. Optimizers, methods for regulating the learning rate η practically and precisely, typically accompany the gradient descent. These optimizers coordinate with the loss function L and the model parameters θ to fine-tune the learning rate η based on the loss function's output.

2.1.3 Distributed Learning Via Gradient Descent

As large-scale distributed learning, of which FL is a part gained popularity, gradient descent-based optimisation techniques have attracted renewed attention. Several nodes may work together while training a complicated model, such as a DNN with millions of variables. These nodes are grouped as compute nodes, which carry out calculations for the network as a whole [26]. Training computations may be accelerated with the use of concurrency approaches like model parallelism and data parallelism. Machine learning models can benefit from model parallelism by having their work broken down into smaller, more manageable work that can then be dispersed over all the available computing nodes. Model parallelism requires frequent communication and synchronization across compute nodes in a cluster, as well as mini-batch data replication at those nodes [27]. Instead, data parallelism keeps the whole model running on each node while spreading the training data evenly across all the machines in a cluster ("sharding"). Data parallelism is available in several modern machine learning structures, including TensorFlow 3 and Pytorch 4. Hybridized parallelism is the amalgamation of the two parallelism approaches to maximise the benefits and minimise the downsides of each, resulting in a system with increased efficiency and scalability.

2.2 Security Measures for Machine Learning

In a distributed learning system, privacy preservation techniques serve two primary purposes:

(i) Ensuring that only legal users have admission to the training dataset, and

(ii) Preventing unauthorized users from accessing the local model parameters that are communicated between nodes as part of an optimization algorithm.

2.2.1 De-identification of Data

De-identification Data (anonymization) is a method through which personally identifiable information (PII) and other sensitive data are obscured or removed from a dataset. Since concealing or omitting information may lessen the usefulness of the dataset, data anonymization must strike a good balance between privacy-guarantee and utility [28]. A privacy assault known as a linkage attack can be launched against a data subject when it is linked with supplementary information from other anonymous datasets. Several approaches have been proposed to thwart linking assaults; they are k-anonymity, l-diversity, and t-closeness. Such privacy safeguards, however, are vulnerable to linkage attacks if the attacker already knows part of the private data. This shortcoming of k-anonymity-based methods necessitates the development of alternate methods, such as differential privacy, that provide strong privacy guarantees.

2.2.2 Homomorphic Encryption

When training a model with collected data in a centralized system, such as the cloud, it is significant to guard the confidentiality of that data by using homomorphic encryption techniques [29]. The encrypted data may be computed without the requirement for the secret key, due to homomorphic encryption. The computation's requester is the only one who can decode the encrypted results. Furthermore, homomorphic encryption guarantees that the result obtained after decryption is identical to the result obtained when using the original, unencrypted dataset. The homomorphic encryption family includes the subsets partially homomorphic encryption, some homomorphic encryption, and fully homomorphic encryption. The Rivest-Shamir-Adleman (RSA) algorithm, one of the most used encryption algorithms, allows for simple mathematical operations like adding and multiplying to be performed [30]. Using FHE, the cipher-text may undergo entirely arbitrary operations, and both the operations and their outputs can be encrypted. In FHE, a decryption function can be used to transmit mathematical computations from the original data or cipher-text without causing any incompatibilities. Although the original plain-text data is never revealed using homomorphic encryption, the enormous processing cost prevents any meaningful calculation from being performed over the cipher-text. Therefore, using homomorphic encryption for massive data training exercises is still not feasible.

2.3 The GDPR

The GDPR replaced the Data Protection Directive, in effect since 1995, 25th May 2018. The GDPR has as its driving principle that "personal data can only be obtained lawfully, under rigorous restrictions, for a valid purpose" (a broader range is shown in "Which?" — Fig. 1). In its 99 paragraphs, the law lays out in great detail the values that organizations must uphold and the technological and administrative standards that they must meet in order to process personal data [31]. As can be shown in Fig. 1, the General Data Protection Regulation ("Global") will have a far-reaching effect on businesses and government agencies across the world that handle the personal information of residents. In accord with data protection legislation, the GDPR makes a clean break between the roles of Data Subject, Data Controller, and Data Processor. The GPDR is an EU regulation with the stated goal of "protecting the rights and freedoms of natural persons with regard to the processing of personal data, in particular the right to privacy". To that end, the GDPR establishes severe requirements for the consent legal basis, which mandate that the Data Controller get the Data Subject's agreement before processing any personal data. Under the conditions of the authorization [32], the Data Controller has total control over the whys and hows of processing personal data.

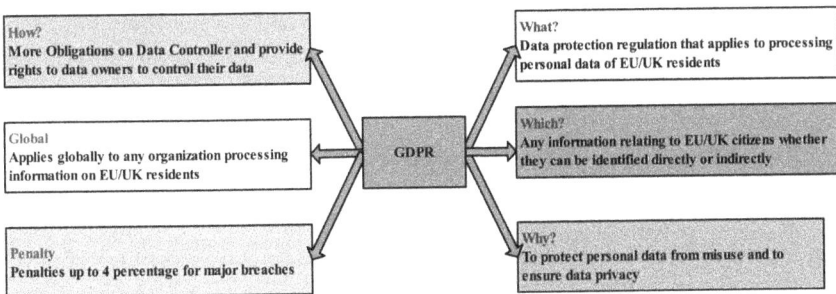

Figure 1. The GDPR regulation.

2.4 *Difficulties in Meeting GDPR Requirements*

Standard ML-based apps and services must employ ways that properly safeguard and manage personal data in accordance with the six data protection principles outlined in the General Data Protection Regulation (GDPR). It may be challenging, if not impossible, to apply these responsibilities in a centralized ML-based system, despite the fact that ML-based systems benefit from a variety of privacy-preserving safeguards [33].

Such ML-based systems collect, aggregate, and analyse massive amounts of data on a single server, making them vulnerable to catastrophic data breaches caused by a single point of failure. It is challenging to provide meaningful analysis of ML algorithms due to the black box-like nature of their execution and decision-making [34]. As a result, the standards of GDPR's transparency, impartiality, and automated decision-making are challenging for most ML-based systems to meet.

In addition, ML-based systems don't always realistically meet the constraints of purpose limiting and data reduction. Most machine learning algorithms are extremely dependent on the quality and amount of the data they are given, thus researchers generally aim to collect as much of it as possible [35]. So, before running such ML algorithms, it is difficult to (1) ascertain the motivations for data collection and (2) ascertain whether data is appropriate, constrained, and relevant exclusively to the declared aims. These limitations put artificial limits on the potential of machine learning (ML) services and applications.

Finally, machine learning algorithms are mostly created to maximise efficiency, while privacy protection methods continue to be limited to basic caveats. To fulfill the stringent necessities of the GDPR, machine learning algorithms will need to undergo a fundamental redesign at the algorithm level. To achieve such a radical change would require enormous, maybe unachievable, technological accuracy and financial backing. It appears that many service providers are weighing the costs and benefits of prioritising efficiency above privacy-guarantee, with the realisation that doing so may render them unable to continue providing their current services.

3. Federated Learning: An Approach to Distributed Collaborative Learning

Due to the difficulties in adhering to stringent data protection requirements on the massive gathering and processing of personal data, the old cloud-centric ML techniques are no longer suitable in many situations. End-user devices (such as smartphones, tablets, and wearables) have growing computational power and Internet access, and so create the vast majority of personal data [36]. Decentralized AI has arisen to move intelligence from the cloud to the edge using mobile edge computing and artificial intelligence/machine learning methods in reaction to the proliferation of such personal gadgets and the associated increase in privacy issues.

Since federated learning enables users to maintain their original personal data on their devices throughout collaborative training of an ML model, it may help mitigate some of the hazards associated with cloud-based ML. It is an interdisciplinary approach to training a centralized ML model using decentralized data, with applications in machine learning, distributed computing, data privacy, and security. When combining and updating models, only training parameters are traded. When it comes to training data, traditionally distributed learning assumes that each compute node's local training datasets are IID and are around the same size. This is not the case with FL. Since FL can accommodate non-IID data sets whose sizes could vary by several orders of magnitude, it is an advance in distributed learning. These unrelated data sets are scattered over a wide variety of mobile devices with varying degrees of network availability and data transmission rates.

3.1 Model Training in Federated Learning

With a model constructed as the minimization of specific objective functions over a training dataset, FL is mainly effective for parameter estimation in ML models. One such statement of the purpose of minimization is:

$$\min_{w \in \mathbb{R}^d} f(w) = \frac{1}{n} \Sigma_{i=1}^{m} f_i(w) \tag{2}$$

whereas the input-output pairs (xi, yi) used for training are of the type $x_i \in R^d$ and $y_i \in R, \forall i \in \{1, 2, ., n\}$. Dataset sample size (n), parameter vector $w \in R^d$ and loss function (fi (w)) are denoted in Equation 2. This idea applies to a broad variety of ANN-related tasks, including linear and logistic regressions, support vector machines, and more. It even extends to sophisticated non-convex challenges in ANN, such as Deep Learning. Minimizing the sum of the losses as a function of the model parameters may be easily computed using the gradient descent technique using back-propagation as an optimization strategy.

These kinds of processes do a high number of quick iterations on a huge dataset that has been uniformly partitioned in data servers, as is done in conventional ML methods. Such algorithms necessitate extremely fast and low-latency links to the training data. The conventional ML approaches to solving this optimisation problem do not apply in FL settings since these conditions are not met. FL training data, in contrast to the balanced and

IID data utilised by conventional methods, is dispersed over millions of separate mobile devices connected by substantially higher-latency, lower-throughput networks. As an added caveat, data from individual devices and access to computers are only made available on an as-needed basis throughout the training process. Optimisation algorithms will need to be successfully adapted to and implemented in federated environments before FL can be put into practise.

3.2 Federated Optimization

Optimization algorithms that work well in federated environments where training data is substantially and unevenly dispersed among local nodes are a cornerstone of FL. The following formula describes the dispersed parameters for the federated optimization. Let K indicates the total number of nodes in the cluster, P_k denote the collection of data points held by node $k \in \{1, 2, ., K\}$, and $n_k = |P_k|$ denote the total no of data points held by node k. Given that each local node stores its own unique set of personal information, we may deduce that, $\mathbb{P}_k \cap \mathbb{P}_l = \emptyset$ if $k \neq l$ and $\Sigma_{k=1}^{K} n_k = n$. For minimization, the following defines the distributed problem formulation:

$$\min_{w \in \mathbb{R}^d} f(w) = \frac{n_k}{n} \Sigma_{k=1}^{K} F_k(w) \tag{3}$$

$$F_k(w) = \frac{1}{n_k} \Sigma_{i \in \mathbb{P}_k} f_i(w) \tag{4}$$

Due to the limited communication capabilities of the local nodes in this federated system, it is essential to decrease the number of iterations utilised by optimization techniques. The SVRG and DAN algorithms for distributed optimization were combined by the authors [25] to create a unique distributed gradient descent (Federated SVRG). The FSVRG updates the parameters at each node after each cycle by computing gradients from P k data at each local node k, and then averaging those weighted values across all k local nodes. The next step is to evaluate the system's ability to forecast whether a post will garner comments on Google+ public postings using a sample of around 10,000 individuals as local nodes. The optimal solution is reached by the FSVRG in 30 fewer iterations than by the original gradient descent approach.

However, in federated contexts, where the amount of data specimen and supplies of data among personal mobile devices vary greatly, the assumption that standard distributed ML algorithms predominantly train on IID data no longer holds. It indicates that in federated systems, non-IID data training results in much worse accuracy and slower convergence than

IID data training. Methods for reducing the data volume of parameter updates sent to the orchestration server to boost the behaviour of FSVRG algorithms in such environments. Updates to ML models may be made more quickly and with fewer variables by using structured updates or sketching updates. Training local nodes in many machine learning models separately as tasks in a single learning goal is one such ambitious federated optimization strategy. It is common for local nodes to generate data with varying distributions, making it natural to use different learning models for each. However, because these models are structurally similar, they can be modelled using a multi-tasking learning (MTL) framework, which allows for the modelling of the similarity between them. This method assures learning convergence and boosts performance while working with non-IID data. The Federated SGD and the Federated Averaging approach were developed by building on earlier federated optimisation studies, allowing for the training of a deep network with a hundred times fewer communications than the naïve FSVRG.

The motivation for these algorithms is the need to do more than only compute gradient steps using the powerful processors found in today's personal mobile devices. In this case, the gradients are calculated by each client, and the local model is likewise computed locally numerous times; the coordination server just aggregates the results. This reduces the number of communications required while still providing a good global model during training. These suggested algorithms work very well under severe constraints, such as low bandwidth and excessive jitter, and delay in communications. Naive FSVRG algorithms are not effective enough in these cases. G-board, the Google keyboard for Android phones, makes use of these algorithms to anticipate what users will type. In this setup, the smartphone runs the Federated SGD to perform gradient descent computations using local information. The information is received by a cloud server, which then processes it. To improve the global model at the start of a new training cycle, the server employs the Federated Average method, picking a random subset of mobile devices and averaging the gradients they supply. After receiving the updated global model, the local nodes apply the necessary changes to their models.

3.3 Cycle of the Federated Learning Process

The network's nodes individually train the model using their own data and report back to the centralized server at regular intervals. The workflow cycle in a centralized FL architecture includes the following four steps:

1. Selection of Participants and Global Distribution of Models: A subset of users who qualify for the training are selected by the server. After that, it updates the global ML model and notifies the trainees of the upcoming training session.

2. *Local Computation*: Participants refresh their in-device ML models and retrain them using the in-device dataset after downloading the server-side ML model. An FL client software is needed on client devices so that training algorithms like Federated SGD and Federated Averaging can be executed on local nodes, global model updates can be obtained, and local machine learning model parameters can be sent to and from the server.

3. *Local Models Aggregation*: To refresh the worldwide machine learning model, the server aggregates enough locally trained models from the contributors to do so (the next step). To avert the server from accessing sensitive machine learning model factors, it is essential to incorporate privacy-protecting measures within the aggregation process. Among these strategies are secure aggregation, differential privacy, and cutting-edge cryptographic protocols.

4. *Revised Global Model*: The server then updates the present global machine learning model using the combined model factors from step 3. Participants in the subsequent round of training will be provided with this revised global model.

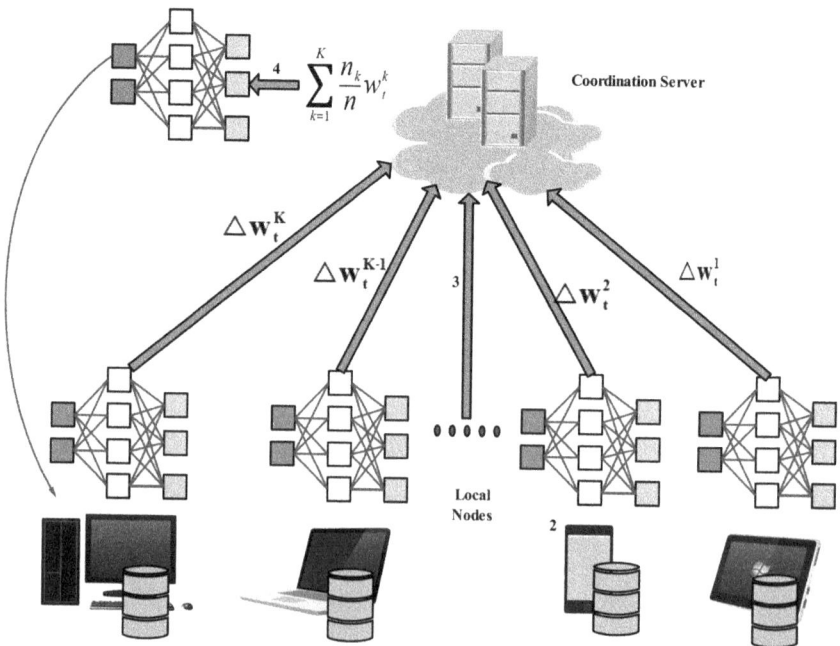

Figure 2. The four-stage cycle of a centralized FL workflow.

It takes several iterations of this 4-stage process to produce a reliable global model. Note that the four steps of the cycle are not required for every workout. An asynchronous SGD strategy may be utilised to rapidly update the local model based on training results without having to wait for updates from other participants. This asynchronous method, which optimises the rate of updates, is often used in the dispersed training of DL models on a big dataset. With cutting-edge tools like aggregation compression, secure aggregation with SMC, and differential privacy, the synchronous approach in FL contexts offers substantial benefits over the asynchronous one, particularly in communication efficiency and security. This coordination necessitates the use of a central server.

4. Centralized FL Framework for Privacy-preservation

While conventional ML approaches require that training data and computation be shared between devices, FL allows for cooperative training of an ML model without compromising privacy. Local ML models retain some aspects of the training data samples in their parameters even if private information is not transferred directly to the coordination server. For instance, the authors [32] show that sensitive information may be extracted from trained models by making use of hidden correlations that are exposed by the training data. Modern deep-learning models, as pointed out by authors [19], hide internal representations of many aspects, some of which are unrelated to the task at hand. Data about the training sets may be inferred from these coincidental features. Due to their reliance on inference, FL systems are susceptible to inference attacks including membership and reconstruction assaults. Coordination servers provide local nodes with access to the global training model in its intermediate stages. There should be research into security models and careful consideration of privacy assurances before adopting a centralized FL framework, as this opens the door for hostile actors to influence the training process by providing arbitrary updates that taint the global model. Additional confidentiality and privacy measures are desired to ensure the security of sensitive data and compliance with stringent data protection standards like the General Data Protection Regulation inside the federated learning system. We see a breakdown of the many attacks on centralized FL and the various privacy protection techniques it employs.

4.1 Coordination Server in Privacy-preservation Solutions

Most modern privacy-preserving FL system solutions are founded on state-of-the-art cryptographic protocols like SMC and differential privacy.

To prevent (i) inference attacks from being launched by server insiders and (ii) model poisoning by Byzantine players, several techniques are implemented on the server. Security against Inference Attacks, Version 4.3.1. During global model training, numerous server-side strategies have been developed to handle inference attacks and prevent the coordinate server from inspecting the factors supplied by a single user. In particular, client-sent parameters may be encrypted using the Secure Aggregate protocol, which is built on top of SMC. SMC is the first step in the protocol since it encrypts the model parameters so that users can average them without revealing their individual contributions. Participants and a coordinate server interact in four rounds: round one involves the distribution of public keys, round two involves the computation of masked inputs on the client side, round three involves ensuring that at least t participants are training the same model, and round four involves unmasking (round 4).

The third round of the protocol is only essential if the server is malevolent, but it may be skipped if the server is simply interested. Clients and a central coordination server must deal with more communication overhead and higher computational complexity as a result of using this protocol. It's worth mentioning that Google's Tensor Flow Federated framework, where Secure Aggregation may be implemented available for use in FL research and practical experimentation. For SGD to run and provide global model updates, all that is needed is for a selection of participants' local model weights to be averaged, as per the Secure Aggregation process. Therefore, the coordination server doesn't have to collect participant-specific local updates. As a result, the server would be unable to conduct inference attacks based on user data. Secure Aggregation protocol, in conjunction with Federated Averaging, allows for the secure execution of SGDs, with increased resilience in the face of errors and reduced communication cost, on a server with restricted trust. Though, this method is only viable in settings with trustworthy individuals. When dealing with Byzantine players, such as when they collude with a rogue server to reveal a client's inputs, there is no assurance that the protocol will be available or accurate.

4.2 *Local Nodes Privacy-preservation Solutions*

To prevent privacy leakage caused by an inside attacker executing inference attacks on the coordination server, Secure Aggregation must be enabled on both the local nodes and the server. Our SMC-based aggregation approach to FL-based multi-party deep learning is made possible by homomorphic encryption of local model parameters from all participants. When enough local models have been collected, an encrypted global model is transmitted to the coordination server. This guarantees that confidentiality is maintained for all global model contributors. The latest model's parameters and the original training dataset may be kept secret by local nodes using the perturbation strategy, which is designed to thwart disclosure attempts by the coordination server and other attackers. Before sharing the model with other nodes, the perturbation technique can be used locally at each node to conceal some parts of the model. Even with these model parameters in hand, the adversary still won't be able to successfully recreate the original training data or glean any useful information.

To rephrase, the perturbation technique has the potential to stop inference assaults on a client's locally trained model. Differential privacy is a popular method for achieving this goal; it entails introducing random variation into either the training data or the model parameters. Differential privacy through SMC was offered as a safe method of integrating independently trained neural networks before FL was developed. Since then, this method has been refined to the point that results from different participants are statistically similar and outside noise has minimal impact on the overall model's precision. Since attackers cannot tell individual records apart during FL training, sensitive data is safeguarded, and inference attacks are averted. Local nodes in FL environments can benefit from differential privacy techniques on two different levels: the batch level, where random noise is generated through sensitivity analysis of parameters based on data points in a mini-batch, and the user level, where the noise is generated by the users themselves.

5. GDPR-Compliance in Centralized FL Systems

FL is an innovative approach to addressing privacy issues in ML-based apps by offloading model aggregation to a third-party server. Compared to centralized ML methods, FL has been shown to better safeguard user privacy. When training an ML model, it allows users to save their training data on their devices. To improve the global model, the most relevant parameters from locally trained models are sent to a coordination server. Even yet, such model parameters still include certain private characteristics that might be utilised to reconstruct or infer significant personal data. Therefore, an FL system is still accountable for adhering to and following the requirements of the General Data Protection Regulation. In this part, we look at each GDPR rule and determine if it is relevant and must be followed in FL circumstances or whether it may be disregarded. The challenges to full GDPR compliance are highlighted and analysed.

5.1 Roles and Obligations

The General Data Protection Regulation (GDPR) establishes the meaning of "Data Subject," "Data Controller," and "Data Processor" within the framework of data protection legislation. Data Processors may not begin working with personal data until the Data Controller has disclosed the GDPR-compliant objectives for and means of processing. In addition, Data Controllers must put in place suitable measures to notify Data Subjects of how their personal data will be transmitted and processed securely and privately. The General Data Protection Regulation guarantees individuals the "right to be informed," "right of access to," and "right to be forgotten". As can be seen in Table 1, data points in FL contexts are not considered discrete samples but rather variables in a regional model.

Table 1. Role of GDPR in various services.

GDPR roles	Centralized federated learning	Traditional machine learning
Data Processor	Service Benefactor	Service Benefactor, Third-parties
Data Controller	Service Benefactor	Service Benefactor
Data Subject	End-users	End-users
Personal Data	Local model factors	Original training data

For instance, in step 1 the service provider instructs the users to train an ML model on their local training data and share such locally trained model, in step (ii), the service provider processes the local model parameters sent from the users, and in step (iii), the service provider disseminates the global models to all users and requests that they update their local models. And in centralised FL environments, service providers may only have access to a global, anonymous, and therefore non-collaborative ML model. Therefore, data processors are the only service providers and the only ones able to supply FL under such circumstances. Since FL concentrates on accumulating local ML models and updating the global ML model, its processing methods are substantially easier to implement.

5.2 GDPR Principles

Figure 3 depicts the six principles established by the General Data Protection Regulation (GDPR) as reasonable guidance for service providers to manage personal data. The duty of care that requires Data Controllers to take on the onus of adhering to the principles and putting in place adequate mechanisms to guarantee compliance is strikingly similar to that of the Data Protection Act of 1998.

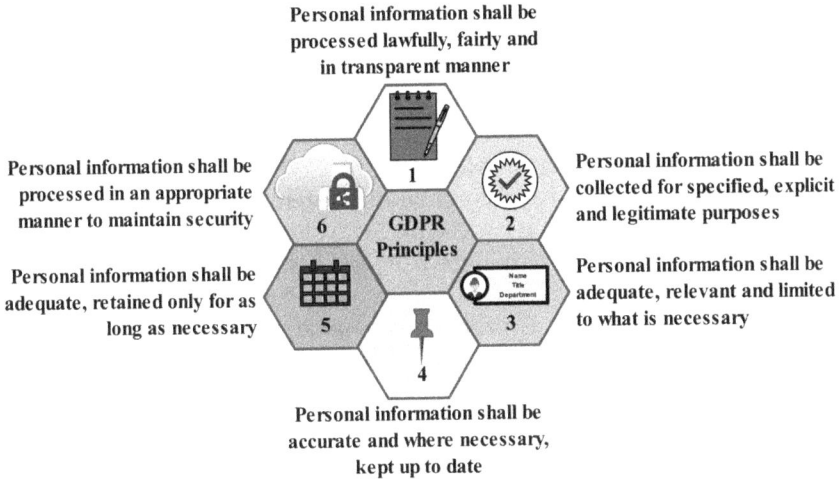

Figure 3. Basic GDPR principles.

5.2.1 Constitutionality, Equity, and Transparent

As the first rule, the FL application's service provider, acting as the Data Controller, must explain why it's end-users must undergo FL training. Six types of legal grounds are specified under the General Data Protection Regulation (GDPR) for processing personal data: There are six reasons, and they are as follows: There are six justifiable reasons for doing something: consent, understanding, compulsion, interest, and public responsibility. Some types of special category data, such as those related to a person's health, genetic composition, or race or ethnicity, may require additional requirements beyond the satisfaction of this legal basis to be legitimately processed. Before beginning any processing of personal data, the most suitable one should be selected and recorded, considering the precise goals and circumstances of the processing.

There is no way for the coordination server to access users' devices in FL systems to get raw training data or locally constructed ML models. Instead, users will become more invested in the FL system by communicating their findings to the coordination server when they are ready to do so. Users of FL client applications should be able to train in any environment they find most convenient, whether locally or by pushing and pulling changes to a server-based ML model. In addition, FL systems only process data for a predetermined purpose, in a manner that customers would reasonably expect, with negligible impact on privacy. In other words, an FL application can be justified by either obtaining the applicant's permission or demonstrating genuine interest.

When employing AI/ML methods like deep learning, it is typical for the Transparency standards not to be satisfied due to a lack of understanding of the bias in data samples and the training process. No preferential treatment is given to FL systems.

If a machine learning system (including FL) is fed biased training data, the system is likely to return biased results. When the trained model is used in a fully automated decision-making system, it might provide unfair and biassed outcomes. One issue with FL systems is secure aggregation, which prevents service providers from accessing the original training dataset and prevents consumers from verifying their locally trained ML model. Therefore, it is problematic for an FL system to carry out training activities openly and guarantee the correctness of any automated judgments made by the system. As a result, all FL solutions become impracticable and fall short of the fairness and openness mandated by the GDPR. These open questions necessitate more study of AI/ML algorithms' interpretability, bias, and openness, and thus call for a loosening of GDPR's oversight boards' stipulations for AI/ML, FL included. One possible approach to meeting this GDPR requirement is to develop ML models and algorithms that promote responsible ML governance by being naturally interpretable.

5.2.2 Purpose Limitation

To adhere to the purpose restriction principle, an FL service provider must make it crystal apparent to customers why and how their personal data and computing power are being sent to the service provider's servers to train a global machine learning model. According to this concept, the service provider is free to use the information for any reasonable purpose. For FL systems to be in full conformity with the principles, they must be able to include acceptable privacy-preserving technologies, like Secure Accumulation and differential privacy, straightforwardly. This is because the coordination server has no way to extract and use the clients' locally trained ML models for anything other than the global model updates. Unauthorized training can be conducted thanks to a hidden back-door model installed by a hostile service provider or Byzantine player. Because centralised FL makes use of safe aggregation, no server-side model anomaly detection solution exists to protect against this sort of attack; hence, achieving entire compliance with the GDPR remains problematic for any FL system.

5.2.3 Data Minimization

The GDPR places an obligation on "Data Controllers," which includes service providers, to collect and handle only sufficient personal data for a specific, explicit, and legitimate purpose. Data reduction is a challenge for traditional centralized ML algorithms since it is difficult to know in advance what data and how much data will be required to successfully train a given ML model. Unlike traditional ML systems, which need users constantly collect and analyse new training data, FL just necessitates that a service provider gets local ML models from participants. When an FL system employs privacy-preserving methods, the coordination server may only collect averages of local model factors from participants for global model enhancement. The aggregation approach further guarantees that the global model does not include any personally identifiable information that may be used by enemies for illicit purposes. Backdoor attacks, which aim to introduce an unauthorized purpose, are possible because of the purpose limitation principle. As a result, the smallest set of parameters for the local ML model collected from the participants is now too large for the intended goal and too little for the unauthorised side job. This implanted side-task is an unresolved difficulty for FL systems since it might be used to reveal the participant's private information.

5.3 Case Study

The GDPR mandates that Data Controllers provide the following rights to Data Subjects wherever possible: There are several rights that individuals have regarding their personal data, including

(1) The right to be informed
(2) The right of access
(3) The right to rectify
(4) The right to erasure (Right to be forgotten)
(5) The right to restrict processing
(6) The right to data portability
(7) The right to object and
(8) Rights in relation to automated decision making and profiling.

5.3.1 Information

The need that the Data Controller to simply and clearly disclose the nature and goal of the FL training, the local machine learning model used, and the expected result of the process makes it challenging to meet this right for Data Subjects. FL is a black-box model, making it impossible to comprehend how it works and to predict its outputs with any degree of certainty, as is the case with many other complex ML processes. Since an FL system's Terms and Conditions may explain why a client's local ML model will be used (i.e., to construct a global ML model), when that data will not be utilised (i.e., after each training cycle), and who will have access to it, it meets the right to be informed (only the co-ordination server).

5.3.2 *Automated Decision Making and Profiling*

The General Data Protection Regulation guarantees individuals the "right not to be subject to a decision based solely on automated processing, including profiling". If an FL customer has reason to believe that the processing used in an automated decision-making system might have legal effects concerning the customer, or similarly significantly affect the customer, the customer has the right to be informed and to have any concerns explained in a way that is understandable and not overly technical. The findings cannot be explained because the operational model is a black box, and the ML (including FL) training procedure is not publicly available (such as a global ML model in FL). This makes it hard to predict how a Data Subject's circumstances, behaviour, or choices might be affected by the outcomes of an ML model, or how the Data Subject's legal status or legal rights could be affected. Therefore, no FL system satisfies the promise made by the GDPR to provide human control over automated decision making. Data Controllers are not required to adhere to the right to be informed when telling Data Subjects that their rights with regard to automated decision making and profiling will not be honoured.

5.4 *Investigation and Demonstration*

Independent governmental agencies known as Data Protection Authorities are set up as supervisory bodies under the GDPR in each member state (DPAs). The burden of proof for complying with data protection laws rests on the Data Controller, while the DPA is responsible for monitoring and evaluating the Data Controller's actions to ensure compliance. These are astute questions to ask: How may DPAs look into and evaluate an FL setup, and how can compliance be proved?

5.4.1 *Competence of DPA's Supervision*

As shown in Fig. 4, DPAs conduct investigations into consumer suspicions or claims of noncompliance and determine appropriate penalties. An organization's compliance audit will reveal whether or not a Data Protection Impact Assessment (DPIA) and Privacy Impact Assessment (PIA) have been conducted, fulfilling a requirement of the General Data Protection Regulation (GDPR). Businesses that fail to comply with the General Data Protection Regulation (GDPR) face severe penalties,

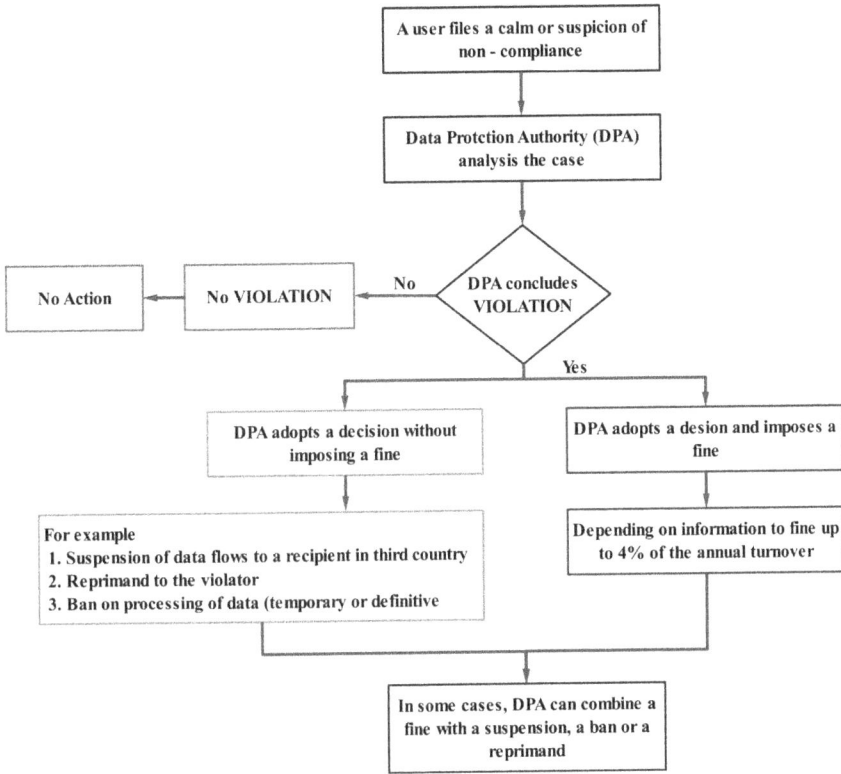

```
┌─────────────────────────────┐
│ A user files a calm or suspicion of │
│      non - compliance       │
└─────────────────────────────┘
               │
               ▼
┌─────────────────────────────┐
│ Data Protction Authority (DPA) │
│      analysis the case      │
└─────────────────────────────┘
               │
               ▼
            ◇ DPA concludes ◇
            ◇   VIOLATION   ◇
```

┌──────────┐   ┌──────────────┐   No   ◇ DPA concludes ◇
│ No Action │ ◄── │ No VIOLATION │ ◄───── ◇   VIOLATION   ◇
└──────────┘   └──────────────┘   ◇   ◇
                                              │ Yes
┌───────────────────────────────┐   ┌──────────────────────────────┐
│ DPA adopts a decision without │   │ DPA adopts a desion and imposes a │
│   imposing a fine   │   │   fine   │
└───────────────────────────────┘   └──────────────────────────────┘

┌──┐   ┌──────────────────────────────────┐
│ For example   │   │ Depending on information to fine up │
│ 1. Suspension of data flows to a recipient in third country │   │   to 4% of the annual turnover   │
│ 2. Reprimand to the violator   │   └──────────────────────────────────┘
│ 3. Ban on processing of data (temporary or definitive │
└──┘

┌─────────────────────────────────┐
│ In some cases, DPA can combine a │
│ fine with a suspension, a ban or a │
│   reprimand   │
└─────────────────────────────────┘

Figure 4. Workflow of the GDPR-compliance inspection and punishment procedure.

including fines of up to 2 crore or 4% of worldwide annual sales, whichever is greater, as well as possible public humiliation, exclusion from relevant markets, or even the suspension of operations. The level of cooperation shown during the investigation; the nature, severity, and duration of the current infringement; the existence of any prior violations; and the existence of any prior violations are all considered when determining the appropriate penalty, in addition to other factors explicitly stated in the GDPR. The law states that these firms urge customers to tick a little box indicating their agreement with the company's privacy policy in exchange for access to the services they provide.

According to data privacy specialists, this practise is in direct contravention of the General Data Protection Regulation. Wikipedia Article 13 provides a list of GDPR-related penalties and notifications (including the reasons for noncompliance). DPAs may request evidence on organisational and technological measures related to the fulfilment of GDPR criteria, as well as reports on independent DPIAs and PIAs done by the Controller. Data protection authorities may also need access to the underlying technology and management procedures for the servers housing any processed personal data. Providers of FL services may only demonstrate compliance with legal requirements by providing evidence of end-user consents and privacy-protecting technological protections such as homomorphic encryption, secure aggregation, and differential privacy. Any FL system would also struggle to answer questions from DPAs requiring access to training data or examination of local model parameters for a single user.

5.4.2 Compliance Demonstration

To create and show GDPR compliance, AI/ML-based service providers should conduct DPIAs and PIAs from the outset of the project and document the procedures properly. Data security and privacy risk assessments are crucial for demonstrating compliance, effectively managing risks, and determining what action has been done to mitigate such risks. However, not every data processing activity requires a DPIA or PIA to be conducted. If the operation "is likely to result in a high danger to the rights and freedoms of natural people," (NS/ > (1)), then the consent of those affected is necessary. DPAs must follow the criteria described in (3), 35(4) of the DPA Act to conduct DPIAs and PIAs. The following procedures for the DPIA/PIA should be carried out by all FL service providers in order to achieve GDPR-compliant and to demonstrate compliance as required by DPAs:

1. The steps involved in processing data, their associated goals, and an explanation and justification of those steps. For instance, it is important to record how the ML model parameters were communicated to the coordination server and how the Data Subject was asked to permit local ML training.

2. An analysis of the relevance and appropriateness of each procedure in light of its goals. For example, while a differential privacy technique is not absolutely necessary, implementing a Secure Aggregation mechanism is.

3. Each operation's potential impact on data security and privacy, as well as any technical steps used to address such concerns, must be evaluated. For instance, in an FL system, inference attacks might cause privacy issues when sending parameters from a local machine learning model to an organization server for a universal machine learning model enhancement. The methods described in the white paper make use of Secure Aggregation and Homomorphic Encryption. FL systems still have vulnerabilities that may be subjected to malicious

reasons, like model destruction through hidden sub-tasks, even with these additional precautions for users' privacy in place. It is critical to prevent such assaults to avoid potential violations of the General Data Protection Regulation.

It is difficult for service providers and DPAs to understand and audit FL since it is built on top of opaque ML models, similar to other AI/ML-based systems. Therefore, it seems that doing a DPIA/PIA on an FL system is a shallow, necessitating significant effort to identify violations of the requirements, allowing for safer operations and more stringent privacy protections.

6. Conclusions

The creation of AI and ML-based applications and services is a major focus for many sectors. However, as the proliferation of personal data goes unchecked, concerns about data privacy have grown. After high-profile data breaches like Facebook's Cambridge Analytica scandal and the SARS-CoV-2 epidemic, questions of algorithmic bias, ethics, political consequences, and legal accountability have garnered significant attention. For "data-hungry" artificial intelligence/machine learning-based systems in particular, FL presents a workable solution to the pressing demand for efficient privacy-preserving approaches. By decoupling data storage and processing at end-user devices from data aggregation on the server side, FL allows consumers to reclaim complete ownership of their data and makes them less susceptible to massive data breaches than they would be in a centralised system. Even though FL is still in its infancy, collaborative computing with decentralised data storage as in FL systems gives enormous benefits, allowing for a variety of AI/ML-based applications to be enabled without having direct access to end-users data. FL systems must adhere to the General Data Protection Regulation and other stringent privacy rules.

Since the parameters of a locally trained ML model hide sensitive elements that may be exploited to infer personal information about end users, they are an example of a local processing outcome that satisfies the GDPR's statutory data protection framework. Therefore, FL systems may violate the GDPR's privacy standards if they are vulnerable to attacks like inference attacks and model poisoning. Systems need to incorporate adequate privacy safeguards such as encrypted transfer learning, differential privacy, and secure multi-party computing (SMC). We discuss the potential security models, risks, and privacy-preserving methods in FL systems and examine their effectiveness in preventing information disclosure. An insightful examination of the GDPR compliance of an FL system is also given. Following the logical rules of the GDPR's six principles, this article analyses in depth the obligations of a service provider and the proper steps to design an FL system that is compatible with the GDPR. Since FL is still in its infancy, a wealth of multi/disciplinary study has only begun to ensure its eventual success and GDPR compliance.

To begin, distributed collaborative learning needs improved encryption and privacy primitives, especially for avoiding model poisoning and inference attacks. There is a need for greater study into the most efficient ways to use privacy-protecting techniques like SMC in FL applications because of their non-trivial performance overheads. Second, there must be an extensive study of FL systems' openness, interpretability, and algorithmic fairness. While much progress has been made on AI and ML in centralized settings, their applicability in decentralized contexts—where training data is spread across several servers—remains an open subject.

It is important to examine how limited sample sizes affect global training model bias and how this bias might be reduced. The agnostic FL framework, for instance, models the target distribution as an indeterminate blend of the distributions, thus it consistently provides good-intent fairness, whereas other FL training approaches use a uniform distribution. This FL structure eliminates the possibility of bias during the instruction phase. Third, for state-of-the-art encryption methods to be further integrated into FL systems, more study into interpretable and unbiased ML models and algorithms that can be applied across encrypted contexts is required. Furthermore, it is essential to investigate the trade-offs inherent in an FL framework with regard to privacy, accuracy, interpretability, and fairness. Users and politicians like the GDPR's regulatory body will be more receptive to the secure decentralized collaborative learning solutions if these prerequisites are satisfied.

References

[1] Eva, G., Liese, G., Stephanie, B., Petr, H., Leslie, M., Roel, V. and Greet, S. 2022. Position paper on management of personal data in environment and health research in Europe. Environment International, 165.

[2] Huang, W., Liu, J., Li, T., Huang, T., Ji, S. and Wan, J. 2023. FedDSR: Daily schedule recommendation in a federated deep reinforcement learning framework. IEEE Transactions on Knowledge and Data Engineering, 35(4): 3912–3924.

[3] Ma, C., Ren, X., Xu, G. and He, B. 2023. FedGR: Federated graph neural network for recommendation systems. Axioms, 12(2).

[4] Sousa, S. and Kern, R. 2023. How to keep text private? A systematic review of deep learning methods for privacy-preserving natural language processing. Artificial Intelligence Review, 56(2): 1427–1492.

[5] Tewari, A. 2023. mHealth systems need a privacy-by-design approach: Commentary on federated machine learning, privacy-enhancing technologies, and data protection laws in medical research: Scoping review. Journal of Medical Internet Research, 25.

[6] Brauneck, A., Schmalhorst, L., Kazemi Majdabadi, M.M., Bakhtiari, M., Völker, U., Baumbach, J. and Buchholtz, G. 2023. Federated machine learning, privacy-enhancing technologies, and data protection laws in medical research: Scoping review. Journal of Medical Internet Research, 25.

[7] Tacconelli, E. et al. 2022. Challenges of data sharing in European Covid-19 projects: A learning opportunity for advancing pandemic preparedness and response. The Lancet Regional Health - Europe, 21.

[8] Liu, Y. et al. 2022. Federated forest. IEEE Transactions on Big Data, 8(3): 843–854.

[9] Paragliola, G. and Coronato, A. 2022. Definition of a novel Federated Learning approach to reduce communication costs. Expert Systems with Applications, 189.

[10] Zuziak, M. and Rinzivillo, S. 2022. Federated Learning as an Analytical Framework for Personal Data Management—A Proposition Paper, 3221.

[11] Eliot, D. and Wood, D.M. 2022. Culling the FLoC: Market forces, regulatory regimes and Google's (mis)steps on the path away from targeted advertising. Information Polity, 27(2): 259–274.

[12] Xu, M. and Li, X. 2022. Federated Learning-based IDS against poisoning attacks. LNICST, 423: 331–345.

[13] Liu, S., Wang, J. and Zhang, W. 2022. Federated personalized random forest for human activity recognition. Mathematical Biosciences and Engineering, 19(1): 953–971.

[14] Truong, N., Sun, K., Wang, S., Guitton, F. and Guo, Y. 2021. Privacy preservation in federated learning: An insightful survey from the GDPR perspective. Computers and Security, 110.

[15] Li, S.C., Chen, Y.W. and Huang, Y. 2021. Examining compliance with personal data protection regulations in interorganizational data analysis. Sustainability (Switzerland), 13(20).

[16] Raja, V.V., Hemamalini, R.R. and Anand, A.J. 2011. Implementation of RETSINA-based congestion control technique in realistic WSN scenarios. Int. J. Comput. Sci. Eng. Technol., 1(10).

[17] Kalloori, S. and Klingler, S. 2021. Horizontal cross-silo federated recommender systems, pp. 680–684.

[18] Lin, Z., Pan, W. and Ming, Z. 2021. FR-FMSS: Federated recommendation via fake marks and secret sharing, pp. 668–673.

[19] Huang, A., Liu, Y., Chen, T., Zhou, Y., Sun, Q., Chai, H. and Yang, Q. 2021. StarFL: Hybrid Federated Learning architecture for smart Urban computing. ACM Transactions on Intelligent Systems and Technology, 12(4).

[20] Iqbal, Z. and Chan, H.Y. 2021. Concepts, key challenges and open problems of federated learning. International Journal of Engineering, Transactions A: Basics, 34(7): 1667–1683.

[21] Geng, J., Kanwal, N., Jaatun, M.G. and Rong, C. 2021. DID-efed: Facilitating Federated Learning as a service with decentralized identities, pp. 329–335.

[22] Raja, V.V., Hemamalini, R.R. and Anand, A.J. 2011. Study of traditional fairness congestion control algorithm in realistic WSN scenarios. Int. J. Eng. Res. Appl. (IJERA), ISSN, pp. 2248–9622.

[23] Jiang, D., Tan, C., Peng, J., Chen, C., Wu, X., Zhao, W. and Deng, L. 2021. A GDPR-compliant ecosystem for speech recognition with transfer, federated, and evolutionary learning. ACM Transactions on Intelligent Systems and Technology, 12(3).

[24] Pedrosa, M., Zuquete, A. and Costa, C. 2021. A pseudonymisation protocol with implicit and explicit consent routes for health records in federated ledgers. IEEE Journal of Biomedical and Health Informatics, 25(6): 2172–2183.

[25] Can, Y.S. and Ersoy, C. 2021. Privacy-preserving Federated deep learning for wearable IoT-based biomedical monitoring. ACM Transactions on Internet Technology, 21(1).

[26] Chen, L.Y., Chiu, T.C., Pang, A.C. and Cheng, L.C. 2021. FedEqual: Defending Model Poisoning Attacks in Heterogeneous Federated Learning.

[27] Cui, X., Lu, S. and Kingsbury, B. 2021. Federated acoustic modeling for automatic speech recognition, vol. 2021-June. pp. 6748–6752.

[28] Cheng, K., Fan, T., Jin, Y., Liu, Y., Chen, T., Papadopoulos, D. and Yang, Q. 2021. SecureBoost: A lossless Federated Learning framework. IEEE Intelligent Systems, 36(6): 87–98.

[29] Zhao, S., Bharati, R., Borcea, C. and Chen, Y. 2020. Privacy-aware Federated Learning for Page Recommendation, pp. 1071–1080.

[30] De Souza, L.A.C., Antonio Rebello, G., Camilo, G.F., Guimaraes, L.C.B. and Duarte, O.C.M.B. 2020. DFedForest: Decentralized Federated Forest, pp. 90–97.

[31] Yi, L., Zhang, J., Zhang, R., Shi, J., Wang, G. and Liu, X. 2020. SU-Net: An efficient encoder-decoder model of federated learning for brain tumor segmentation, 12396 LNCS, pp. 761–773.

[32] Tan, C., Jiang, D., Mo, H., Peng, J., Tong, Y., Zhao, W. and Xu, Q. 2020. Federated acoustic model optimization for automatic speech recognition, vol. 12114 LNCS, pp. 771–774.

[33] Vaishnavi, R., Anand, J. and Janarthanan, R. 2009. Efficient security for desktop data grid using cryptographic protocol. In 2009 International Conference on Control, Automation, Communication and Energy Conservation, pp. 1–6.

[34] Haffar, R., Domingo-Ferrer, J. and Sánchez, D. 2020. Explaining misclassification and attacks in deep learning via random forests, vol. 12256 LNAI, pp. 273–285.

[35] Yang, Q., Liu, Y., Cheng, Y., Kang, Y., Chen, T. and Yu, H. 2020. Federated Learning. Synthesis Lectures on Artificial Intelligence and Machine Learning, 13(3): 1–207.

[36] Song, T., Tong, Y. and Wei, S. 2019. Profit Allocation for Federated Learning, pp. 2577–2586.

Chapter 6

A Federated Learning Approach for Resource-Constrained IoT Security Monitoring

P Sakthibalan,[1] M Saravanan,[1,]* V Ansal,[2] Amuthakkannan Rajakannu,[3] K Vijayalakshmi[4] and K Divya Vani[5]

1. Introduction

"Internet of Things" (IoT) denotes a network in which everyday objects are digitally and physically connected to one another and the outside world [1]. Markit predicts that by the year 2030, there will be 125 billion

[1] Assistant Professor, Department of Electronics and Communication Engineering, Annamalai University, Annamalai Nagar, Chidambaram, Tamil Nadu, India - 608002.
[2] Department of Electrical and Electronics Engineering, NIT Goa, Farmagudi, Ponda, Goa-403401.
[3] Associate Professor, Department of Mechanical and Industrial Engineering, National University of Science and Technology, Muscat, Sultanate of Oman.
[4] Assistant Professor, Department of Electrical and Communication Engineering, College of Engineering, National University of Science and technology, Sultanate of Oman, Muscat.
[5] Assistant Professor, Department of Electronics and Communication Engineering, St. Martin's Engineering College, Secunderabad, Telangana, India.
Emails: balan1109@gmail.com; ansal.v@gmail.com; amuthakkannan@nu.edu.om; vijayalakshmi@nu.edu.om; sarav20021989@gmail.com
* Corresponding author: saravananm180982@gmail.com

Internet of Things devices in use worldwide. Smart homes, intelligent automation, smart cities, and cyber-physical systems are just a few of the many applications for IoT devices. Many kinds of embedded technology, including sensors, processors, and communication hardware, were used by these gadgets to gather and disperse information. The information gathered by these gadgets is shared with other similar devices on the edge network. By pooling resources, we can enhance data management and monitoring, as well as communication between machines and people and AI-based analysis. Despite this, these gadgets are increasingly being used as entry points for cybercriminals and other forms of online terrorism. As an illustration, the widespread Distributed Denial-of-Service (DDoS) attack powered by the Mirai infection on vulnerable Internet of Things (IoT) devices results in a catastrophic outage [2]. However, the processing capacity, memory, and processors of these IoT gadgets are quite limited. As a result, artificial intelligence (AI) methods optimised for desktop computers and servers are inapplicable to IoT gadgets due to their limited hardware and software resources. Therefore, the processes to tackle these difficulties in an FL setting where they are exacerbated must be efficient and effective in terms of both resources and outcomes.

ML methods, and Deep Neural Network (DNN) in particular, have been shown to have interesting applications in cyber security monitoring in recent research [3–5]. In contrast, a DNN-based cyber security scheme is unsuitable for application in the context of security monitoring because of the dispersed and constrained nature of IoT devices. Companies are concerned about data security when using centralised methods to train AI-based strategies (e.g., data centres). Despite the FL approach's potential, it may not be used by a wide variety of IoT and cyber-physical devices because of limitations in local storage, network throughput, processing power, and memory. The extensive list of model parameters can make it challenging to train a federated DNN model on a low-powered IoT device (millions in some cases). The following RQs are investigated in an effort to provide a workable federated DNN-based solution.

RQ1: Can state-of-the-art DNNs be trained under FL conditions, yielding a model well-suited to Network of Things security monitoring even when resources are scarce? Clarification can be found in Section 3.2.

RQ2: Will the resulting REFDL be able to reliably detect attacks on IoT networks? Subsection 5.

Using a Federated Averaging (FedAvg) DNN and eight Internet of Things (IoT) benchmark datasets, we construct a REFDL model for our experiments. Classification performance is enhanced in both synthetic and real-world testbed federation environments with the final Resource Efficient Federated Deep Learning, and it does so with reduced memory consumption. Federated integration of models helps protect private information while training models locally on IoT devices.

This chapter's structure continues as follows. Studies of relevance are reviewed in detail in Section 2. The proposed approach and FL technique employed are described in Section 3, and the evaluation procedure is outlined in Section 4. For the findings and discussion, please refer to Section 5. Afterwards, the report wraps up with suggestions for future study in Section 6.

2. Related Work

Here, we provide some related research on the topic of deep learning for keeping an eye on the safety of the Internet of Things, and we do the same for FL and its uses in Internet of Things settings.

Deep Neural Network in IoT. There has been a lot of discussion in the academic world about using AI to keep an eye on the security of the IoT. These methods often employed deep neural networks [6]. To classify network traffic, authors used DNN to analyze information from the Internet of Things [7]. Authors performed the same analysis with data from smart cities enabled by the Internet of Things [8]. Convolutional neural networks with compact structures were introduced by the authors for use in a low-resource Internet of Things (IoT) environment (CNN) [9]. By proving its worth on widely used testbeds like CIFAR-10 and Imagenet, we can see that it is a powerful tool. Major obstacles to the approach's success include the absence of validation of Internet of Things benchmark datasets and the failure to take memory utilisation into account. Quantized convolutional neural networks were used by literature to extract information from radar sensor data (CNN) [10]. To complete classification tasks on very low power devices connected to the Internet of Things, authors use DNN and, more particularly, Fully Connected Neural Networks (FCNN) [11]. To achieve this goal, we will not compromise accuracy for simplicity in the model. Model complexity was likely not considered while choosing the FCNN design, which could limit the applicability of the methodology. Quantization of weights and bias factors was also considered by the majority of the described optimization methods. Effective and trustworthy DNN techniques for centralised IoT security monitoring were developed by authors [12]. However, the approach we propose in this paper may function in distributed settings, using less memory and accelerating the completion of tasks. The method's memory requirements are decreased, and its precision is increased using simulated micro-batching, pruning, and variable regularisation. This is helpful for situations where there are limited resources available for the task of dispersed learning.

FL in IoT Environments. The original FedAvg FL strategy proposed by the authors, offered training a single local model on numerous clients without requiring them to exchange their local data with a server [13]. Model convergence employing diverse client-side data in non-decentralized and non-identical distributions appears promising thanks to this method. As a result, scholars from many different fields have taken many different approaches to studying FL techniques. FL is becoming widely used in the field of monitoring the security of the IoT. The authors described utilising FL to identify intrusions in the Internet of Things networks [14]. Literature has discussed how to work around the restrictions of FL software on IoT gadgets [15]. FL's attack detection capabilities for industrial IoT devices were described in detail by authors [16]. Literature investigation is substantially improved with the addition of raw sensor reading data [17]. Federated Learning training on edge devices can be made more efficient with the help of model pruning [18]. The authors created a Federated Learning architecture that may be easily converted to mobile devices to lower the overall cost of communication [19]. A zero-day flaw in an Internet of Things network was discovered using FL [20]. These implementations make use of the FL data privacy and disregard any restrictions imposed by the underlying hardware or software. As a means of enhancing FL and reducing the load on IoT network memory resources, authors [21–23] implemented model parameter reduction and data parallelism (micro-batching). However, none of these methods consider the possibility of optimizing FL to lessen the time and memory costs of the many DNN variants. We use raw data on network traffic from a variety of IoT devices to fine-tune the federated training process and get around this problem. Our low-overhead REFDL method was rigorously tested with both simulated and real embedded devices in a wide range of conditions to ensure that it maintains state-of-the-art accuracy while requiring a minimum of memory and processing time.

3. Methodology

This will be demonstrated by determining the REFDL by optimising the BFDL on a small sample of real-world and IoT benchmark datasets.

We show that the REFDL can be generated using a carefully optimised version of the BFDL algorithm. When applied to non-IoT datasets, the efficient REFDL can detect IoT attack actions and provide accurate categorization.

3.1 *Baseline Federated Deep Learning*

Baseline Federated Deep Learning combined a traditional DNN (FCNN/ CNN) model with the FedAvg algorithm. The integrated DNN is an artificial neural network with many layers of neurons that can accurately represent the data you feed it. These neurons represent the computational units that may communicate the outcomes of operations performed on their activation function and input. In the FCNN, neurons are linked to their weights and biases in a sequentially. In this sense, the weights and biases are similar to computer memory. In Algorithm 1, we find the network architecture, activation functions, weight, and bias values, and BFDL (Mn) baseline model. Parameters like weight and bias can be tweaked to fine-tune the error function Mn produced over training data Dtr. Following these guidelines will result in a "master" FedAvg model that includes information from all your client models. Using backpropagation and the Stochastic Gradient Descent (SGD) algorithm, Algorithm 1 describes how to train an Mn-size (or larger) network. Using Algorithm 1's Device UPDATE feature, the server can update the master model and re-distribute it to the clients after every communication round. In the third algorithm, clients iteratively use their data to update the gradient descent weights they submit to the server. This is settled upon to develop a universal reference model by minimizing the cost function in Equations 1 and 2. Algorithm 1's lines 11 and 12 are then used to make educated guesses about the algorithm's execution time and memory requirements. After the local weights have been updated, this data represents the device-level training resource logs. In the third method, the weights of the models are transmitted to the coordinating server. When integrating global models, the server will often use the return weight midpoint. By adding a Dtr-learning function to the global model, previously unseen data can be accurately mapped. For classification purposes, the Baseline Federated Deep Learning technique employs supervised DNN (FCNN and Convolutional Neural Network), with Mn receiving a data stream (Dtr) and producing a probability class vector ($\hat{Y}$). As shown in Equation 3, a threshold value t is used to round the results to the nearest integer. Example of safe (1) or dangerous (0) traffic against a sample of the Internet of Things data or an image dataset is shown in this output.

Algorithm 1. Baseline BFDL Training

Input: Labelled data D*tr*, Iteration number T, Batch size S
Output: Baseline model M*n*
function Base (D*tr*[ ])   Training baseline model
 for $i = 1$ to T **do**
          Mini-batch $B = \{(x_1,y_1),...,(x_m, y_m)\} \subset D_{tr}$
          $F_p(B)$ Forward propagation with B
          $\mathcal{E}_i \leftarrow L$ L= Base loss
          $B_p(B)$ Backward propagation
          function Device Update((d))   Run on device d
               $B_s \leftarrow$ (data P_d in batches of size B)
               for batch $b \in B_s$ **do**
                    $w \leftarrow$ local weights update   device local weights update computation
                    Estimate m_i Execution memory at epoch i
                    Estimate t_i Execution time at epoch i
                    M*n* = Trained model that estimate E*i*, m_i, t_i
               end for
          end function
 end for
 return w to server in Algorithm 3 *Calls* to coordinating server in Algorithm 3 for weights averaging
 return (M*n*, E*i*, m_i, t_i)
end function

$$J(W,b) = \frac{1}{m}\Sigma_{i=1}^{m} L(\hat{Y}^i, Y^i) \qquad (1)$$

$$L(\hat{Y}^i, Y^i) = -(Y \log \hat{Y} + (1-Y)\log(1-\hat{Y})) \qquad (2)$$

$$\text{Output} = \begin{cases} 0 \text{ if } \hat{Y} \leq t \\ 1 \text{ if } \hat{Y} > t \end{cases} \qquad (3)$$

3.2 Resource Efficient Federated Deep Learning (REFDL)

It can be difficult to train a DNN model for FL that uses fewer resources, as was noted above, and this is especially true for IoT security monitoring [24, 25]. To design and implement an optimal architecture, we need to think about both the frequency of FL communications and the specifics of the DNN itself. The difficulty of this method increases as more dimensions are added to the dataset.

Algorithm 2. The Proposed Strategy for Obtaining Resource Efficient Federated Deep Learning

Input: Penalty term = λ, (Dtr, T, B, L, in Algorithm 1)
Output: An Efficient Model of Me
function Efficient (D*tr*[ ])
  for j = 1 to T ; **do**
      Micro-batch $M = \{(x_1,y_1),...,(x_m,y_m)\} \subset B$
      $F_p(M)$ M = Forward propagation
      $\mathcal{E}_t = L$ L = Initialized loss
      Memory and time can be initialised based on εt, therefore mt and tt can be estimated.
      $\mathcal{E}_j \leftarrow \mathcal{E}_t + \lambda \sum_{j=1}^{W} \dfrac{\left(w_j^2/w_0^2\right)}{\left(1 + w_j^2/w_0^2\right)}$
      B_p(M) M = Backward propagation
      function Device Update$((d))$ Run on device d
          $M_t \leftarrow$ (data P_d in batches of size M)
          for batch $b \in M_t$ **do**
              $w \leftarrow$ local weights update   device local weights update computation
              if (Ej $\leq$ Et) **then**
                  $\lambda = \lambda + \Delta\lambda$
                  Estimate m_j Execution memory at epoch j
                  Estimate t_j Execution time at epoch j
                  if $((m_j < m_t) \wedge (t_j < t_t))$ **then**
                        $m_{tr} = m_j$ m_{tr} = Efficient memory
                        $t_{tr} = t_j$ t_{tr} = Efficient time
                        Me = Trained model that estimate Ej,m_{tr},t_{tr}
                  end if
              end if
          end for
      end function
  end for
  return w to server in Algorithm 3 Calls to coordinating server in Algorithm 3 for weights averaging
  return (Me,Ej,m_{tr},t_{tr})
end function

To that purpose, we employ BFDL's reference architecture to generate more efficient variants of the basic model (REFDL). Using Dtr, the REFDL model's efficient Me is optimized during training, as explained in Algorithm 2. Micro-batching is shown in Line 3 of Algorithm 2, which is useful for breaking down large datasets into manageable chunks for quick on-device model training. Micro-batching, unlike mini-batching, is compatible with most datasets, including those produced by the IoT. A penalty (weight removal) approach with a threshold value of w0 was used to regularize Equation 4 and thereby reduce the number of network nodes [26]. To efficiently perform local weight updates in the network, it is necessary to identify the groups of weights that are relevant to the problem at hand. In particular, we need to find out which of the baseline model's huge weights are actually important. For weights above w0 with a complexity cost of around 1, a regularization with the penalty parameter λ is necessary. Algorithm 2 instead finds the optimal amount of weights to drop, considering the DNN architecture and any additional constraints.

Specifically, in line 7, the regularization considers a situation in which the initialized model generates a larger error value ε_i. We used the parameters to get a reduced error value εj, which resulted in better performance. When the initialized values in line 6 are compared to the calculated computational memory footprints and execution time, the client device model returns the minimal memory constraint in lines 15 and 16. In the third algorithm, the device models that use the fewest resources are sent back to the central server along with their respective weights. The coordinating server in a federated setup updates the weights used in model aggregation by taking an average of the weights provided by the client models. The client's time and energy spent on computing can be cut down by using this method while developing the REFDL aggregate model. When the model finally converges, the cumulative savings in time and energy from each client device over the course of each federated round could be substantial.

Algorithm 3. Method of Coordinating Algorithms 1 and 2

> **Server Executes:**
> **function** Server Weights Update
>   initialize weight w;
>   **while** $t \le n$ **do**   *n* federated round
> $R \leftarrow$ random set of $max(C.K, 1)$ *C.K* fraction of clients *K*
>   **for** $k \in R$ in parallel **do**   *k* client index
>   Weight device update *Federated* model weight update for Algorithms 1 or 2
>   **end for**
>   Averaged weights update
>   **end while**
>   **return** Averaged updated weights
> **end function**

4. Evaluation

The criteria by which the BFDL and REFDL approaches are judged are outlined below. Moreover, it provides access to the datasets that were employed to test the methodology.

4.1 Datasets Used

The IoT dataset is a collection of multiple representative sample data from nine commercial IoT devices, and it includes a large quantity of botnet and benign network traffic flows. Both the BASHLITE and the Mirai strains are rather widespread, thus each unit has one of them. To be more precise, we randomly select eight categories of devices that have the most demands in terms of the Internet of Things. Just a few examples are as follows: This package includes the following products: an Ecoobee thermostat, a Danmini doorbell, an Ennio doorbell, a Provision PT-838, a Provision PT-737E, a SimpleHome XCS-1002WHT, a Samsung SNH-1011-N, and a SimpleHome XCS-1003WHT. Each device may keep track of 115 attributes vector, which is more than enough information to mathematically represent network flows. This includes numerous assault varieties such as syn, ack, scan, trash, UDP, TCP, udpplain, combo, and regular. We, therefore, present the N-BaIoT dataset as the foundation for the device-centric IoT security monitoring technique. Our BFDL and REFDL models were trained and evaluated using federated data from commercial devices linked to the N-BaIoT.

Multiple traffic flows from a simulated SCADA system are included in the WUSTL dataset [27–29]. This data collection may be suitable for studying the practicability of AI algorithms for use in surveillance. There are 7,037,983 samples of raw data to look at. To assess the efficacy of our approach, we take into account the frequency of occurrence of 471,545 attacks and 6,566,438 control instances.

4.2 Experimental Establishment of a Workforce of Virtual Employees

All procedures were implemented in Python 3.7 and tested on a personal computer outfitted with four 2.10 GHz Intel Xeon E5-2695 processors and 16 GB of RAM. To estimate memory consumption, we created a profile of integrated memory usage [30]. For the simulated on-device training, we employed the PyTorch 1.4.0 [28] and PySyft 0.2.9 frameworks [31]. Making a virtual worker is a breeze with the Pysyft framework. For the BFDL and REFDL, we simulated the FL setting by using these digital employees. These workers are emulations of genuine virtual machines that can function independently in the same Python application as their data [32, 33]. Each client's virtual worker model was updated computationally and sent to a coordinating server worker, who was part of our federation training program. A federated client model has five unique layers: an input, an intermediate, an output layer, and two further layers. Topologies were chosen for each dataset to optimize performance metrics and reduce the number of processes required [34]. The findings of the parameter adjusting approach indicate that the examined experimental circumstances are suitable for binary classification. When comparing the performance of BFDL to the suggested REFDL method, the underlying architectural settings are kept constant. The deployed model architecture for each FL method is shown in Table 1 for easy comparison. The preferred topology for the Wustl dataset was reported based on the tuning method.

Table 1. Data on each device's normal and assault distribution and design.

Device	Normal	Attack	Inputs	Outputs	Architecture
Ennio Doorbell	39,200	317,500	115	1	83-128-128-83
Danmini Doorbell	49,612	967,860	115	1	83-128-128-83
Ecobee Thermostat	13,124	824,775	115	1	83-128-128-83
SimpleHome XCS-1003-WHT	20,535	842,319	115	1	83-128-128-83
Samsung SNH-1011-N	52,161	323,084	115	1	83-128-128-83
Wustl	6,567,443	472,556	6	1	26-128-128-26
Provision PT-838	98,523	729,873	115	1	83-128-128-83
SimpleHome XCS-1002-WHT	46,596	817,482	115	1	83-128-128-83
Provision PT-737E	62,163	767,108	115	1	83-128-128-83

lr = 0.001 was employed in both the primary and secondary model training procedures. To construct the REFDL method model, we used values of 0.01 for λ, $\Delta\lambda$, and the threshold w0 across 4 microbatches. The Relu activation function can be used in the fully connected layers, while a sigmoid can be used in the last layer. For FedAvg training, both BFDL and Resource Efficient Federated Deep Learning employ an SGD optimizer. Optimal convergences were achieved by training each federated model in a total of 128 batches over the course of 4 epochs and 30 rounds of worker-to-worker communication. The coordinated worker is informed of the typical weights once the client's model training is complete. This worker must compile the necessary weights for the global model. The underlying code for this answer can be found on GitHub [35, 36].

4.3 Establishing an Experimental Testbed and Conducting Tests

The PySyft python framework version 0.2.9, which comprises a client and server class coupled by WebSocket (WS), was utilised to examine the viability of federated communication in the REFDL and BFDL environments. Our distributed FL training environment is well-suited to low-powered devices, and we take advantage of the fact that PyTorch might be used as a library for PySyft. The configurations are meant to function in a decentralized fashion similar to the client-server communication scenario they simulate. It can help construct virtual and realistic testbed environments in this situation. We thought of using a laptop and 4 Gigabyte Brix (GB-BXBT-2807) in a network-realistic testbed environment. For the sake of simulating low-frequency communications in a wireless network, the individual laptop takes the role of the coordinating server. Model weights are aggregated on the server and sent out to users. In Algorithms 1 and 2, the client's devices train a local model on the client's dataset using the server model weights and then send those weights back to the server. Therefore, the client side, which includes the edge devices, bears a greater share of the communication burden than the server computer. Ubuntu 20.04.4 LTS is the version of Ubuntu preinstalled on GB-BXBT-2807 clients. The PySyft framework and all necessary dependencies are pre-installed on each client. The source code for implementing a federated network testbed is available to the public.

This was accomplished by contrasting the simulated runtime of BFDL and REFDL with their real execution time using four workers (Alice, Bob, Charlie, and Jane) and their distributed training data (shown in Fig. 1). The enhanced tuning approach yields a 64-mini-batch size for usage with a 50-round federated communication round with two epoch iterations. For efficient FedAvg SGD education, a sample size of 1000 is used for the test batch with a learning rate (lr) of 0.01. Both Algorithm 1 and Algorithm 2 use real-time models, which consist of an input, four hidden (128-128-128-128), and an output layer. Model convergence can be accomplished efficiently and effectively with the help of the selected

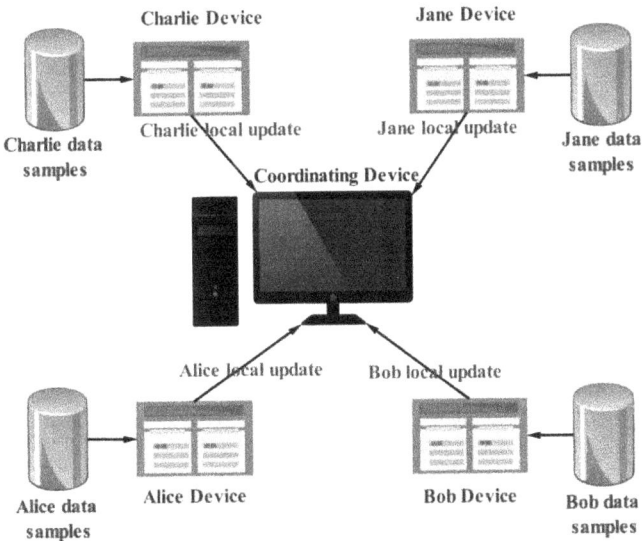

Figure 1. Model-training testbed for BFDL and REFDL using gigabyte-class devices.

architecture. Studying the CNN DNN variation in real-world scenarios with clients using the MNIST picture dataset allowed us to evaluate the efficacy and generalizability of REFDL. There are two convolutional layers in the CNN design (Conv-2). A 5 square kernel is used in the first 2D convolutional layer, which takes a single input and generates 20 convolutional features (1, 20, 5, 1). Using a 3 square kernel, the second 2D convolutional layer can produce 50 convolutional features from its 20 input layers (20, 50, 5, 1). There are two layers of real-time architecture, the first of which is represented by (800) (4450) (128), and the second by (128, 10). There were no hiccups when applying Max-2d Pool's algorithm to the input image. Comparing the convolutional fully connected hidden layers to Table 1 version yields similar results.

5. Results and Discussion

The paper reflects on the experimental outcomes. It describes the testbed and simulation analyses that compared the enhanced REFDL and the original BFDL FedAvg models.

5.1 Virtual Workers Simulation Results

Scientists assessed the effort and time needed to train BFDL and Resource Efficient Federated Deep Learning federated algorithms on nine widely used IoT datasets utilising remote client PCs. Tabulated in Table 2 below is the amount of storage space and processing speed needed for each data collection. With REFDL training, you may reduce your system's memory usage and improve performance. However, REFDL and BFDL maintained the same level of accuracy on all the test data sets. This could be because all the models use a very similar network architecture and because the tested datasets are highly skewed in favour of the test records. For a comparison with a balanced dataset with a small sample size of test records, see Table 6.4 and Fig. 10.

The percentages of memory and time saved using Resource Efficient Federated Deep Learning, as shown in Table 2, are displayed in Figs. 2 and 3. The outcomes show that large amounts of memory are conserved across all datasets. As far as the client's processing time is concerned, REFDL is more effective. It's an improvement over BFDL in that it requires fewer inputs, has a higher potential for rapid learning, and produces more fruitful performance outcomes. Since it uses fewer resources, it's a better option for keeping an eye on the safety of the Internet of Things. In particular, for resource-limited edge devices in a distributed network, on-device learning is essential.

Table 2. As compared to REFDL, BFDL's federated model training requires more memory.

Dataset	Model	Memory (MB)	Time (mins)	Test (accuracy %)
Danmini Doorbell	BFDL	3.792	0.098	96.24
	Resource Efficient Federated Deep Learning	0.857	0.084	96.24
Ecobee Thermostat	BFDL	3.745	0.093	94.45
	Resource Efficient Federated Deep Learning	0.815	0.082	94.45
Ennio Doorbell	BFDL	4.147	0.092	89.87
	Resource Efficient Federated Deep Learning	0.805	0.076	89.87
Provision PT-838	BFDL	3.436	0.092	89.06
	Resource Efficient Federated Deep Learning	0.814	0.076	89.06
Provision PT-737E	BFDL	3.463	0.094	93.48
	Resource Efficient Federated Deep Learning	0.853	0.081	93.48
Samsung SNH-1011-N	BFDL	3.792	0.099	87.01
	Resource Efficient Federated Deep Learning	0.858	0.081	86.06
Simple Home XCS-1002	BFDL	3.494	0.090	94.65
	Resource Efficient Federated Deep Learning	0.816	0.072	94.65
Simple Home XCS-1003	BFDL	3.914	0.085	97.73
	Resource Efficient Federated Deep Learning	0.801	0.071	97.73
Wustl	BFDL	3.002	0.095	94.26
	Resource Efficient Federated Deep Learning	0.816	0.076	94.26

Table 3 displays the outcomes of the deployed BFDL method and its optimized equivalent REFDL in comparison to training methods. Multiple model iterations employing both hidden layers (L) and virtual workers (VW) were analyzed for their effects on training time, memory needs, and accuracy. As demonstrated with the XCS-1003 dataset, the REFDL uses less time and memory than its predecessor. With four hidden layers, federated models such as BFDL and REFDL achieve marginally higher accuracy (4L). Therefore, efficient federated learning in distributed environments is affected by the number of hidden layers. However, clients with more computing power can have a much larger impact on the global

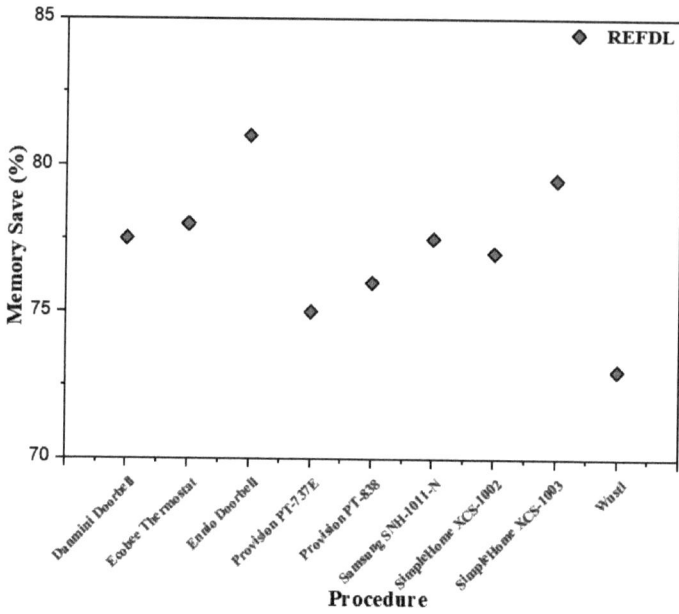

Figure 2. Memory resources for training models in REFDL's federated format, indexed against datasets.

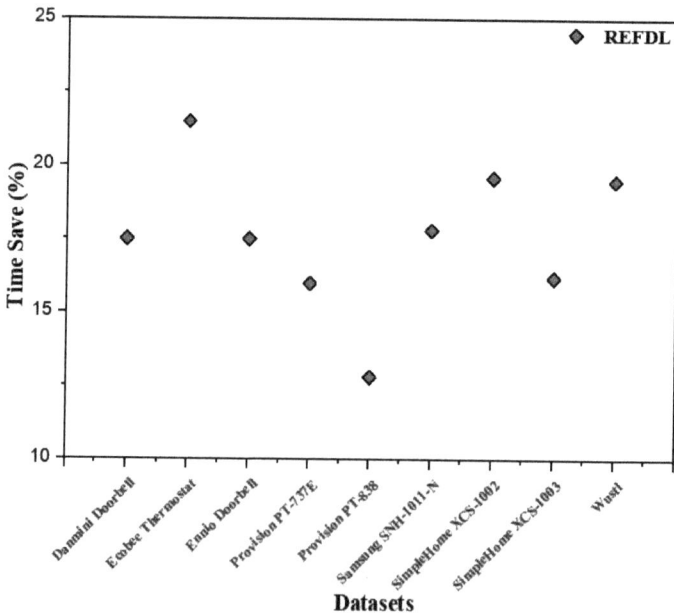

Figure 3. Time and resources were conserved while training models via the REFDL federation.

Table 3. Performance evaluations on the Simple Home XCS-1003 dataset compared to the FL training process.

Process	Model	Time (Min)	Memory (MB)	Test (acc %)
2VW-3L	BFDL	0.038	1.906	97.72
	Resource Efficient Federated Deep Learning	0.027	0.550	97.72
2VW-4L	BFDL	0.046	2.698	97.73
	Resource Efficient Federated Deep Learning	0.036	0.052	97.73
4VW-3L	BFDL	0.067	2.971	97.72
	Resource Efficient Federated Deep Learning	0.060	0.294	97.72
4VW-4L	BFDL	0.085	3.914	97.73
	Resource Efficient Federated Deep Learning	0.071	0.801	97.73

model than those with fewer resources. To further complicate matters, the global model's precision may be impacted by the clientele's skewed distribution of income and education levels. In subsequent research, it would be fascinating to investigate these restrictions.

Savings in memory and processing time due to REFDL are shown in Figs. 4 and 5 concerning Table 3. The outcomes show that REFDL performs better at minimizing resources (memory and time) compared to each training method. Results show that using constructions from the PySyft virtual worker, which mimic real-world virtual machines and run independently within the same Python script, improves REFDL's performance. Specifically, it shows how the REFDL model's employment of 4 hidden layers (4L) and 2 virtual workers (2VW) may exhibit the potential model for lower memory consumption. It also shows that memory savings may be achieved by increasing the number of virtual employees using a three-hidden-layer network architecture.

We evaluated REFDL on the MNIST image dataset using an FCNN and CNN integration in the FL scenario to demonstrate its generalizability and effectiveness outside of the cybersecurity sector (see Table 4). We can understand whether the method can be utilized as a general-purpose option for on-device learning by evaluating its performance on datasets that are unrelated to the Internet of Things. In particular, to reap the rewards of CNN's dominance over human experts in image classification and its capacity for efficient resource management. Through this analysis, we compared the efficacy of BFDL and REFDL across all federated training situations using the PySyft WS (network) simulated workforce. As

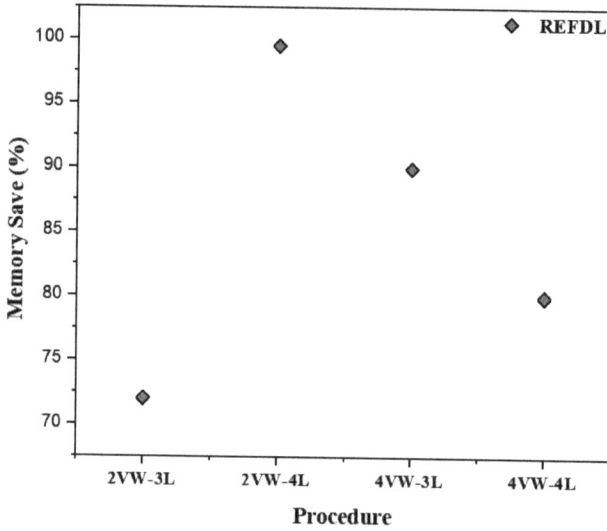

Figure 4. Model training data for the XCS-1003 dataset in REFDL federated memory.

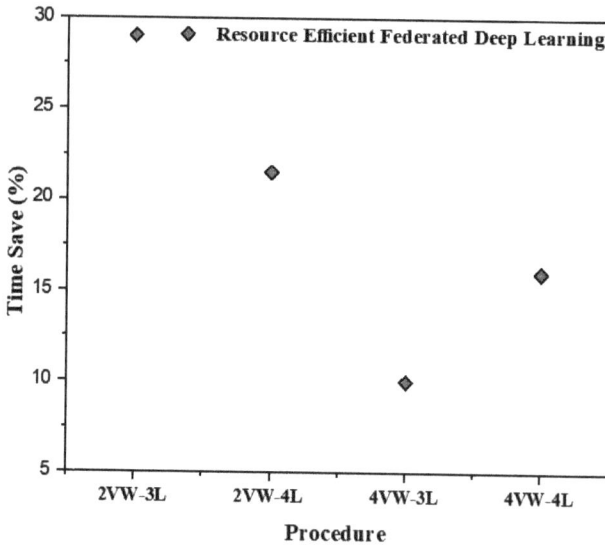

Figure 5. Training using the XCS-1003 dataset reduces the time and resources efficient federated deep learning models.

opposed to PySft virtual workers, which operate as a construct within the same Python programme, this method evaluates REFDL's performance by simulating a network in which both clients and servers reside on the same machine. The REFDL outperforms the BFDL in accuracy with

Table 4. A study comparing the effectiveness of simulated federated training on the MNIST dataset using BFDL and REFDL.

Process	Federated Learning	Time (Min)	Test set (acc %)	Time (save %)
FCNN-MNIST	BFDL	1.393	34.64	N/A
	Resource Efficient Federated Deep Learning	1.346	91.03	3.374
CNN-MNIST	BFDL	1.583	90.59	N/A
	Resource Efficient Federated Deep Learning	1.457	98.28	7.960

all DNN (CNN and FCNN) variants. Because the MNIST dataset has a relatively balanced distribution for both training and testing examples, it may be responsible for the superior performance it has achieved. And the time needed to finish a training session is cut down significantly. The importance of regularization in protecting accuracy when dealing with DNN fluctuation is demonstrated by the outcomes. This motivates additional research in practical contexts.

5.2 *Network Workers Testbed Results*

The amount of space required to train Resource Efficient Federated Deep Learning and Baseline Federated Deep Learning on four GB-BXBT-2807 edge devices using data from the Ennio Doorbell and the Samsung SNH Internet of Things datasets was calculated with the help of a mobile network simulator. Data in Table 5 for memory specifications are a mean across all four devices. The findings demonstrate that the REFDL is more suited for real-time IoT attack detection with fewer memory requirements than the BFDL. The number of federation clients and the feature distribution choices determine the precise memory and CPU (runtime) reductions. Increased cost reductions may result from a federation's broad client base processing complex data.

Table 5. Training memory utilisation for federated models using Resource Efficient Federated Deep Learning and Baseline Federated Deep Learning.

Dataset	Model	Memory (MB)	Test (accuracy %)	Memory (Saving %)
Ennio Doorbell	BFDL	33.965	89.00	Not Applicable
	Resource Efficient Federated Deep Learning	31.981	89.00	5.84
Samsung SNH	BFDL	32.519	86.10	Not Applicable
	Resource Efficient Federated Deep Learning	30.550	86.10	6.05

Data from the Ennio Doorbell and Samsung SNH Internet of Things datasets were used to train both BFDL and Resource Efficient Federated Deep Learning, as shown in Fig. 6. Real-time depiction of the average predicted convergence is also provided. When it comes to spotting IoT assaults, the REFDL doesn't need as much real-time as the BFDL does. The findings validate REFDL's use for reducing computing demands in limited settings. Figure 7 shows that more distributed edge devices can increase efficiency.

The outcome proves that REFDL saves more money than virtual WS connections when used in actual networks. It demonstrates REFDL's potential to reduce energy consumption in settings where limited resources are shared across numerous client devices.

We used the MNIST data set to compare REFDL with BFDL in terms of how much time it saves. According to Fig. 8, REFDL is more effective than BFDL at all phases of training. When compared with the MNIST-FCNN, the MNIST-federated CNN's training procedure is more laborious and needs more data storage space. One possible solution for learning directly on the device is the FCNN DNN variant of Resource Efficient Federated Deep Learning, especially if minimizing development costs is a top goal. REFDL is the preferred approach for implementation in an IoT resource setting.

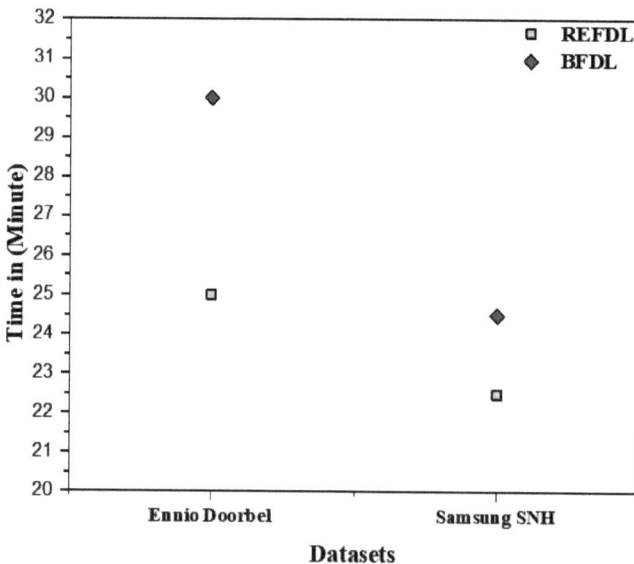

Figure 6. Comparison of REFDL and BFDL's training execution times on different datasets using a federated model.

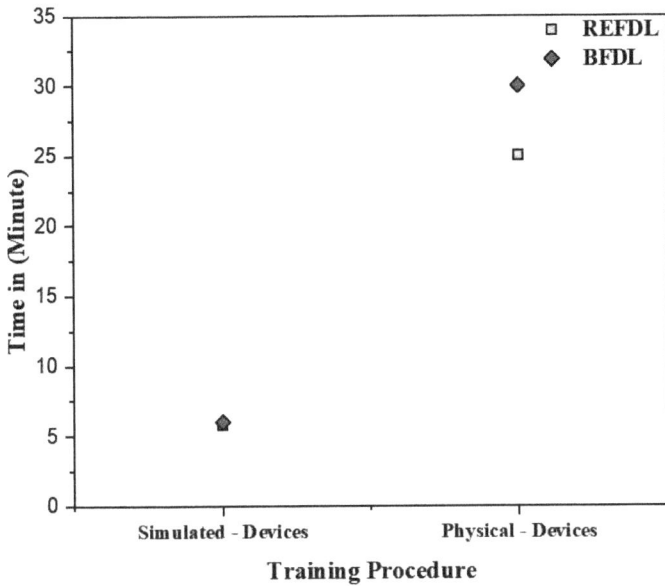

Figure 7. Examining the performance of REFDL and BFDL on the Ennio dataset in a synthetic and realistic network context.

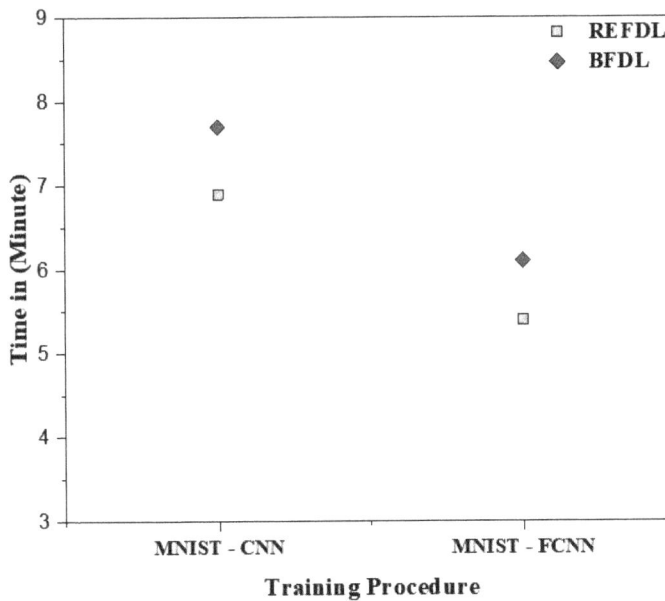

Figure 8. Training times for federated models on the mnist dataset using REFDL and BFDL.

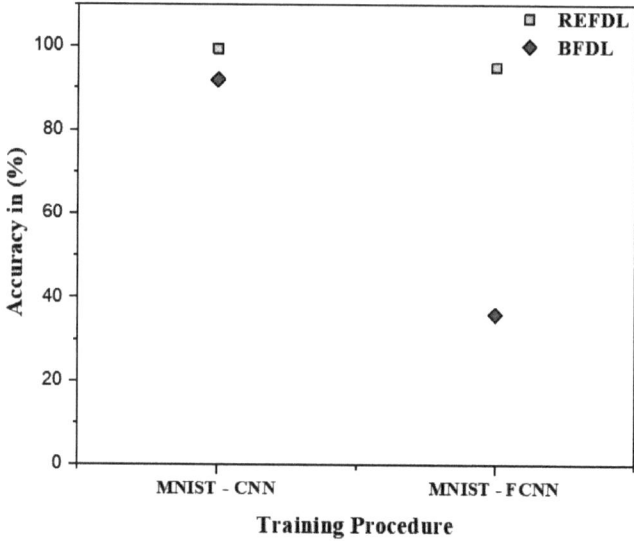

Figure 9. The mnist dataset is utilised to estimate the federated model accuracy of REFDL and BFDL.

A comparison of REFDL and BFDL's convergence accuracy on the MNIST dataset to those of other DNN versions is shown in Fig. 9. Therefore, it may be assumed that REFDL will always function better than BFDL. By combining multiple variations of the CNN and FCNN (DNN) models, it can correctly categorize image samples. This finding points to the value of optimization processes in the development of a globally deep federated model. It further highlights how effectively using CNN in the FL technique can boost accuracy. This is helpful since it makes use of the differences in on-device learning efficiency between the various DNN models.

To evaluate REFDL's efficiency and speed of learning in a federated environment, we run it through the FCNN-MNIST process using different numbers of epoch iterations on the GB-BXBT-2807 testbed. We can rapidly evaluate the relative merits of several federated methods there. Figure 10 shows that a better level of precision can be achieved by the REFDL with only one local epoch and 50 communication cycles. All subsequent epoch iterations maintain this pattern of ever-increasing precision. In particular, the integrated FCNN model speeds up learning across edge devices, demonstrating REFDL's applicability. The restriction that REFDL must have at least one epoch is a plus. Especially in a context like the IoT, where memory constraints are already in place.

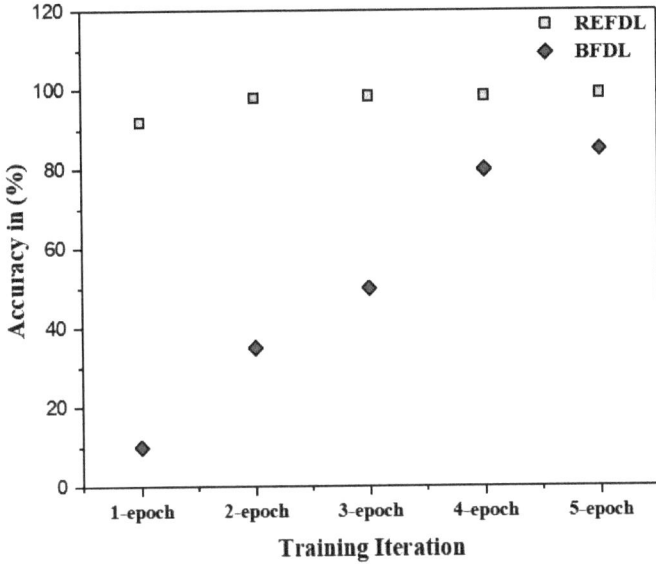

Figure 10. The effectiveness of a federated model on the mnist dataset at different epochs comparing REFDL and BFDL.

6. Conclusion

In this article, we explored the feasibility of training Federated Learning in low-resource environments like the Internet of Things, where its distributed Machine Learning model can simultaneously safeguard the privacy of a large volume of customer data and facilitate on-device learning at the network's peripheral nodes. To furnish effective and practical protections against the Internet of Things. In this study, we combine the BFDL algorithm FedAvg with a few hand-picked model optimization strategies to create a federated REFDL model that is both powerful and economical with its use of data. Eight IoT datasets and one picture dataset were used to evaluate REFDL's efficacy, simplicity, and efficiency on both a simulated and a real-world GBBXBT-2807 testbed. It is superior to other methods in its ability to identify Internet of Things assaults while utilising very few resources. In a federated training setting, it can classify image data more accurately than either a fully connected network or a convolutional network trained with the same data. In addition, a more accurate FL model can be generated by REFDL in less time than its competitors. Inspired by these encouraging findings, we want to do additional research into the use of more nodes/client devices in computational networks during deployment, focusing on both wired and wireless environments. To be more specific, in a realistic network scenario with a greater range of connected edge devices than those explored in this research, we aim to investigate the REFDL's resilient capabilities to increase its security resilience against adversarial assaults. In this way, we can examine the impact of our proposed method on resource usage and security monitoring over a wide set of geographically scattered edge devices.

References

[1] Tabassum, N., Ahmed, M., Shorna, N.J., Sowad, M.M.U.R. and Haque, H.M.Z. 2023. Depression detection through smartphone sensing: A Federated Learning approach. Int. J. Interact. Mob. Technol., 17(1): 40–56. doi: 10.3991/ijim.v17i01.35131.

[2] Ahmed, U., Lin, J.C.W. and Srivastava, G. 2022. Mitigating adversarial evasion attacks by deep active learning for medical image classification. Multimed. Tools Appl., 81(29): 41899–41910. doi: 10.1007/s11042-021-11473-z.

[3] Zhang, C., Zhu, Y., Markos, C., Yu and Yu, J.J.Q. 2022. Toward crowdsourced transportation mode identification: A semisupervised Federated Learning approach. IEEE Internet Things J., 9(14): 11868–11882. doi: 10.1109/JIOT.2021.3132056.

[4] Marulli, F., Balzanella, A., Campanile, L., Iacono, M. and Mastroianni, M. 2021. Exploring a Federated Learning approach to enhance authorship attribution of misleading information from heterogeneous sources. In Proceedings of the International Joint Conference on Neural Networks, vol. 2021-July. doi: 10.1109/IJCNN52387.2021.9534377.

[5] Mei, G., Guo, Z., Liu, S. and Pan, L. 2019. SGNN: A graph neural network based federated learning approach by hiding structure. In Proceedings—2019 IEEE International Conference on Big Data, Big Data 2019, pp. 2560–2568. doi: 10.1109/BigData47090.2019.9005983.

[6] Zhang, Liu, X., Xie, X., Zhang, J., Niu, B. and Li, K. 2022. A cross-domain federated learning framework for wireless human sensing. IEEE Netw., 36(5): 122–128. doi: 10.1109/MNET.001.2200231.

[7] Abdelbasset, M., Hawash, H., Sallam, K.M., Elgendi, I., Munasinghe, K. and Jamalipour, A. 2022. Efficient and lightweight convolutional networks for IoT Malware detection: A Federated Learning approach. IEEE Internet Things J., pp. 1. doi: 10.1109/JIOT.2022.3229005.

[8] Gogineni, V.C., Werner, S., Huang, Y. and Kuh, A. 2022. Communication-efficient online federated learning strategies for kernel regression. IEEE Internet Things J., pp. 1. doi: 10.1109/JIOT.2022.3218484.

[9] Wang, S., Hosseinalipour, S., Gorlatova, M., Brinton, C.G. and Chiang, M. 2022. UAV-assisted online machine learning over multi-tiered networks: A hierarchical nested personalized Federated Learning approach. IEEE Trans. Netw. Serv. Manag., pp. 1. doi: 10.1109/TNSM.2022.3216326.

[10] Imani, H., Anderson, J. and El-Ghazawi, T. 2022. ISample: Intelligent client sampling in Federated Learning. In Proceedings—6th IEEE International Conference on Fog and Edge Computing, ICFEC 2022, pp. 58–65. doi: 10.1109/ICFEC54809.2022.00015.

[11] Zhang, D.Y., Kou, Z. and Wang, D. 2021. FedSens: A Federated Learning approach for smart health sensing with class imbalance in resource constrained edge computing. In Proceedings—IEEE INFOCOM, vol. 2021-May. doi: 10.1109/INFOCOM42981.2021.9488776.

[12] Akubathini, P., Chouksey, S. and Satheesh, H.S. 2021. Evaluation of machine learning approaches for resource constrained IIoT devices. In 2021 13th International Conference on Information Technology and Electrical Engineering, ICITEE 2021, pp. 74–79. doi: 10.1109/ICITEE53064.2021.9611880.

[13] Alghamdi, R. and Bellaiche, M. 2023. A cascaded federated deep learning based framework for detecting wormhole attacks in IoT networks. Comput. Secur., 125. doi: 10.1016/j.cose.2022.103014.

[14] Huang, C., Xu, G., Chen, S., Zhou, W., Ng, E.Y.K. and Albuquerque, V.H.C.D. 2022. An improved Federated Learning approach enhanced internet of health things framework for private decentralized distributed data. Inf. Sci. (Ny)., 614: 138–152. doi: 10.1016/j.ins.2022.10.011.

[15] Gupta, R. and Alam, T. 2022. Survey on Federated-Learning approaches in distributed environment. Wirel. Pers. Commun., 125(2): 1631–1652. doi: 10.1007/s11277-022-09624-y.

[16] Zhang, Z., Guan, C., Chen, H., Yang, X., Gong, W. and Yang, A. 2022. Adaptive privacy-preserving federated learning for fault diagnosis in internet of ships. IEEE Internet Things J., 9(9): 6844–6854. doi: 10.1109/JIOT.2021.3115817.

[17] Khalil, M., Esseghir, M. and Merghem-Boulahia, L. 2022. A Federated Learning approach for thermal comfort management. Adv. Eng. Informatics, 52. doi: 10.1016/j.aei.2022.101526.

[18] Govindwar, G.D. and Dhande, S.S. 2022. A review on Federated Learning approach in Artificial Intelligence. In 2022 6th International Conference on Computing, Communication, Control and Automation, ICCUBEA 2022. doi: 10.1109/ICCUBEA54992.2022.10010798.

[19] Gkillas, A. and Lalos, A.S. 2022. Missing data imputation for multivariate time series in industrial IoT: A Federated Learning approach. In IEEE International Conference on Industrial Informatics (INDIN), vol. 2022-July, pp. 87–94. doi: 10.1109/INDIN51773.2022.9976093.

[20] Tekchandani, P., Pradhan, I., Das, A.K., Kumar, N. and Park, Y. 2022. Blockchain-enabled secure big data analytics for Internet of Things smart applications. IEEE Internet Things J., pp. 1. doi: 10.1109/JIOT.2022.3227162.

[21] Uddin, R. and Kumar, S. 2022. SDN-based Federated Learning approach for satellite-IoT framework to enhance data security and privacy in space communication. In 2022 IEEE International Conference on Wireless for Space and Extreme Environments, WiSEE 2022, pp. 71–76. doi: 10.1109/WiSEE49342.2022.9926943.

[22] Yin, Y., Li, Y., Gao, H., Liang, T. and Pan, Q. 2022. FGC: GCN based Federated Learning approach for trust industrial service recommendation. IEEE Trans. Ind. Informatics, pp. 1–11. doi: 10.1109/TII.2022.3214308.

[23] Shubyn, B., Mrozek, D., Fabry, L., Maksymyuk, T., Amhoud, E.M. and Gazda, J. 2022. Federated Learning techniques for 5G mobile networks. In Proceedings —16th International Conference on Advanced Trends in Radioelectronics, Telecommunications and Computer Engineering, TCSET 2022, pp. 653–657. doi: 10.1109/TCSET55632.2022.9767080.

[24] Alazzam, M.B., Alassery, F. and Almulihi, A. 2022. Federated deep learning approaches for the privacy and security of IoT systems. Wirel. Commun. Mob. Comput., 2022. doi: 10.1155/2022/1522179.

[25] Huang, J., Tong, Z. and Feng, Z. 2022. Geographical POI recommendation for Internet of Things: A Federated Learning approach using matrix factorization. Int. J. Commun. Syst. doi: 10.1002/dac.5161.

[26] Alam, T. 2022. Federated Learning approach for privacy-preserving on the D2D communication in IoT. Lecture Notes in Networks and Systems, 322: 369–380. doi: 10.1007/978-3-030-85990-9_31.

[27] Hsu, Y.L., Liu, C.F., Samarakoon, S., Wei, H.Y. and Bennis, M. 2021. Age-optimal power allocation in industrial IoT: A risk-sensitive Federated Learning approach. In IEEE International Symposium on Personal, Indoor and Mobile Radio Communications, PIMRC, vol. 2021-September, pp. 1323–1328. doi: 10.1109/PIMRC50174.2021.9569536.

[28] Fadlullah, Z.M. and Kato, N. 2021. On smart IoT remote sensing over integrated terrestrial-aerial-space networks: An asynchronous Federated Learning approach. IEEE Netw., 35(5): 129–135. doi: 10.1109/MNET.101.2100125.

[29] Balasubramanian, V., Aloqaily, M., Reisslein, M. and Scaglione, A. 2021. Intelligent resource management at the edge for ubiquitous IoT: An SDN-based Federated Learning approach. IEEE Netw., 35(5): 114–121. doi: 10.1109/MNET.011.2100121.

[30] Hatzivasilis, G., Ioannidis, S., Fysarakis, K., Spanoudakis, G. and Papadakis, N. 2021. The green blockchains of circular economy. Electron., 10(16). doi: 10.3390/electronics10162008.

[31] Zhu, Y., Xu, J., Xie, Y., Jiang, J., Yang, X. and Li, Z. 2021. Dynamic task offloading in power grid Internet of Things: A fast-convergent Federated Learning approach. In 2021 IEEE 6th International Conference on Computer and Communication Systems, ICCCS 2021, pp. 933–937. doi: 10.1109/ICCCS52626.2021.9449265.

[32] Liu, Y., Garg, S., Nie, J., Zhang, Y., Xiong, Z., Kang, J. and Hossain, M.S. 2021. Deep anomaly detection for time-series data in industrial IoT: A communication-efficient on-device Federated Learning approach. IEEE Internet Things J., 8(8): 6348–6358. doi: 10.1109/JIOT.2020.3011726.

[33] Zhang, W., Lu, Q., Yu, Q., Li, Z., Liu, Y., Lo, S.K., ... and Zhu, L. 2021. Blockchain-based federated learning for device failure detection in industrial IoT. IEEE Internet Things J., 8(7): 5926–5937. doi: 10.1109/JIOT.2020.3032544.

[34] Perry, D., Wang, N. and Ho, S.S. 2021. Energy demand prediction with optimized clustering-based Federated Learning. In 2021 IEEE Global Communications Conference, GLOBECOM 2021—Proceedings. doi: 10.1109/GLOBECOM46510.2021.9685647.

[35] Stergiou, K.D., Psannis, K.E., Roumeliotis, M., Kokkonis, G. and Ishibashi, Y. 2021. Optimising decision making on communication systems: The Federated Learning approach. In 2021 IEEE 9th International Conference on Information, Communication and Networks, ICICN 2021, pp. 119–123. doi: 10.1109/ICICN52636.2021.9674002.

[36] Duy, P.T., Hao, H.N., Chu, H.M. and Pham, V.H. 2021. A secure and privacy preserving Federated Learning approach for IoT intrusion detection system. Lecture Notes in Computer Science (including subseries Lecture Notes in Artificial Intelligence and Lecture Notes in Bioinformatics), vol. 13041 LNCS, pp. 353–368. doi: 10.1007/978-3-030-92708-0_23.

Chapter 7

Efficient Federated Learning Techniques for Data Loss Prevention in Cloud Environment

A Peter Soosai Anandaraj,[1,*] *S Sridevi,*[2] *R Vaishnavi,*[3]
M Meenalakshimi[3] and *RV Chandrashekhar*[4]

1. Introduction

Data security has emerged as the main type of protection in the big data era, as practically all information is sent across networks. There are still issues with privacy and the leakage of sensitive information, despite the significant advancements in network information security protection mechanisms [1]. Machine learning, the most popular data analysis and processing method in the big data age, is extensively employed in a variety of industries because of advancements in computer computational capacity.

[1] Associate Professor, Department of Computer Science and Engineering, Veltech Rangarajan Dr. Sagunthala R&D Institute of Science and Technology, Chennai, 600062, India.
[2] Professor, Department of Computer Science and Engineering, Veltech Rangarajan Dr. Sagunthala R&D Institute of Science and Technology, Chennai, 600062, India.
[3] Assistant Professor, Department of Computer Science and Engineering, Veltech Rangarajan Dr. Sagunthala R&D Institute of Science and Technology, Chennai, 600062, India
[4] Cyber Liasion Manager, IT Operations, Ford Motor Private Limited, Chennai, 600119, India.
* Corresponding author: anandsiriya@gmail.com

However, there are two issues with machine learning development. On the one hand, it can be challenging to maintain data security, and concern over privacy protection is rising. However, as big data has emerged as a new trend and businesses have tightened data protection measures to prevent data leaks, data sharing has decreased. The difficulties machine learning has encountered transferring data for training are what have led to data silos. Using a cutting-edge machine learning architecture known as Federated Learning, clients who are wary of one another can still benefit from the collaborative training model without disclosing their confidential information sets in the open [3]. Federated learning is also known as the Decentralised ML Framework. Due to the constrained network capacity, substantial communication expenses among the cloud server and clients have emerged as the primary issue. Additionally, model inversion attacks can be carried out using the same model parameters.

A combined model is stored on a cloud server, while the training data are scattered across several devices. Customers may benefit from the collaborative training technique with federated learning, a ground-breaking machine learning design that safeguards the confidentiality of their personal data. It is also known as the Decentralized ML Framework [5]. The training data is spread across multiple devices, and the combined model is stored on an online computer. However, high transmission expenses between the cloud server and clients, as well as model inversion assaults, have surfaced as the main problem.

A potential method that supports distributed collaborative learning without revealing the original training data is federated learning (FL) [2]. FL offers a straightforward answer in this case. That is, by keeping the data local, A combined model is stored on a cloud server, while the training data are scattered across several devices. Customers may take advantage of the collaborative training approach with FL, a revolutionary ML design that protects the privacy of their private information. This is accomplished by looking for data breaches and ATO (Account Takeover) fraud. Furthermore, it examines credit scores and learns a user's footprint to avoid fraudulent KYC without uploading data to the cloud [7]. FL opens the path for Fintech risk mitigation. It develops unique and creative techniques for its customers and enterprises [4]. It establishes confidence between the two parties. It also allows them to progress their connection. There is a tremendous increase in information as technology advances. As a result, additional privacy restrictions are in place to protect such information. Many businesses have begun to use federated learning. They train their algorithms on different datasets without transferring data. Federated learning tries to safeguard data acquired through several channels. It also keeps critical information close at hand [6]. FL is a technology that enables machine learning on-device without sending the user's private data to a central cloud. As a result, federated learning can aid with personalisation. As well as improving device performance in IoT applications. Recently, this research has discussed the present obstacles to adopting FL approaches, as well as its improvements in real-time systems such as medical AI, IoT, Edge Systems, and self-governing businesses that can adapt FL in their expanding domains in the future when put to practical use.

Despite the obstacles of data imbalance, resource allocation, and privacy concerns, this study focuses on techniques of Federated learning techniques in Cloud computing and the core challenges faced by FL which may include effective and efficient communication between federated networks, management of systems heterogeneity, and statistical heterogeneity within the same network [8].

The information is then utilised to develop insightful inference models or to provide insights. These factors make this tactic no longer workable. First off, privacy is a growing concern for the data owners. To address consumers' privacy concerns in the age of big data, several data privacy regulations have been adopted, including consumer privacy rights in the US [11] and the GDPR of Europe [10].

On end devices, DNN models are jointly trained using FL. The two procedures that typically make up the FL exercise operation are regional simulation on the final devices and the worldwide collection of changed attributes on the FL server. To generalise FL's local model training, we initially give an overview of the DNN prototype experiment in this section. Even though FL may be used to train ML models generally, we concentrate on training DNN models in this part because most of the articles we discuss in the next section look at federated training of DNN models. The fact that DNN models are simple to aggregate and exceed traditional ML techniques makes them superior in many situations [16]. The organic adoption of FL at mobile edge networks should consequently be aided by the expanding DL and growing processing power and data volume gained by dispersed end devices.

FL was developed in response to data owners' privacy concerns [21]. FL allays customers' privacy worries by letting users jointly train an identical framework and maintaining their data on their devices. Because of this, FL could be a solution that makes it possible for mobile edge networks to train ML models. For a summary of the various FL possibilities, including parallel FL and straight up FL.

2. Related Works

Federated learning (FL), a ground-breaking technique, addresses the problem of safeguarding personal information via the security measures mechanism by enabling consumers to collectively develop a global model while maintaining local control over the data. The platform aggregates the models until convergence while customers train their local models. To encourage users to provide enormous amounts of high-quality data, the server employs a reward structure in this process.

Despite some works having applied FL on the Internet of Things (IoT), medical, industrial, and supplementary sectors, FL application is still in its infancy and there are still several problems that remain unresolved [9]. Today, improving FL model quality is one of the hot research topics and difficult problems. The methods for raising FL model quality are objectively and in-depth investigated in this work. Since practitioners frequently worry that achieving privacy protection necessitates compromising learning quality, we are particularly interested in the study and application developments of FL as well as the impact comparison between FL and non-FL [10]. We evaluate the present FL-related literature methodically. The information and insights provided in this review might be useful to both academics and business people. We examine important factors affecting FL model quality, trends in academic and industrial FL application research, and the effectiveness of FL versus non-FL algorithms for learning. Based on the findings of our investigation, we make some recommendations for enhancing the calibre of FL models.

Massive volumes of data are produced nowadays by a great number of networked devices, including mobile phones, wearable technologies, and autonomous automobiles (Big Data). The need to store and analyse data locally is growing because of the rapidly rising processing capability of these devices and privacy concerns, which are moving computing from the cloud to the edge. To fully exploit big data's promise, artificial intelligence (AI) is required: Deep Learning has demonstrated to be quite successful at extracting knowledge from difficult data. Deep neural network designs have been shown to perform better than humans in classifying photos from the well-known ImageNet dataset. Since all data must be gathered in one place, which is frequently a cloud data centre, traditional machine learning algorithms may be violate user privacy and data confidentiality laws. Technology companies are now required to abide by user privacy rules in many different regions of the world when processing user facts. One such example is the European Union's General Regulation for the Protection of Data.

Federated learning is a viable approach for using the enormous and diverse datasets and rising processing capacity of the devices on the border of the network. This method also protects the privacy of the data being used. Financial and healthcare apps must be private owing to the sensitivity of the data they handle and the challenges of sharing it. Even if there are a few instances of FL being utilised effectively in business settings, several research issues must be cleared out before these designs are often employed in routine applications [11]. FL has become a cutting-edge paradigm in response to concerns about the security of personal data as well as the increasing computational power of gadgets like smartphones, wearables, and driverless vehicles [13]. Due to the increased need for personal storage of data and the move of ML computations to end gadgets while minimising data transport additional costs, scientists have attempted to adapt FL architectures in a few disciplines [12]. Federated learning, which is like a group presentation or report, entails the remote sharing of data among numerous people to collectively train a single deep learning model and progressively improve it. The model, which is frequently a foundation model that has previously been trained, is provided to each participant from a cloud datacentre. Once the model has been updated using their personal data, it is summarised and encrypted. The modified models are sent back to the cloud where they are averaged, encrypted, and integrated into the original model. Up until the model is fully trained, collaborative training is carried out iteration after iteration.

Which training methods should be prioritised more in a federated learning infrastructure—synchronous or asynchronous—is a crucial design choice. Asynchronous training has been employed in many successful deep learning research, such as Dean et al. (2012), but recently, synchronous massive batch training has been gradually increasing, even in data centres (Goyal et al., 2017; Smith et al., 2018) [14]. The McMahan et al. (2017) Federated Averaging technique employs a similar approach. Additionally, a few methods for improving FL privacy guarantees, like differential privacy (McMahan et al., 2018) and Secure Aggregation (Bonawitz et al., 2017), essentially call for some sort of synchronisation on a collection of fixed devices to ensure the learning algorithm's server-side component only must use a simple aggregate of updates from numerous users [15].

Large-scale cloud-based systems have been described and are being utilised in practise, and significant work has been done on distributed machine learning. A lot of systems, such as those mentioned by Dean et al. (2012) and Low et al. (2012), support several distribution strategies, such as model parallelism and data parallelism. Since virtualization technology the nodes have significantly higher bandwidth and reliability than mobile devices, our solution imposes a more organised approach suitable for the area of mobile devices. Rather than focusing on an ad hoc distributed compute architecture, we opt for a synchronous FL protocol. According to Brisimi TS et al. (2018), the goal of this study is to provide a distributed (federated) method for predicting hospitalisations for patients with heart conditions over a year based on their medical histories as documented in their Electronic Health Records (EHRs). Every patient's record might be kept in the EHR systems of other institutions or on their smartphone. To create a global hospitalisation prediction model, several stakeholders (agents) must collaborate in all cases. The topic will be framed as a binary supervised classification problem, and a distributed soft-margin l1-regularized (sparse) Support Vector Machines (sSVM) technique will be developed. SVMs are considered since they are effective classifiers and have a high degree of predictability for hospitalisations. To allow for the interpretation of predictions, sparse classifiers can also provide very little predictive data. We can optimise from a system viewpoint for the particular use case because of this domain expertise. The parameter vector, the parameter server—used by Li et al. (2014), Dean et al. (2012), and Abadi et al. (2016)—is a particularly well-liked method in the datacenter because it lets several staff members cooperate on a single big-picture model. In that field of study, emphasis is placed on developing a server architecture that can successfully manage vectors with sizes ranging from 109 to 1012. Workers asynchronously access and change the global state through the parameter server. Since we need a precise rendezvous between a group of devices and the FL server to perform a synchronous update using Secure Aggregation, our method is essentially unable to work with such a global state. They are now particularly interested in discovering ways to solve the problems associated with healthcare data: (1) data reside in various locations (e.g., hospitals, doctors' offices, home-based devices, patients' smartphones); (2) data availability is growing, making scalable structures essential; and (3) collecting data in one repository is impractical or undesirable due to scale and/or data privacy issues. The ability to utilise anonymization techniques (such as k-anonymity [13]) is made possible by centralising all data storage, but it also creates a single point of attack or failure and makes it possible for a data breach to disclose identifying information for many individuals. A central data repository must also deal with information governance challenges, such as gaining authorization for processing and storing, and make large infrastructure expenditures. Our version's capacity to support a range of configurations, from totally centralised to totally decentralised, is one of its primary applications. Our intriguing healthcare challenge is structured as a binary classification problem. Two layers of information processing are possible: that of the patient, for instance, when they use their smartphones, and that of the hospital, which manages patient data. A central data repository must also deal with information governance challenges, such as gaining authorization for processing and storing, and making large infrastructure expenditures. Our version's capacity to support a range of configurations, from totally centralised to totally decentralised, is one of its primary applications. Our intriguing healthcare challenge is structured as a binary classification problem. Two levels of information processing are possible: that of the patient, for instance, using a mobile, and that of the hospital, which analyses patient data.

The term "federated learning," which Google first used in 2016, was introduced at a time when the use and exploitation of personal data was gaining international attention. Users of Facebook and other comparable websites are now more aware of the dangers of disclosing personal information online thanks to the Cambridge Analytica incident. Additionally, it sparked a larger conversation about how many people are being watched online, frequently without their knowledge. At the same time, the public's confidence in businesses' ability to secure personal information has been eroded by a string of high-profile data breaches. When California, the epicentre of digital platforms fuelled by advertising revenue, ratified the GDPR, a comprehensive data privacy policy for Europe, other states soon followed suit. Since then, legislation governing digital privacy has been debated or passed in Brazil, Argentina, and Canada. Numerous individuals remotely share their data to jointly train a single deep learning model, iteratively improving it, like a team presentation or report. Each participant receives the model, which is frequently a pre-trained foundation model, from a cloud datacentre. Before summing and encrypting the model's new setup, they train it using their data. Updates to the model are sent back to the cloud, where they are averaged, encrypted, and combined with the central model. Up until the model is fully trained, the collaborative training procedure is performed iteration after iteration.

3. System Design

Federated learning, which is like a group conversation or report, involves the remote sharing of data among numerous people to collectively train a single deep learning model and progressively improve it [15]. A cloud data centre provides a distinct model, often a simple model that has already completed training, to each customer. They use their data to train the model, which is then summarised and encrypted. The updated model is transmitted back to the server, where it is averaged, encrypted, and integrated with the base model. Iteration after iteration of collaborative training continues until the prototypical is completely competent.

Three different versions of this dispersed exist, decentralised training approach. The primary model in horizontal federated learning is trained on comparable datasets [17]. While using the same feature space and a variety of samples, horizontally partitioned federated learning (HFL) uses data that is scattered across silos. The data are complementary in vertical federated learning; movie and book evaluations, for instance, are integrated to forecast a person's musical tastes. Finally, federated transfer learning involves training a foundation model that has already been trained to accomplish one job, such as detecting automobiles, on a different dataset to do another task, such as recognising cats. Foundation models are now being included in federated learning by Baracaldo and her co-workers. The technique of learning from data that has been dispersed across silos and involves a variety of feature spaces and samples is known as federated transfer learning (FTL).

Traditionally, data is collected from several sources and then consolidated for machine learning models. A database lake, a data warehouse, or a lakehouse-style hybrid of the two might serve as this focal point. You select an algorithm, like a decision tree, or a group of algorithms, like the neural networks discussed above, and train it on the collected data. The created model can then be executed on the main server or spread among devices.

It is a simple, widely used method, but it has significant drawbacks, especially when combined with contemporary IoT infrastructures where fresh data is continuously produced at the edge of computer networks. The following are the primary issues that centralised ML is now experiencing.

Network Latency: Real-time applications need models that can react extremely quickly. If it is carried out on the server, data transfer delays are always a possibility.

Problems with connectivity: Once more, effective data transfer between many devices on either side and the primary computer on the other requires a dependable Internet connection.

Concerns about data privacy: For instance, pictures and transcripts are possibly helpful for developing intelligent apps, but they could also include private information that should not be made public.

In one potential usage, banks may train an AI model to spot fraud before using it for different purposes. In Google's original Federated Learning use case, the data is dispersed among end user devices and remote data is used via Federated SGD and averaging to enhance a central model [20]. If improving centrally managed models is a system's main goal, it may be referred to as "model-centric" or "model-centric" in a federated learning context. Data generated locally is still decentralised. Each client stores just their data; the data of other clients cannot be read. Data is not dispersed randomly or consistently. The two federated learning models that are generally used are cross-silo and cross-device. Utilising databases with information from various users are necessary for learning in cross-silo environments. For example, a healthcare analytics server can analyse medical records by fusing data from hospitals and clinics. Contrarily, cross-device settings focus on learning across user devices that include data that was generated by a single user. Cross-device was the first use of federated learning, in which Google trained next-word prediction models using user data from GBoard.

How to design a federated learning system in six steps

Step 1. Model Framework Selection:

Selecting the model architecture that will assist your implementation of the underlying model is the first step. You must be able to use some federated learning elements in the application format with the model framework that you select. Field (such as imaging, NLP, or tabular data), team technical competence, and framework compatibility with existing infrastructure are other criteria for choosing.

Step 2. Explore and identify the Network mechanism:

The next step is to decide the networking strategy to use. This is the communications architecture that the participants of the federated learning network use to exchange instructions. Depending on whether the application is being utilised internally inside an enterprise or externally across organisations, the networking framework and core model framework selection is made. Decisions in the latter case must be endorsed by all organisations concerned. We now have the necessary foundational elements to begin building a federated learning scheme.

Step 3. Build Integrated Service:

The construction of centralised services is required at this level. This service will oversee organising participant communication and supervising the training process. From a functional perspective, this service likely needs to guarantee statelessness to help with load balancing, which necessitates the choice of a storage mechanism to hold intermediate data passing between clients, the inclusion of authentication and authorization mechanisms, as well as the support structure to keep it reliable. To fulfil the requirements of the federated learning system and have the capacity to conduct training sessions, a system must be set up and kept current.

Step 4. Client System Design:

The moment has come to think about a possible client system design. This system must be able to communicate the parameters of the model with the central service and perform customer-side courses of study. The client system will also need local variables across additional networked gadgets to update the local model. To create and deploy a reliable client application, you should try to address the following queries.

Step 5. Setting up the Training process and operationalizing the Model management system:

The federated learning system needs to know which private client data should be utilised to instruct the regional models for a specific time slot. The central service or another individual must deliver this data. It is crucial to keep the meta data, which is frequently handled by the primary service, regarding the available data. Clients must additionally register the meta data about the datasets that are accessible to other clients. Similar metadata indicating which collections should be used for a lesson for each client will need to be retrieved from the central server through a query. Following a training session, a machine learning framework is produced. For the model that was trained to be used by the proper users, the federated learning system must handle model metrics and access.

Step 6. Last but certainly not least, addressing Security and Privacy concerns:

The final model is available locally to one or more people. The above framework was built by spreading weights across numerous parties and may be inverted to disclose more about the core data used for training. Therefore, choosing suitable hazards over the framework itself will be crucial. The possibility of recreating a training set, which may contain sensitive intellectual property, or the capacity to re-identify someone who took part in a particular training set are two examples of these dangers.

Applying different confidentiality to the data weights before sending them to the central service is one way to reduce these dangers. You may strike a compromise between the utility of the final design and the level of risk you deem appropriate by setting the privacy budget.

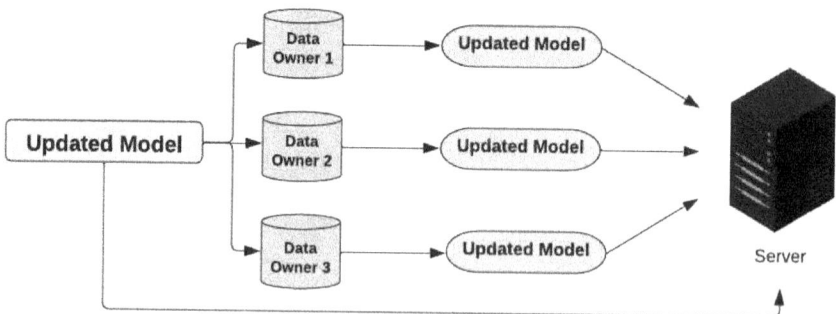

Figure 1. General Federated Learning architecture.

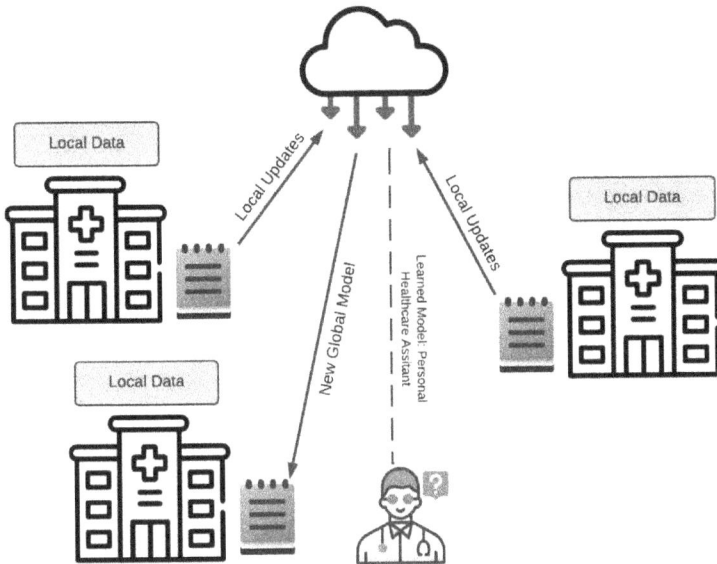

Figure 2. Federated Learning architecture in a clinic.

Figure 2 explores how a FL design might be used in a medical environment. Local data updates are aggregated, and a training model is constructed using ML. Sadly, there are still some substantial obstacles preventing FL from being fully integrated in other contexts, particularly regarding the data. Even the data itself can be difficult to manage since there is typically a lot of diversity within the information, including material, framework, and download-formats.

4. To Improve Efficiency by Using Fl Techniques

There are a few commonly used strategies and algorithms that needs to be understood for understanding FL.

Categories of Federated Learning:

(a) Centralized Federated Learning: For centralised federated learning to function, a significant server is required. It plans the initial client device choice and gathers model updates during training. Only certain edge devices and the central server exchange data.

(b) Decentralized Federated Learning: Decentralised federated learning can be coordinated without a central server. Instead, only the connected edge devices are given access to the updated models. The final model is built on an edge device by merging the local apprises from all the linked verge procedures.

(c) Heterogeneous Federated Learning: The usage of hybrid customers, including cell phones, notebook computers, or Internet of Things (IoT) devices, is required for heterogeneous federated learning. Different hardware, software, processing power, and data kinds could be present in these devices.

Federated Learning algorithms:

(a) Federated stochastic gradient descent (FedSGD)

Traditional SGD computes gradients using a subset of data samples from the complete sample set known as mini batches. These little batches may be compared with multiple customers that each have local data in a federated environment. Clients can access the primary model using FedSGD, and they can independently generate the gradients using local data. To determine the gradient descent step after receiving these gradients, the central server aggregates them depending on the number of samples on each client.

(b) Federated averaging (FedAvg)

The FedSGD algorithm has been extended by federated averaging. Multiple local gradient descent updates can be run by clients. The local model's adjusted weights, not its gradients, are shared with the essential global system. The weights (model parameters) from the customers are then around the global computer. Federated Averaging is a refinement of FedSGD, where aggregating the proportions is the same as aggregating a gradient if all clients start from the same baseline.

(c) Federated Learning with dynamic regularization (FedDyn)

By adding a consequence to the forfeiture process, regularisation in traditional machine wisdom approaches aims to improve generalisation. In federated learning, the local losses from many devices must be combined to determine the overall loss. Because of the variety of clients, limiting global forfeiture is different from minimising local loss. The FedDyn technique responds to data properties like data volume or transmission cost to create the regularisation term for local losses. Dynamic regularisation modifies local losses, allowing them to converge to the global loss. HeteroFL was developed as a countermeasure to well-known Federated Learning approaches that assume the attributes of the local models correspond to those of the main model. However, that doesn't happen very often. HeteroFL can give a single global model for inference after training over a multitude of local models.

The possibility of a single point of failure is eliminated by this strategy, but the network architecture of the edge devices is completely responsible for the model's accuracy. Although this approach appears forthright and produces precise models, the central server presents a bottleneck problem—network issues can halt the entire process. Model-Centric and Data-Centric federated machine learning are the two basic forms. Starting with the more prevalent model-centric approach nowadays.

FL just seeks to provide an answer to the vital question, "Can we train the model without transferring data to a central location?" In general, FL would maintain privacy while enabling simple information viewing and exchange.

- Private FL Systems: A small number of significant computational and data-intensive components are present in private FL systems. Private FL systems must find a way to migrate calculations to data centres while still adhering to strict regulations.
- Public FL Systems: Despite having many entities, public FL systems only have a small amount of data and processing power.

An important technology for integrating federated learning into interactive shadow-edge design.

i) Communication:

Cross-device FL necessitates customers to execute several communication rounds in the directive for the prototypical to congregate, even though clients can connect with edge servers more effectively. Clients may have to spend money on training for this, and the resulting communication delay is undesirable [14]. For the carrying out of federated learning in shadow-edge collaborative design, it is therefore generally acknowledged that ensuring successful communication is a vital technology. Our study has led us to three successful strategies for achieving successful interaction:

- End computing
- Aggregation control
- Model compression
- Privacy and security

The two most common assumptions in most cloud-based FL frameworks are: (i) The clients and edge servers used in FL are all trustworthy devices. They completely adhere to the FL manager's demands. (ii) No individual has a copy of other customers' data. It is not possible to build a trustworthy FL structure in the cloud-edge collaborative design without considering these two suppositions, which may lead to privacy leakage and a failure to protect against forbidden attacks, upsetting the social system and having a negative effect on customers. The cloud-edge cooperative paradigm may not fully trust the edge servers and clients, and malevolent attackers can easily participate in FL. Private and confidential solutions are crucial for the successful deployment of FL.

ii) Personalization:

Customers opt to engage in FL with the overall objective of improving the model when the regional collection is too small to train a method with great accuracy. This puts a danger to privacy and consumes communication and computing resources. However, in some instances, such as when an individual contributing to cross-silo FL has a large dataset, the level of accuracy of the locally generated model may be better than that of the combined model [18]. Model diversity means the potential for various customers to have diverse necessities for the model. In a word prediction task, for example, multiple individuals typing the same "I like..." would undoubtedly expect various prediction outcomes. Basic customising strategies can help to some extent with model heterogeneity issues. The list of several customised federation learning (PFL) methods that may be applied to cloud-edge collaborative architecture is provided below:

- Meta-Learning
- Transfer Learning

Opportunities for interactive shadow-perimeter architecture with federated learning

- FL for computation offloading
- FL for edge caching
- Vehicular networks
- Medicine
- Cyber security

federated learning deployment difficulties and prospective study goals in interactive shadow-perimeter architecture

iii) Outlier:

In edge federated learning, a collection of edge devices is selected for every training session. The client selection approach was used to develop an exclusive model for every customer. However, each of these methods assumes that every customer can continue to connect to the perimeter machines. Since these devices require little vitality and their connectivity surroundings are constantly changing, perimeter devices are prone to becoming disconnected during the instructional process. These disengaged customers are commonly referred to as outliers. A great deal of the previously completed research on handling outliers focuses on sustaining training while there are just a few, but they are unable to handle huge outliers, and the efficacy of federated learning will be significantly reduced.

iv) Aggregation:

Model aggregation, one of the most significant FL operations, has a direct impact on training quality. The aggregation, which is frequently simplistic in FL environments, is unable to manage complex cloud-edge collaborative architecture. Although (i) Managing impaired refreshes have received some attention. (ii) Avoid making forced upgrades. Further investigation into robust FL aggregation is still required, including consideration of accessibility collection along with additional safeguard accumulation.

v) Incentive mechanism:

In viable FL, handler modelling using local client records might suggest extremely correct content, which has financial advantages. It is difficult to devise incentive systems to persuade users to participate in FL, nevertheless, given the learning process incorporates their personal information as people pay greater attention to it. The design of FL incentive structures currently uses techniques from game theory and economic theory, but their efficacy is only guaranteed in simulated tests because subjectivity controls the actual world.

vi) Migration:

Because it is difficult to determine a perimeter device's exact location when using mobile edge computing, it frequently connects to the perimeter server that is nearby, which causes issues. When a new perimeter server connects to the edge device, the training is interrupted, and the network's performance degrades significantly. This is due to the lack of a copy of the previous training on the new edge server. A critical topic for research is the continuance of education. Many studies have been conducted on service migration in perimeter computing. The model in perimeter federated learning is substantial, nevertheless. The performance of federated learning is adversely affected by the significant delay in transferring model copies between perimeter servers, necessitating effective service migration in a cloud-edge collaborative setting.

vii) Asynchronicity:

The synchronous aggregation option of federated learning is used in most current research. Users' power reserves and network quality change in a real-world network environment. The edge nodes find it difficult to continue training, especially with traditional synchronous approaches, because of resource limits. A lot of work has been put into developing efficient asynchronous federated learning algorithms to overcome this problem.

viii) Privacy Issue:

The privacy protection features of the FL system, however, might not work with all devices due to resource constraints. Therefore, novel approaches need to be created that offer strong privacy guarantees along with being effective at communication, cost-effective at computation, and able to handle decreased participants. Numerous studies focus on the problems associated with peripherals that have insufficient processing power. It is not viable to utilise advanced encryption and communication security in this environment due to the potential impact on the computational capabilities of these devices.

ix) Scalability:

The FL method must be very scalable. In an IoT ecosystem, many customers may take part in prototypical preparation. In the FL process, it's also important to make sure the system is scalable. Due to different factors, such as weak indication, short battery life, or poor quality, as with handheld devices, some clients may lose connection to the server when there are many others. It is essential to make sure that such a system is capable of handling certain circumstances. Federated learning, which protects data privacy, may be used to train ML models utilising the large and diverse datasets and the expanding processing capacity of the devices at the periphery of the network. Privacy is essential since financial and healthcare applications are by their very nature very sensitive and sharing this data is usually not viable [19]. A lot of study must be done before these designs are extensively employed in commonplace applications, although a few instances of FL functioning successfully in commercial environments. Due to concerns over the security of private information as well as the increasing computing power of devices like cell phones, devices that are worn, and self-driving cars, FL has emerged as a cutting-edge framework. Experts have made multiple attempts to build FL systems because of the rising requirement for local storage of information and the migration of ML computations to end devices while reducing data transit overhead.

5. Conclusion

Although federated learning is still in its infancy, it will undoubtedly benefit many industries in terms of safeguarding data privacy. Connectivity is a significant obstacle in federated learning networks since the information collected on each device stays private. It is vital to develop communication-efficient strategies to cut down on the total number of communication cycles. The minimum user involvement that occurs from just a subset of the devices being in operation at any given time must also be considered by FL solutions. The future of cyber infrastructure is frequently envisioned as having federated learning applications and hybrid architectures. Even though both shadow-edge collaboration and federated learning are now prominent scholarly issues, there hasn't been much study on how to integrate them into the shadow-edge collaborative design. By carefully analysing the essential skills, encounters, and uses of implementing federated learning in shadow-edge joint design, this research aims to narrow the knowledge gap. It also offers suggestions for further research. Federated learning is more innocuous than straight centralised ML and is capable of handling data silos, however, it is not always successful. System diversity, model diversity, and data diversity are the biggest problems for FL. When the easily accessible local data do not accurately reflect the global distribution, a statistical disparity emerges. FL participants regularly experience different network, energy, processing, and storage capacity conditions. This is what is meant by system variety. Since most FL settings are simultaneous, convergence might be overdue, and some devices might not have the resources to quickly recover local modifications. Model diversity frequently happens in B2B relationships in FL because various clients may provide various model requirements based on various assumptions. Furthermore, many of the original FL algorithms created a security threat to the sincerity of their customers. Due to the aforementioned issues, some customers could feel that the FL-trained model is inferior to their local model, which would deter them from participating in the preparation. Given the aforementioned, the creation of federated learning is essential for Internet of Things (IoT) applications that strongly rely on handling issues like statistical diversity, system diversity, model diversity, and safe administration, where statistical variation is more common. However, these FL techniques are extremely beneficial to the research community in preventing data loss in cloud environments.

References

[1] Salem, A., Wen, R., Backes, M., Ma, S. and Zhang, Y. 2022. Dynamic backdoor attacks against machine learning models. *In*: 2022 IEEE 7th European Symposium on Security and Privacy (EuroS &P). IEEE, pp. 703–718. https://doi.org/10.1109/EuroSP53844.2022.00049.

[2] Sandhu, A.K. 2021. Big data with cloud computing: Discussions and challenges. Big Data Min Analytics, 5(1): 32–40. https://doi.org/10.26599/BDMA.2021.9020016.

[3] Angel, N.A., Ravindran, D., Vincent, P.D.R., Srinivasan, K. and Hu, Y.C. 2021. Recent advances in evolving computing paradigms: Cloud, edge, and fog technologies. Sensors, 22(1): 196. https://doi.org/10.3390/s22010196.

[4] Chen, Y., Liu, B. and Hou, P. et al. 2021. Survey of cloud-edge collaboration. Comput. Eng., 43(02): 242.

[5] Alqudah, N. and Yaseen, Q. 2020. Machine learning for traffic analysis: A review. Procedia Comput. Sci., 170: 911–916. https://doi.org/10.1016/j.procs.2020.03.111.

[6] de Magalhães, S.T. 2020. The european union's general data protection regulation (gdpr). World Sci. https://doi.org/10.1142/9789811204463_0015.

[7] Deng, S., Zhao, H., Fang, W., Yin, J., Dustdar, S. and Zomaya, A.Y. 2020. Edge intelligence: The confluence of edge computing and artificial intelligence. IEEE Int. Things J., 7(8): 7457–7469. https://doi.org/10.1109/jiot.2020.2984887.

[8] Li, T., Sahu, A.K., Talwalkar, A. and Smith, V. 2020. Federated Learning: Challenges, methods, and future directions. IEEE Signal Process. Mag., 37(3): 50–60. https://doi.org/10.1007/978-3-030-85559-8_13.

[9] Lim, W.Y.B., Luong, N.C., Hoang, D.T., Jiao, Y., Liang, Y.C., Yang, Q., Niyato, D. and Miao, C. 2020. Federated Learning in mobile edge networks: A comprehensive survey. IEEE Commun. Surv. Tutorials, 22(3): 2031–2063. https://doi.org/10.1109/comst.2020.2986024.

[10] Yao, H., Gao, P., Zhang, P., Wang, J., Jiang, C. and Lu, L. 2019. Hybrid intrusion detection system for edge-based iiot relying on machine-learning-aided detection. IEEE Netw., 33(5): 75–81. https://doi.org/10.1109/MNET.001.1800479.

[11] Legacy, C., Ashmore, D., Scheurer, J., Stone, J. and Curtis, C. 2019. Planning the driverless city. Transp. Rev., 39(1): 84–102. https://doi.org/10.1080/01441647.2018.1466835.

[12] Hassan, N., Yau, K.L.A. and Wu, C. 2019. Edge computing in 5g: A review. IEEE Access., 7: 127276–127289. https://doi.org/10.1109/ACCESS.2019.2938534.

[13] Xiao, Y., Jia, Y., Liu, C., Cheng, X., Yu, J. and Lv, W. 2019. Edge computing security: State of the art and challenges. Proc. IEEE, 107(8): 1608–1631. https://doi.org/10.1109/jproc.2019.2918437.

[14] Bonawitz, K., Eichner, H., Grieskamp, W., Huba, D., Ingerman, A., Ivanov, V., Kiddon, C., Konečný, J., Mazzocchi, S., McMahan, B., Van Overveldt, T., Petrou, D., Ramage, D. and Roselander, J. 2019. Towards Federated Learning at scale: System design. arXiv preprint arXiv:1902.01046. https://doi.org/10.48550/arXiv.1902.01046.

[15] Brisimi, T.S., Chen, R., Mela, T., Olshevsky, A., Paschalidis, I.C. and Shi, W. 2018. Federated Learning of predictive models from federated electronic health records. Int. J. Med. Inform., 112(1): 59–67. https://doi.org/10.1016/j.ijmedinf.2018.01.007.

[16] McMahan, B., Moore, E., Ramage, D., Hampson, S. and y Arcas, B.A. 2017. Communication-efficient learning of deep networks from decentralized data. *In*: Artificial Intelligence and Statistics. PMLR, pp. 1273–1282. https://doi.org/10.1109/icde51399.2021.00040.

[17] Jordan, M.I. and Mitchell, T.M. 2015. Machine learning: Trends, perspectives, and prospects. Science, 349(6245): 255–260. https://doi.org/10.1126/science.aaa8415.

[18] Gaff, B.M., Sussman, H.E. and Geetter, J. 2014. Privacy and big data. Computer, 47(6): 7–9. https://doi.org/10.1109/MC.2014.161.

[19] Dillon, T., Wu, C. and Chang, E. 2010. Cloud computing: issues and challenges. *In*: 2010 24th IEEE International Conference on Advanced Information Networking and Applications. IEEE, pp 27–33. https://doi.org/10.15373/2249555x/mar2014/181.

[20] Chu, W. and Park, S.T. 2009. Personalized recommendation on dynamic content using predictive bilinear models. *In*: Proceedings of the 18th International Conference on World Wide Web, pp. 691–700. https://doi.org/10.1145/1526709.1526802.

[21] Hayes, B. 2008. Cloud Computing. ACM, New York. https://doi.org/10.1007/978-1-4842-8236-6_2.

Chapter **8**

Maximizing Fog Computing Efficiency with Federated Multi-Agent Deep Reinforcement Learning

G Anitha[1],* and *A Jegatheesan[2]*

1. Introduction

The development of the Internet of Things (IoT) has led to a rise in the adoption of V2X connections, allowing for more effective communication and processing. Agents in vehicles can coordinate with other connected devices and pool their idle processing power to benefit time-critical programmes [1]. Decision-making in automotive systems is notoriously difficult owing to the highly dynamic and nonlinear behaviour of the car network, brought on by the ever-changing nature of the resource-sharing environment [2]. Vehicle applications are time-sensitive, mobile, and subject to fast changes in queue state, all of which make task offloading difficult. Solutions based on machine learning have the potential to reduce

[1] Research Scholar, Institute of Computer Science and Engineering, Saveetha School of Engineering, Saveetha Institute of Medical and Technical Sciences, Chennai, India.
[2] Professor, Institute of Computer Science and Engineering, Saveetha School of Engineering, Saveetha Institute of Medical and Technical Sciences, Chennai, India.
Email: jegatheesanphdcse@gmail.com
* Corresponding author: gani3086@gmail.com

the aforementioned problems [3]. One common approach, grounded in game theory and reinforcement learning, seeks to optimise decisions by continuously gathering data from the world around them [4]. In other words, they can't account for the vast state space that corresponds to the many possible queue statuses at each node. Deep reinforcement learning is used in more complex solutions for the optimisation of decisions, although it is computationally costly and has a sluggish convergence rate [5].

To address delayed convergence caused by environmental unpredictability, limited information sharing, and a centralised design, policy optimization-based approaches are designed to give a close approximation of the job offloading strategy in a wide state space [6]. Solutions discussed here are centralised in nature, as they glean information from centralised fog servers, and knowledge transfer among vehicular agents is limited [7]. Federation learning is a revolutionary distributed learning paradigm that has the potential to improve decision making and speed up convergence [8].

By allowing each device to do its learning, federated learning not only achieves rapid convergence but also achieves reduced communication overhead across the shared-resources ecosystem [9]. When it makes sense to train locally, a distributed learning method called federated learning (FL) can be quite useful [10]. The approach to learning makes use of the increasing computing and storage capacity of today's smart automobiles. Due to storage and processing limitations, it is impractical to train a model with a high number of input parameters on a centralised data server [11]. Transporting raw input data over the network using this centralised method results in prohibitively expensive communication expenses. In particular, data that cannot be transferred via a network due to privacy concerns must be retained locally [12].

In FL, trained models and their parameters are stored in different places, but everyone works together to build a single universal prediction model [13]. Vehicles frequently collect data and utilise it to develop autonomous local models as part of the learning process in the data-parallel FL technique [14]. The next step is to retrain the local model using a global model constructed by combining the parameters of all the previously trained local models. The repetitive process of the learning algorithm ensures eventual convergence [15]. The fact that FL relies on a centralised aggregator makes it less distributed than it could be; yet, complete decentralisation reduces the efficiency of the learning model [16].

Malicious vehicles tampering with local training data or selfish vehicles not supplying their local models can also cause the strategy to fail [17]. In the context of autonomous vehicles, however, the approach is practical and may be put into place to provide safer and faster vehicular services. Using constant feedback from the environment, reinforcement learning (RL) is an AI technique that intelligently learns the best policy to maximise reward [18]. This is similar to how individuals learn about their surroundings through experience rather than direct observation [19]. Traditional RL would work, but it would take too long to completely explore the complicated environment before it could converge on an ideal policy. Scaling RL in such systems, which often have enormous and continuous state-action space, is a significant challenge. It is suggested that Deep Reinforcement Learning (DRL) can enhance generalisation and adaptation to novel situations, making it well-suited for such vast environments [20, 21]. A Reinforcement learning agent engages in interactions with the complicated environment and the deep network to approximately learn the value function or policy for future actions [22]. Applications of DRL in areas as diverse as control, resource management, gaming, robotics, and autonomous driving have been demonstrated [4].

Specifically, Reinforcement Learning excels at sequential decision-making in the face of ambiguity. This is because, unlike more conventional optimization problems with explicit rules, off-loading constraints tend to be more implicit and diversified in highly dynamic mobility circumstances [23]. For instance, vehicle-to-everything (V2X) enables a wide range of services, each with its own set of requirements and challenges [24]. Another factor contributing to the growing popularity of RL-based offloading decision models is the rapid expansion of the use of technological innovations in automobiles. To achieve a goal, these models typically use reinforcement learning to acquire a better approach [25]. Interestingly, this training can be done with distributed algorithms, thus the approach can be developed through distributed collaboration, boosting system-level performance with local knowledge [26]. To accomplish efficient task offloading decisions, we present a DRL-based federated offloading architecture that uses multi-tiered collective learning to offload jobs [27]. By allowing automobiles and fog servers to share information, the framework enhances their capacity to learn. The most important results of this study are:

- To make smarter decisions about offloading work, it's important to investigate multi-agent V2X scenarios in which agents communicate with one another and the surrounding vehicular environment.
- Make use of several distributed learning environments to create a multi-tiered federated offloading system that can learn offloading

decisions at the edge. With less data being shared between levels, this increases both the privacy and efficiency of communication channels.

- Using a hybridized strengthening learning paradigm called asynchronous advantage actor-critic (A3C), researchers may build a two-tiered offloading system. Vehicle agents are used to learn local models, while the fog-servers are used to train the global model. By simultaneously operating numerous separate learning environments, A3C can capture a larger fraction of the state space. Quicker convergence with better offloading choice knowledge is the result of this state-space exploration and multi-tiered learning strategy.

Table 1. The notations and meanings.

P_t	Transmission power
$\mathcal{P}$	Parked vehicles
$\mathcal{V}$	Set of vehicles
d	Distance
$\mathcal{F}$	Fog servers
C	Channel capacity
l	Path loss
N_0	Channel noise
B	Bandwidth
s_j	Task size
λ	Arrival rate
c_j	Computing requirements
μ	Service rate
Z_i	Residence time
L_i	Residence time
s_i	Input size
f_i	Clock frequency
Q	Queue length
D_i^1	Transmission delay
D_i^2	Computation delay
D_i^3	Queuing delay
ψ_i	Frequency reuse cost
Ψ_2	Leased computation cost
T	Action reward
R_i	Agent reward

2. Literature Review

Recent efforts to make effective use of distributed computer nodes across a vehicle network are discussed here.

2.1 *Conventional Task Offloading*

The choice to unload a job is based on the vehicle's speed and the direction it is travelling. In other words, cars headed in the same direction can compete for the offloading job. For instance, both independent and collaborative fog-servers are provided with mobility-aware and location-based work offloading techniques [28]. Timely completion of tasks is achieved through efficient utilisation of available computing and communication resources. This method prioritises device mobility when choosing resources above queue status. To maximise resource utilisation in a V2V resource-sharing network, context-aware opportunistic offloading is proposed based on the vehicles' directions, locations, speeds, and queue conditions. The authors [29] offer an approach to adaptive offloading based on the multi-armed bandit theory and the offloading delay of fog nodes. To reduce stress on the edge fog servers, explain how load balancing may be achieved using a game-theoretic approach in a FiWi-enabled fog-computing network. Centralised network management, a major tenet of SDN, is the key to this remedy [30]. An approach to offloading work is provided in the form of a game with missing information in which the existing resources are pooled and shared. It uses a greedy pruning algorithm to make resource selections that are both efficient and environmentally friendly [31]. The evolutionary method is also used to optimise pool resource allocation. Notably, these meta-heuristics approaches rely on centralised network management and put more emphasis on device-specific factors when allocating resources. These techniques perform badly as the total and queued number of requests queued increases. In addition, AI-based job offloading can analyse large state spaces and adapt to the dynamic conditions around a vehicle. For instance, waiting times for programme execution are decreased using a predictive model in [29]. By adopting ensemble learning, a fleet of UAVs can efficiently compute high-demand computing jobs. To allocate fog servers to queries from mobile nodes, they use an auction-based mechanism [32]. This method uses a DNN model to reduce the time and power consumed by mobile nodes. Another similar paper uses DNNs to schedule tasks and offload computation to the edge [33]. The review shows that time and energy are better allocated. Notably, the limited information exchange and huge state space present in a vehicular setting where DNN-based approaches are used negatively affect their decision-making ability.

2.2 *Reinforcement Learning-based Offloading*

A deep Q-learning-based intelligent job offloading technique with centralised control over information and network data throughput has been proposed in recent work. Similarly, a deep Q-network-based computing resource allocation approach to free Q-learning from its previous limitations [34]. An efficient decision model for offloading work is calculated by factoring in the speeds and computational capabilities of the vehicles in question. For cases with a sizable state space and a sluggish convergence rate because of numerous policy explorations and associated calculations, Q-learning becomes inefficient. An optimal offloading policy is learned by the fog device thanks to online learning performed by a service offloading architecture described [35]. DRL is used to formulate a stochastic resource optimization problem in the context of a "when and where to schedule" paradigm [36]. The model takes in a feature set representative of the states of the task queue and uses DNN to estimate an assignment strategy. The authors [37] of a similar study propose a priority-aware DRL approach that encourages vehicles to pool their unused processing power. The answer specifies a dynamic pricing approach determined by the relative importance of tasks, the mobility of nodes, and the accessibility of resources. A DRL-based offloading technique is similarly optimised [38], where task queue status, mobility, and task dependency are all part of the joint objective function. The method improves performance by cutting down on power usage and lag time while offloading tasks. Experts in deep reinforcement learning investigate the task offloading paradigm [39]. In this research, the authors use a SAC (soft actor critic)-based, model-free version of deep reinforcement learning to maximise entropy and expected utility. However, due to centralization in the learning of task allocation decisions, the given labour is highly concentrated. Our suggested method involves learning two distinct models at the vehicles and the base stations, as opposed to the prior work's singular focus on learning these task offloading decisions at the base stations. DRL approaches generally fail to deliver satisfactory outcomes in continuous state space contexts because of the dynamism of the environment and the lack of communication. This work may be well-suited to distributed and federated DRL, which ensures a richly varied training environment and speeds up the process of reaching the global optimum.

3. System Model

Using a network of cars, some of which are in motion (denoted V), others are parked (denoted P) and some of which are in the fog (denoted F), we examine a resource-sharing vehicular environment (denoted F). The only source of jobs with fluctuating computational demands is the moving automobiles themselves; these can be processed locally or offloaded to other nodes.

3.1 Transmission Model

The route loss l for a wireless channel is defined as where d is the distance between the transmitting and receiving devices.

$$l = a + \beta \log_{10}(d) + b \tag{1}$$

where β is the path attenuation index, a represents the initial offset, and b represents the shadow fading effect. Therefore, the upper bound along the path of communication can be expressed as

$$C = \mathbf{B}\log_2\left(1 + \frac{P_t \cdot l}{N_o}\right) \tag{2}$$

where C is the maximum data rate that can be transmitted over a radio channel with bandwidth. N_0 is the channel noise, and Pt is the transmission power of the vehicular nodes. It is important to note that, we assume the wireless channel is quasistatic and time-invariant. This indicates that the channel state experienced by all nodes is consistent, with equal transmission power and interference. Keep in mind that the transmission delay, a crucial component of the overall delay that we aim to minimise, considers the channel capacity. Our model considers distance, pathloss, and adapts to new values of the transmission delay to teach itself which jobs are best offloaded.

3.2 Task Model

Tasks generated by the various vehicle apps all have the same priority and are completely independent of one another. There are two tuples for each incoming job j ∈ j, whereas sj represents the task size in MBs and cj represents the computation requirements in Mbits.

3.3 Queuing Model

Researchers investigate a multi-server First In First Out queuing model with an infinite population using a Poisson process to characterise the system's workload and an arrival rate of λ. The entire system capacity is represented by the formula $\forall(\mathcal{V}, \mathcal{F}) = \sum_{i=0}^{T} \mathcal{K} \cdot (c \cdot s_i)$, whereas K is the size of the entire calling cluster, c is the number of parallel computing nodes, and si is the amount of the data being input. For a given agent i we may write its residence time, Z_i, as

$$Z_i = \Sigma_{x \in Q} \frac{c_x}{f_i} \tag{3}$$

where, fi is the clock frequency and $\Sigma_{x \in Q} c_x$ is the queue length of queue Q.

3.4 Task Execution Model

The transmitting node consumes some bandwidth while unloading a task. The formula for the sum of all transmission delays, D_i^1, for car i ∈ V is as follows:

$$D_i^1 = \psi_i \, \mathbf{B} \cdot (D_i^\uparrow + D_i^\downarrow) \tag{4}$$

Here, up-link, $D_i^\downarrow$ and down-link, $D_i^\uparrow$ represent the down and up-link transmission delays, respectively ψ_i represents the cost of frequency reuse. Task j's uplink communication delay is written as $D_i^\uparrow = \Sigma_{j \in J} 1_j \cdot \left(\frac{S_j}{C_i} \right)$. Whether or whether a job is offloaded is represented by the indicator random variable, $1_j(\cdot)$, which follows the same pattern. Due to its negligible size, the output delay was disregarded. In addition, the time lag in calculations, D_i^2, for a given vehicle i is expressed as

$$D_i^2 = (\psi_2 f_i) \cdot D_i^* \tag{5}$$

where, f_i compute node's clock frequency and, ψ_2 is the unit leasing cost of the computing resource. The amount of time it takes to do a computation is denoted by the notation, $D_i^* = \Sigma_{j \in J} \frac{c_j}{f_i}$. Like the execution delay, the queuing delay is the amount of time a task spends in the queue before being processed. Therefore, the queue time, D_i^3, is represented as

$$D_i^3 = (\psi_2 f_i) \cdot \Sigma_{j=1}^{J} Z_j \tag{6}$$

whereas, Z_j denotes the residence time, or the amount of time that task j spends waiting in the system queue before being processed. In queue management, the unit leasing cost of the computing resource is $\psi2$. Similarly, all delays add up to what we call total service time, D_i.

$$D_i = D_i^1 + D_i^2 + D_i^3 \tag{7}$$

3.5 Offloading Model

The duties generated by the vehicles might be carried out locally or distributed over the resource-sharing vehicular network, as was indicated earlier. The computer power needed for the task and the proximity of surrounding resources like fog-servers and parked vehicles are two of the main considerations. As shown in the Equation (8), a source vehicle i for a given task j employs a sophisticated decision model to assess whether to distribute the burden between nearby nodes.

$$\delta_i = \{(0, 1): \delta = f(x_1, x_2, x_3, \cdots, x_n)\}, \forall j \in J \tag{8}$$

where the current vehicle condition is represented by the collection of features $(x_1, x_2, x_3, \cdots, x_n)$. The task offloading algorithm learns a function $f(\cdot)$ that relates the state of agent δ_i to an offloading strategy.

4. Problem Formulation

To make better offloading decisions, vehicles communicate with other vehicles and their surroundings. Consider an agent I that, given its current state s-i, decides to take one action a_i in exchange for a reward r-i to progress to state s_i'. The following form can be used to describe the agent's interaction with its surroundings:

4.1 State Space

The current vehicle status is a metaphor for available processing power. Due to the dynamic nature of computing demand in automotive systems, vehicles must be able to meet computing requirements based on the requirements for computing and the availability of computing resources, which can vary substantially. Our approach considers the many variables that can affect jobs and vehicles. When planning the system's current state, we considered the computational needs of each task and the resources available to each node. We have also factored in the transmission delays and vehicle data rates that control the workload allocation. The state space of a vehicle is a map between the vehicle and the collection of critical parameters that appropriately characterise the dynamics of the environment in which the vehicle operates. The corresponding state profile Si is built as $S_i = \{z_i, \Delta_i, V_i, W_i, f_i, (s_j, c_j)\}$ for an agent i carrying out task j.

- Z_i: includes local, fog-server, mobile, and stationary vehicles and their respective dwell times.
- Δ_i: is the range between the vehicle carrying out the mission to any available resources, such as fog servers or other vehicles.

- V_i: includes moving and stationary vehicles, as well as fog-servers, as a collection of speeds in relation to one another.
- W_i: consists of many methods of resource offloading, such as local servers, fog servers, mobile devices, and stationary vehicles.
- f_i: equals the processing power of the agent that generated the task.
- (s_j, c_j): denotes characteristics of the task, such as the amount of input data and the needed cycle rate.

It is important to note that the performance metrics should be measured in real time so that environmental changes can be accounted for. Capturing the full dynamics of these networks is difficult since their topology changes rapidly and there is a wide variety of possible states. To determine the most appropriate offloading action for the current situation, we use a state building module that generates the environment state in real time.

4.2 Action Space

The set of possible actions in a vehicle-based resource-sharing network is defined by an offloading action space. This area represents the local, fog-server, and mobile/stationary vehicle compute resources. The action space, Ai, for a vehicle agent i performing task j is represented as follows:

$$A_i = \{a_1, a_2, a_3, a_4\} \tag{9}$$

in which case (a-1) the job is queued for execution locally, (a-2) it is offloaded to an adjacent moving automobile, (a-3) it is offloaded to a neighbouring fog-server, and (a-4) it is offloaded to a vehicle in the immediate vicinity that is parked.

4.3 Reward Function

An incentive function can boost the efficiency of a learning algorithm. Here, we refer to a subset of resident time optimization that, by selecting the best action from the given action space, maximises reward across multiple repetitions. Let's assume that a vehicle agent I receives reward, ri for performing an action, ai at a destination node $k \in \{i, x_i\}$, where, $r_i = -\left(\dfrac{1}{\lambda}\right) Z_k$, whereas λ is the task arrival rate and Z-k is the amount of time k needs to spend at its final destination.

Therefore, we may define Ri, the sum of all rewards earned by agent i as

$$R_i = \Sigma_{t=0}^{T} \frac{r_i(t) - R_i}{M_i} \tag{10}$$

where, if there are T timesteps in the simulation, then Mi is the total number of tasks allocated to agent i and ri is the reward for action ai at timestep t.

In conclusion, the agents' learning process begins with the input of the state vector. The agents' enhanced ability to comprehend the condition of the environment thanks to the sharing of local and global models also leads to better decision making. In the beginning, agents make decisions at random and then learn to make better ones based on the rewards they obtain for taking certain actions. Because the reward is proportional to the amount of time spent at a node, picking those with shorter dwell times results in larger payouts. Therefore, given the accessibility of cars in the neighbourhood, vehicular agents will choose the action with the highest reward, even if this means spending less time there.

4.4 *Multi-agent Asynchronous Advantage Actor Critic (A3C) Algorithm*

The actor-critic network is an example of a policy gradient implementation, which is a method for optimising a policy by gradually refining its parameters over time. Both classical Q-learning and deep Q-nets (DQN) select the optimal next step based on the current state, but they couldn't be more different in how they go about doing it. The best policy, or mapping from states to actions, can be learned directly via policy gradients. The actor is the one who really does something about the policy, while the critic provides feedback to help the actor make better choices. When the cumulative return, R-t is removed from a baseline function, the variance or advantage is reduced as, $R_t - b_t(s_t)$. The typical benchmark is the learning value estimate ($V\pi(st)$).

In a multi-agent deep reinforcement scenario, N agents collaborate within a car ecosystem. This creates a simple Markov decision process (MDP) in which tuples like s,a,r.s'. represent agents. The current state of the agent is denoted by the variable s, the action chosen by the agent by the variable a, the reward for the action by the variable r, and the anticipated state of the agent by the variable s. Supposing that at time t, N agents take cues from their environment and act autonomously, with transition probability p, they are rewarded and go on to states. An agent's goal is to discover a mapping from states to actions $\pi(s) \rightarrow a$ that maximises the expected discounted future reward $V^{\pi}(s_t) = \mathbb{E}(R_t | (s_t, a_i))$. Where, Rt is the discounted cumulative future return calculated as, $R_t = \sum_{k=0}^{\infty} \gamma^k r_{t+1}$ with a discount factor $\gamma \in (0, 1)$.

During model training, the A3C approach makes use of numerous concurrent threads, each of which benefits from unique and improved learning experiences. Here, many agents are doing parallel exploration of different regions of the state space of moving vehicles. Uncorrelated updates from many asynchronous interactions allow for faster, more thorough probing of the state space. By subtracting the reward function for success from the estimated value function, $V\pi$ (s_t), the method arrives at an advantage function for time instant t, where s is the time instant.

$$A(s_t, a_i; \theta, w) = r_t + \gamma V(s_{t+1}; w) - V(s_t; w) \tag{11}$$

An action's profitability relative to other possible actions is represented by the benefit function. Using the formula, the actor network does self-updates,

$$\begin{aligned} d\theta \leftarrow d\theta + \nabla_{\theta'} \log \pi_{\theta'}(a_i | s_t; \theta') A(s_t, a_i; \theta', w') \\ + \beta \nabla_{\theta'} H(\pi(s_t; \theta')) \end{aligned} \tag{12}$$

where the measure of uncertainty in the policy option is the entropy parameter. An increase in entropy suggests the agent's behaviour is becoming more chaotic and unpredictable. The agent will converge to a sub-optimal behaviour without the entropy term, giving more weight to the acts that result in the highest reward. While is a regularisation term that adjusts the balance between exploiting and exploring by adjusting the entropy strength, the latter is controlled by the exploitation parameter β. The equation also serves to modernise the critic model.

$$dw \leftarrow dw + 2A(s_t, a_k; w, w_c)\nabla_{w'} A(s_t, a_k; \theta', w_c)$$
$$dw \leftarrow dw + \frac{\partial A(s_t, a_k; \theta', w')^2}{\partial w'} \tag{13}$$

Here, the critic update keeps an eye on the benefit function to ensure that the difference between expected and desired rewards is kept to a minimum.

4.5 System Consumption Problem

When agents produce work, it may be put to the local queue or sent to neighbouring nodes. The function for choosing whether to unload or enqueue locally is intricate. Optimizing decisions can make better use of available computational resources, decreasing the total time spent in a building. Multiagent systems aim to One of the goals of deep reinforcement is to reduce the value of the future in such a way that

$$P1: \text{minimize } V^\pi(s = s_t) = \Sigma_{i=0}^{T} E(R_t \mid (s_t, a_i)) \tag{14}$$

$$\text{badhbox } \Sigma w_j \le W, j \in E$$

$$C3: A(s_t, a_i; \theta, w) \approx 0; \forall T \text{ s.t } \pi^*(s) = a^* \tag{15}$$

The expected payoff for pursuing policy from state s-t. is given by Equation (14), as previously described. It agrees with the estimated value that faithfully depicts the elapsed times of queued tasks. By reducing this value estimate, we can cut down on the number of time tasks spent in the queue, which is the ultimate goal. Decision tuples C1 containing local and offload actions are indicated by the notation, a-i.=0,1. To reduce their residence time, agents favour offloading operations that contribute to the aim function. Because of their lack of context, agents are more likely to explore and act erratically from the outset. The agent's actions grow increasingly predetermined as it gains experience and comes closer to the goal. Workload w must be less than or equal to C2, where C2 is the entire capacity W. Cumulative reward and approximate value function have just a small gap, as shown by the C3 benefit function. To achieve this goal, the agent applies the learned weights θ and w to find the best policy π* that will lead to the best possible actions, a*.

Since there is less opportunity to exchange information in a vehicle, picking the best offloading strategy for activities might be difficult, leading to longer wait times and slower convergence. However, as described in [39], this is possible with federated learning by simply trading gradients between the upper and lower levels of a multi-tiered learning architecture. The system can now scale more quickly and more reliably because of this. We did the same thing with the A3C algorithm, which uses parallel simulation to improve performance. More of the state space can be explored in less time, allowing for better decisions to be made as a result.

5. Proposed FL Architecture for Computation Offloading

This study presents a multi-agent DRL-based federated task offloading framework. The suggested method establishes a multi-tier framework in which regional and global models communicate and collaborate to determine when and where to delegate work. As was previously established, each vehicle agent learns a local model by interacting with its surroundings and so captures the underlying dynamics of the vehicular environment. In addition, the agents' local models are regularly updated thanks to data sent from a centralized global model stored on an edge device. As was previously explained, the vehicle nodes are responsible for learning and storing the models. Federation learning reduces network latency by allowing vehicles to train locally on individual nodes. Training is done on the vehicle nodes, and only the most essential data is transmitted across the resource sharing environment, both of which improve privacy. Because only gradients are communicated between networks, vehicle privacy is protected, and the trained model is more resistant to changes in the data. Since less information is being shared, the trained model is more stable, and the data is less likely to be altered.

In Fig. 1, the two-tiered federated architecture's workflow in action was shown. Two levels, one over the other. Lower tiers host the vehicle nodes, whereas higher tiers are supported by a federation of fog servers. As the actor and critic network of the lower-tier vehicles learns about policy and value, they take over. The upper-tier fog servers collect the data from the cars in the range and aggregate the actor and critic updates. The offloading decision model benefits from the two tiers exchanging up-to-date information.

In a distributed and multi-tiered learning system, N mobile agents compete for limited computational resources. Each vehicle conducts its investigation of the state space within its own A3C instance of the complete vehicular environment, as we have many A3C instances running in parallel. Each agent in a highly dynamic resource-sharing system works to create a policy π that will allow it to reduce the amount of time it spends on each activity. In contrast to existing DRL methods, which rely on experience replay or have a centralised architecture, we employ a distributed approach, wherein learning occurs across numerous levels simultaneously. Improved deliberation in a continuously convergent state space setting has important implications for a wide range of delay-tolerant vehicle services. As was indicated up top, there are two tiers to the learning process.

- Lower Tier Learning: Each vehicle agent uses its instance of an actor-critic network when it connects to the larger network. The agent can use the network to learn a customised version of the local model. To begin, the agent imports the weights (θ_i, w_i) from the global actor-critic network into its local network, which then resets to their original values. Thereafter, it starts learning in an online manner by interacting with its environment. Then it begins learning online through interactions with its surroundings. In other words, when a new task is received, the agent's actor network will perform an offloading action, ai according to a policy $\pi(s) \rightarrow a$, and then receive a reward ri. By calculating a state value, the local critic network provides the actor with constructive criticism $V^\pi(s_i)$. Based on this predicted value, the actor network will investigate other offloading options. An offloading

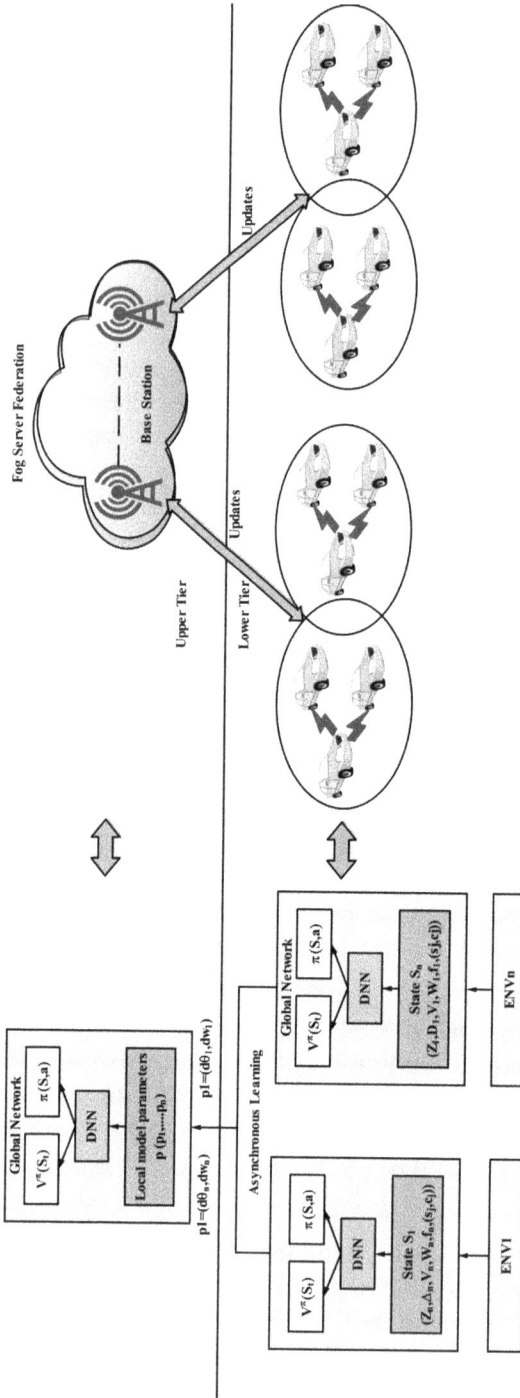

Figure 1. A multi-level infrastructure for federated learning. There is two-way communication between the lower-tier cars and the associated fog node below.

model is mastered jointly by the actor and critic networks. Until a final state is established, the agents interact with their surroundings in accordance with local policy for finite amounts of time, t-max. The agents update the global network based on their calculated local value and policy loss following the conclusion of each episode.

- **Upper-tier Learning:** The locally learned offloading model's weights are sent to a global actor-critic network at the end of each episode using the notation, $(\theta'_1, \theta'_2, \cdots, \theta'_n, w'_1, w'_2, \cdots, w'_n)$. If the ending condition is met or the game is over, the episode concludes. In this project, we deploy the worldwide web via a consortium of fog-servers. Keep in mind that the task of learning the global network has been handed off to a single federate, which might be a fog-server with low residence time $\left(\underset{i \in F}{\arg\min} Z_i \right)$. Inter-tier communication is handled by the remaining federates. The global network updates the global model $(\theta_1, \theta_2, \cdots, \theta_n, w_1, w_2, \cdots, w_n)$ by adding the weights of the local models. Owing to the asynchronous nature of the approach, the global network doesn't need to receive regular local updates. The method swiftly converges by reducing the predicted discounted future utility (Rt | (st, ai)), which is proportional to the residence duration of the jobs in the queue, because of the rich communication and collaborative information exchange between the two tiers.

Because of this correlation between task residence duration and reward, vehicles make offloading judgments. As was previously said, cars select an action to delegate chores to nearby neighbours, and the reward they earn is contingent upon the action they select. The offloading vehicle monitors the vehicles' health throughout the run and verifies their total reward via a time-based calculation of the duration of each task. The truck that spends the least amount of time at each job node is the one that gets the most money and gets to the end faster. The solution is thus able to make offloading decisions for varying arrival rates based on the trained model's weights and probability distribution. Distributed learning of the task of offloading model by vehicles, as we have seen, resolve the optimization problem at the foundational level. Since we are using the distributed A3C technique, we are simultaneously executing several simulation scenarios. The agents operating the vehicles are also engaged in conversation within their virtual worlds. Each car learns from its surroundings in real time, making unloading decisions that are optimal for its unique situation. The higher tier must aggregate the parameter updates sent up from the lower tier and send them back down again. By compiling the latest information for all the cars in a simulation environment, the upper layer is assisting the agents in developing a more varied training experience.

Algorithm 1 Proposed A3C Based Task Offloading Solution

```
 1: procedure MAIN(self)
 2:     if Training Model !exists then
 3:         x ← LowerTierLearning()                              ▷ Agents Online Learning
 4:     end if
 5: end procedure
 6: procedure LOWERTIERLEARNING(V₁, V₂, V₃, ...., Vₙ)
 7:     Get Concurrent Threads
 8:     GlobalParameters ← (θ, w)
 9:     LocalParameters ← (θ', w')
10:     dθ = dw = 0                                              ▷ Zero Gradient initialization
11:     θ' = θ and w' = w                                       ▷ reset using global weights
12:     x ← AgentExperience()
13: end procedure
14: procedure AGENTEXPERIENCE(self,s,a,r,s')
15:     s = sample state for an agent vᵢ
16:     for t ≤ tₘₐₓ do:
17:         while (episodeᵢ != End) do:
18:             Actor ← π(s; aᵢ)                                ▷ Actor Takes Action
19:             rₜ(s, s')                                       ▷ Gets reward and move to the next state
20:             Update t = t + 1
21:         end while
22:         Critic ← V(Sₜ; w) ;                                 ▷ Critic Estimate
23:         Gradient Computation w.r.t θ' : dθ ← FromEquation ??
24:         Gradient Computation w.r.t w' : dw ← FromEquation 14
25:         x ← UpperTierUpdate()                               ▷ Global Model Update
26:     end for
27: end procedure
28: procedure UPPERTIERUPDATE( dθ, dw )
29:     Globalₐ ← dθ ; Globalc ← dw                            ▷ Asynchronous Local to Global updates
30: end procedure
```

There are many benefits to the proposed multi-tier architecture when working with multiple agents. To begin, the architecture makes use of agent interactions across various locally dispersed settings to achieve a diverse learning experience. Second, agents can learn with greater convergence thanks to asynchronous learning, which also increases their exposure to the underlying resource-sharing environment. Third, a fog federation is used to accomplish the learning at the higher tier, which allows for a more comprehensive picture of the environment to be constructed through the fusion of information from several sources. Lower tiers, in contrast to higher ones, are harder to see since they only interact with things directly below them. Using global policy rather than just local knowledge, collaborative learning occurs through faster convergence to global minima through interactive learning. Due to its restricted connectivity and perceptual range, a vehicular network necessarily has certain restrictions. As a result, judgments are only made at the lowest possible level, which might lead to sub-optimal outcomes due to restricted information flow and a distorted view of the road. Higher-level fog servers can consolidate information gleaned from vehicles and lower-level fog servers to better inform offloading decisions. Cars can efficiently map the vast universe of possible vehicle states when they can both teach and learn from one another. Fourth, the models are trained locally, reducing the amount of data transferred to the global network at the top tier, which maximises the use of the available communication channels. Last but not least, the suggested approach makes efficient use of communication resources through individual effort and restricted collaboration to share information across a shared infrastructure. This approach of shared learning improves decision-making, which in turn reduces the time spent on each activity inside the vehicle's shared resources network. At the foundational level, where the cars learn about the distributed work offloading model, the optimization problem is solved. Since we are using the distributed A3C technique, we are simultaneously executing several simulation scenarios. The agents operating the vehicles are also engaged in conversation within their virtual worlds. Each car learns from its surroundings in real time, making unloading decisions that are optimal for its unique situation. Thus, our method benefits from the combined effects of the two processes. To improve oversight and allow for a more comprehensive exploration of the state space at a specific instance when multiple simulation environments are being performed in parallel, the first process gathers and reshares the changes at the higher layer.

6. Performance Evaluation

To test the proposed framework in a realistic traffic environment, we combine the Manhattan Road and mobility model with the SUMO traffic simulator. The necessary software and simulation settings are detailed in Table 2. In our simulation environment, there are 100 automobiles and 9 fog nodes. Our future method makes use of multithreading to run simulations in several distinct scenarios simultaneously. We have been using an Intel Core i7 2.4GHz Lenovo machine with 8GB of RAM. Two of the system's threads are dedicated to simultaneously processing two simulation scenes. Using a simulation of generative adversarial reinforcement learning, we train our model by simulating a network of nodes, each of which can do a different type of computation. TensorFlow 1.13.1 and Python 3.7 are used to create our federated learning method. To give the learning algorithm a more well-rounded education, the simulated traffic scenario is run at a range of arrival rates, from $60 \leq \lambda \leq 200$ per second. The network is built using an actor-critic network and a four-layer DNN architecture to create a policy for offloading tasks locally. Using softmax activation, the 32 neurons in the input layer gather information about the surrounding environment, while the 4 neurons in the output layer make one of four offloading options. There are a total of eighteen (18) neurons in the two covert layers. One example of a hyperparameter is the ability to customise the density of hidden neurons. We looked at the model's performance

Table 2. Variables used in simulations.

Variables	Values
Vehicle/fog transmission range	100 m/200 m
OS	Windows 8.164 – bit
Transmit power/white Gaussian noise	1 W/10^{-3} [23]
Vehicle/fog clock frequency	0.6–1.0GHz/2.0GHz
NN layers/Activation Function	(32,16,8,4)/tanh/Softmax
Learning Rate/Discount Factor	$10e^{-5}$/0.955 [42]
Simulation time/area	1000 time steps/900 × 900 m
Mobility model/vehicle speed	Manhattan/Random
#Vehicles/#fog/#parked	100/9/20
Task cycles/data size	0.2–1.0 Mbits/1–6 MBs
System	Intel Core i7 2.4GHz 8GB
V2V/V2R/V2P bandwidths	1.0/2.0/1.0MHz

with varying numbers of hidden-layer neurons. The network makes use of the Adam optimizer and the tanh activation function.

The algorithm's capacity for learning will suffer if the number of hidden layers is not optimal. Underfitting happens when there are few hidden layers in the input layers, while overfitting happens when there are many. Learning the task offloading parameters requires as little work as possible, therefore picking the right number of layers is important. While there is no universal rule, hidden layers are often a moderate size between the input and output layers. The number of hidden layers ought to be relatively low, especially in comparison to the number of input layers. The tanh function has faster convergence than the sigmoid function because it is symmetric around 0. The simulation converges more quickly on the tanh because of its steeper gradient. As was previously mentioned, the vehicle state space is being simultaneously explored by many agents. This means that the A3C algorithm's distributed nature allows for numerous simulation environments to operate at once in independently controlled threads. The cars can experiment with many strategies at once and see how they pan out thanks to the availability of multiple simulation scenarios. The distributed structure of the technique allows us to probe a wider region of the state space. To complement the cars' online learning, base stations aggregate locally learnt models.

6.1 Online Learning

The online learning data generating process is at the base of our suggested approach. To train a job offloading model, the A3C method gathers fresh data from the environment. Since work offloading is still a novel concept, there is no solid dataset that exhaustively covers the driving environment. Vehicle networks have a very dynamic architecture, leaving little possibility for unrestricted data sharing between users. Creating a dataset that considers all facets of the automotive system is difficult for this reason. A vehicle system is also an environment with a vast or continuous state space, so a sizable sample size is needed to get reliable results. Therefore, the suggested method is an online learning-based answer that produces data in real-time and discovers a preference for offloading tasks. To learn offloading judgments, it is fundamentally distinct from experience replay type techniques, in which a buffer of past events is mined at random.

We utilise the following two algorithms as a basis for comparison:

Stochastic Selection algorithm (SA) – randomly chooses a destination node from a pool of computing nodes within the range of its data transfer. Fog servers, stationary cars, and moving vehicles are all equally represented in the pool from which the resources are drawn.

Greedy Algorithm (GA) – eagerly selects from the available processing power within its region of data transmission, a nearby destination node.

7. Results and Discussion

7.1 Residence Time

It includes the time spent waiting for a work to begin and the time spent gathering the necessary materials. Prison terms are always reduced when better decisions are made. In Fig. 2, we compare the proposed method to two other, more traditional ones for analysing residency duration patterns across a range of arrival rates. All approaches exhibit nearly the same behaviour when the arrival rate is low. However, the pattern gets more noticeable as the arrival rate >=140. Our method appears to have a 58% improvement at =200. Smaller queues and less overall delay time are the results of the suggested method's more careful decision making.

7.2 End-to-end Delay

Determines the total time lost by the network because of activities such as task transmission, waiting in line, and processing at the final node. The proposed method predicts a 57.11% drop at $\lambda = 200$. The best decision-making in a shared-resources environment is indicated by a decrease in the overall end-to-end latency, whereas the worst is disclosed by an increase in both the delay and the queue lengths.

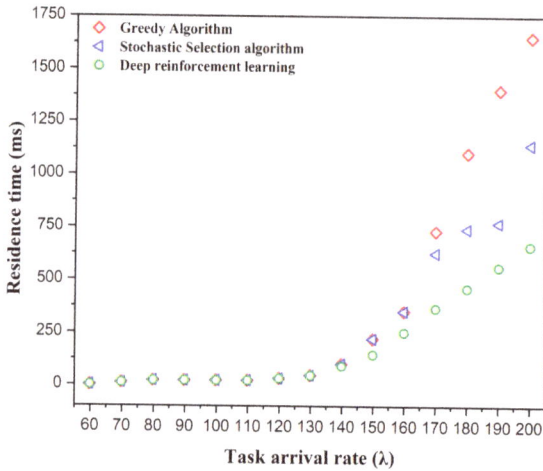

Figure 2. Mean residence time for a variety of task arrival rates.

7.3 *Transmission Delay*

Is the lag caused by the finite data rate of the wireless connection? Remember that this variable impacts the total time of your stay. In comparison to the GA and soft actor critic approaches, the overall transmission delay for the DRL technique is shown in Figure 3. While GA's offloading to nearby neighbours means it has the shortest transmission latency, it also causes the longest residence time. When comparing the proposed scheme to GA and SA, we see a decrease of 8.5 percent and an increase of 105.6 percent, respectively. By carefully considering each node's offloading preferences,

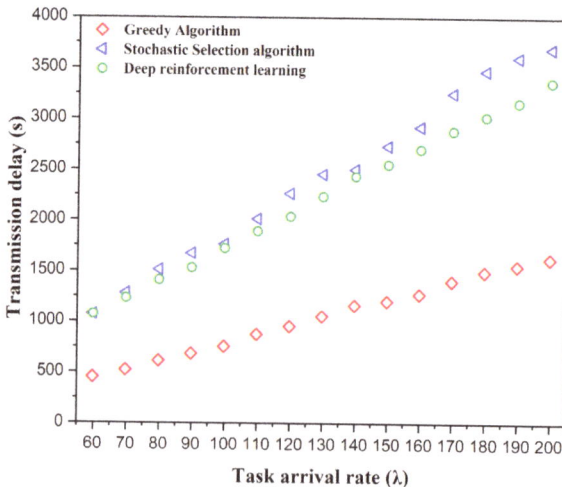

Figure 3. Total transmission delay.

the DRL method makes efficient use of the idle resources of neighbouring nodes.

7.4 System Cost

Is the total of both the offloaded and local executions. Several offloading options, including the proposed approach, and their average system costs. Clearly, the largest cost in an aware decision model is the workload that both GA and soft actor critic must bear. Specifically, the cost of SA is up 16% and the cost of GA is up 56% when $\lambda = 200$. The proposed approach increases the total system cost of the resource-sharing setting by decreasing residence time, as explained. This improvement is attributable to a 58% reduction in residence time for $\lambda = 200$.

7.5 Delivery rate

Is the fraction of the original work that was offloaded, computed, and returned to the original nodes. Figure 4 shows that both the GA and SA sachems' deliveries have significantly decreased. There is a 45.9% decrease in GA and a 20.4% decrease in SA at $\lambda = 200$. This significant decrease in deliveries might be attributed to the uneven distribution of workloads, which results in longer queues and more failures. As more work is shifted to the nodes around it, GA experiences a higher rate of failure due to growing queues.

At a task arrival rate of $\lambda = 200$, as shown in Fig. 5, the total number of tasks given is displayed. When compared to GA and SA, which

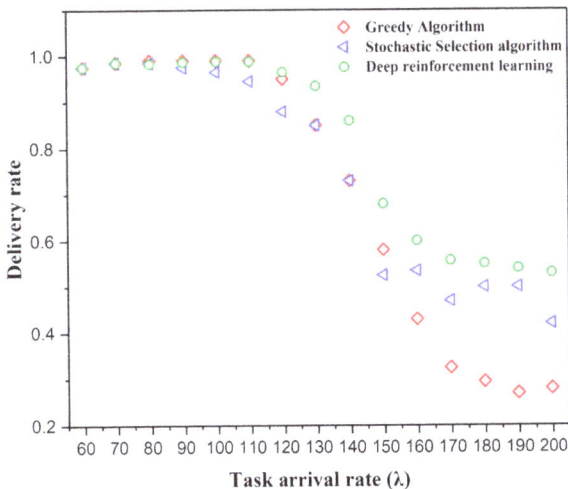

Figure 4. Variable arrival rate and result delivery ratio.

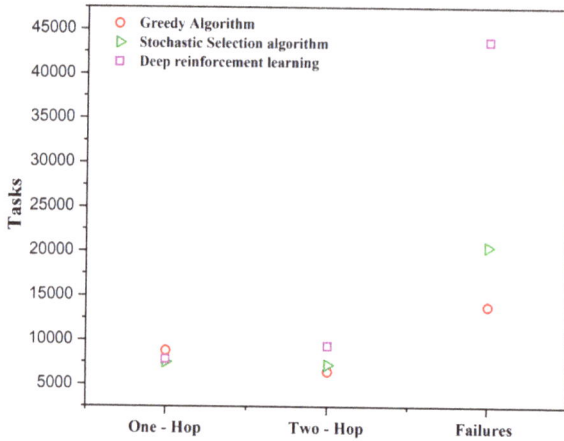

Figure 5. Rate of task arrival statistics with a value of 200.

complete roughly 17000 and 14000 tasks, respectively, the proposed system completes around 15000 tasks. The proposed method also has a 9000 single-hop delivery rate out of a total of 15000 delivered assignments. Remember that delivery is the sum of one hop and two hops. BP makes better use of the local nodes' resources than SA and GA do, which both use 8000 for first-hop delivery. The proposed approach has much fewer failures than the GA and SA (by 69% and 53% respectively). So, as the number of arrivals increases, the SA and GA method reduces the number of deliveries and increases the number of successful attempts.

7.6 Workload Distribution

Figure 6 displays the system's task assignment and completion rates. The data at $\theta\lambda = 200$ shows that the offloading system allocates tasks in the most effective way possible. The proposed method increases the proportion of locally executed jobs by 30% compared to GA and by 4.1% compared to SA. The proposed method employs nearby resource allocation optimally by assigning 15000 tasks to each destination. The suggested results demonstrate that delays and result delivery are decreased across the resource sharing framework and that the GA technique increases its use of neighbouring mobile nodes by 65%. However, efficiency has decreased, as measured by the lowest delivery costs, according to the results.

7.7 Utilization Ratio

Specifies the fraction of incoming tasks that were successfully processed by the available compute nodes. The proposed DRL-based method is

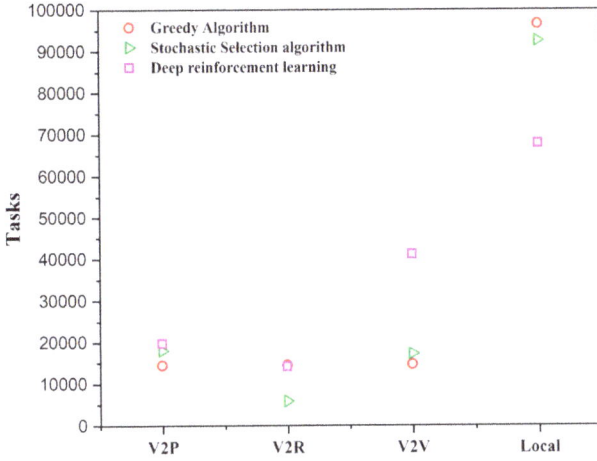

Figure 6. Statistics for tasks with an arrival rate of $\lambda = 200$.

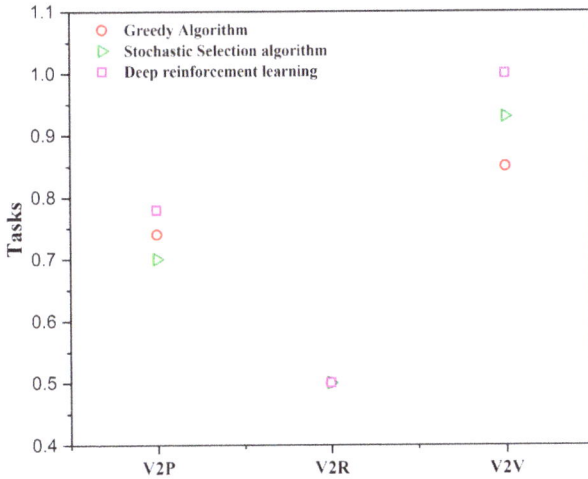

Figure 7. Task arrival rate $\sigma = 200$(task/s) utilisation ratio.

seen to make good use of the shared resources available. Figure 7 shows the distribution of the three possible scenarios: V2V, V2F, and V2P. The proposed method clearly outperforms GA and SA in terms of task utilisation efficiency. The proposed method improves resource utilisation in all three scenarios: V2P (5.35% points) and vehicle-to-vehicle (8.5% improvement).

Remember, these are the numbers for when the system is at capacity. Loss of actor support and critic support networks are shown in the upper and centre plots of Fig. 8. Initially, the actor network probes its

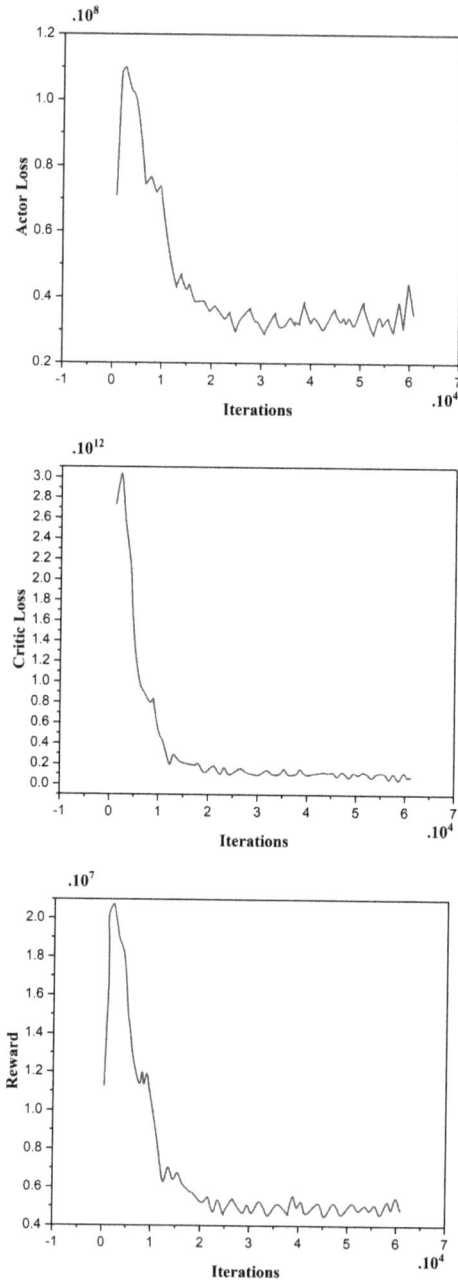

Figure 8. The expected rate of convergence for the proposed deep reinforcement learning task offloading structure, with a task arrival rate of $\lambda = 200$. The cumulative loss for the actor network is shown at the top, the cumulative loss for the critic network in the middle, and the cumulative reward at the bottom.

surroundings by acting arbitrarily and noting the resultant reward. The actor revises its offloading strategy in light of the total reward and then chooses actions that will lead to more future rewards with shorter residence durations. As observed in the bottom rewards plot, closing the gap between the goal reward and the value function improves the value estimate and, by implication, the critic-network loss. This might serve as both a guide and a critique of the actor network as it reconsiders its approach.

8. Conclusion

Using deep reinforcement learning, this study presents a framework for federated multi-agent-powered resource allocation. Fast convergence is achieved in the proposed model for delay-sensitive resource sharing by learning the decision model at many levels simultaneously. Our experimental results demonstrate that our multi-agent DRL framework may improve performance in a high-stakes environment by increasing the speed with which solutions are delivered, decreasing end-to-end latency, and making the most efficient use of available resources. The utilisation ratio is increased by 5.35% points in V2P and 8.5% in V2V because of these advancements to the suggested work, and 9000 tasks are now supplied in a single hop. In addition, a 30% improvement in local execution speed over GA has been achieved. In the future, we hope to expand this work by comparing more advanced DRL approaches. The framework will be hosted in a public GitHub repository for easy access by everybody. In addition, further methods can be implemented to enhance the collaborative system's performance in terms of sampling efficiency, transmission latency, and other metrics. The vast quantity of gradient updates communicated in the context of a resource-sharing vehicle may be a burden on the shared network. Data transferred between layers remains private with local training, but the changes are exposed to gradient spoofing attacks.

References

[1] Ponmalar, A., Uma, M., Jyothi, B.R., Aishwariya, K. and Dharshini, S. 2021. Transparent solar panel using cloud. In 2021 6th International Conference on Communication and Electronics Systems (ICCES), pp. 870–874.

[2] Anand, J. and Perinbam, J.R.P. 2014. A Survey on energy efficient biomedical wireless sensor networks. Am. Int. J. Res. Sci. Technol. Eng. Math., 3(7): 212–216.

[3] Anand, J., Yaseen, M.M. and Radha, K. 2014. Multi-session packet scheduling approaches over multi-hop wireless networks. Int. J. Adv. Res. Comput. Commun. Eng., 3(7): 1–7.

[4] Shabir, B., Rahman, A.U., Malik, A.W., Buyya, R. and Khan, M.A. 2023. A federated multi-agent deep reinforcement learning for vehicular fog computing. Journal of Supercomputing, 79(6): 6141–6167.

[5] Jamil, B., Ijaz, H., Shojafar, M. and Munir, K. 2023. IRATS: A DRL-based intelligent priority and deadline-aware online resource allocation and task scheduling algorithm in a vehicular fog network. Ad Hoc Networks, 141.

[6] Wei, Z., Li, B., Zhang, R., Cheng, X. and Yang, L. 2023. Many-to-Many task offloading in vehicular fog computing: A multi-agent deep reinforcement learning approach. IEEE Transactions on Mobile Computing, pp. 1–16.

[7] Du, J. and Jiang, C. 2023. Priority-aware computational resource allocation. Wireless Networks (United Kingdom), pp. 271–305.

[8] Chen, N., Zhang, P., Kumar, N., Hsu, C.H., Abualigah, L. and Zhu, H. 2022. Spectral graph theory-based virtual network embedding for vehicular fog computing: A deep reinforcement learning architecture. Knowledge-Based Systems, 257.

[9] Hazarika, B., Singh, K., Biswas, S. and Li, C.P. 2022. DRL-Based resource allocation for computation offloading in IoV networks. IEEE Transactions on Industrial Informatics, 18(11): 8027–8038.

[10] Wei, Z., Li, B., Zhang, R., Cheng, X. and Yang, L. 2022. Dynamic Many-to-Many task offloading in vehicular fog computing: A multi-agent DRL approach, pp. 6301–6306.

[11] Sarkar, I. and Kumar, S. Delay-aware intelligent task offloading strategy in vehicular fog computing.

[12] Shi, J., Du, J., Wang, J. and Yuan, J. 2022. Federated deep reinforcement learning-based task allocation in vehicular fog computing, vol. 2022-June.

[13] Hazarika, B., Singh, K., Biswas, S., Mumtaz, S. and Li, C.P. 2022. SAC-based resource allocation for computation offloading in IoV networks, pp. 314–319.

[14] Son, D.B., An, V.T., Hai, T.T., Nguyen, B.M., Le, N.P. and Binh, H.T.T. 2021. Fuzzy deep Q-learning task offloading in delay constrained vehicular fog computing, vol. 2021-July.

[15] Fu, F., Kang, Y., Zhang, Z., Yu, F.R. and Wu, T. 2021. Soft actor-critic DRL for live transcoding and streaming in vehicular fog-computing-enabled IoV. IEEE Internet of Things Journal, 8(3): 1308–1321.

[16] Shi, J., Du, J., Wang, J. and Yuan, J. 2021. Deep reinforcement learning-based V2V partial computation offloading in vehicular fog computing, vol. 2021-March.

[17] Lan, D., Taherkordi, A., Eliassen, F. and Liu, L. 2020. Deep reinforcement learning for computation offloading and caching in fog-based vehicular networks, pp. 622–630.

[18] Shi, J., Du, J., Wang, J., Wang, J. and Yuan, 2020. Priority-aware task offloading in vehicular fog computing based on deep reinforcement learning. IEEE Transactions on Vehicular Technology, 69(12): 16067–16081.

[19] Fu, F., Kang, Y., Zhang, Z. and Yu, F.R. 2020. Transcoding for live streaming-based on vehicular fog computing: An actor-critic DRL approach, pp. 1015–1020.

[20] Zhao, J., Kong, M., Li, Q. and Sun, X. 2020. Contract-based computing resource management via deep reinforcement learning in vehicular fog computing. IEEE Access, 8: 3319–3329.

[21] Chen, X., Leng, S., Zhang, K. and Xiong, K. 2019. A machine-learning based time constrained resource allocation scheme for vehicular fog computing. China Communications, 16(11): 29–41.

[22] Lee, S.S. and Lee, S. 2019. Poster abstract: Deep reinforcement learning-based resource allocation in vehicular fog computing, pp. 1029–1030.

[23] Liu, Y., Jiang, L., Qi, Q. and Xie, S. 2023. Energy-efficient space-air-ground integrated edge computing for internet of remote things: A Federated DRL approach. IEEE Internet of Things Journal, 10(6): 4845–4856.

[24] Jo, S., Kim, U., Kim, J., Jong, C. and Pak, C. 2023. Deep reinforcement learning-based joint optimization of computation offloading and resource allocation in F-RAN. IET Communications, 17(5): 549–564.

[25] Tian, A., Feng, B., Zhou, H., Huang, Y., Sood, K., Yu, S. and Zhang, H. 2023. Efficient federated DRL-based cooperative caching for mobile edge networks. IEEE Transactions on Network and Service Management, 20(1): 246–260.

[26] Liu, S., Yang, S., Zhang, H. and Wu, W. 2023. A Federated Learning and deep reinforcement learning-based method with two types of agents for computation offload. Sensors, 23(4).

[27] Xu, Y., Bhuiyan, M.Z.A., Wang, T., Zhou, X. and Singh, A.K. 2023. C-FDRL: Context-aware privacy-preserving offloading through federated deep reinforcement learning in cloud-enabled IoT. IEEE Transactions on Industrial Informatics, 19(2): 1155–1164.

[28] Nguyen, L.N., Sigg, S., Lietzen, J., Findling, R.D. and Ruttik, K. 2023. Camouflage learning: Feature value obscuring ambient intelligence for constrained devices. IEEE Transactions on Mobile Computing, 22(2): 781–796.

[29] Huang, Y., Li, M., Yu, F.R., Si, P. and Zhang, Y. 2023. Performance optimization for energy-efficient industrial internet of things based on ambient backscatter communication: An A3C-FL Approach. IEEE Transactions on Green Communications and Networking, 1.

[30] Song, X., Xu, B., Zhang, X., Wang, S., Song, T., Xing, G. and Fang, L. 2023. Everyone-centric heterogeneous multi-server computation offloading in ITS with pervasive AI. IEEE Network, pp. 1–7.

[31] Ndikumana, A., Nguyen, K.K. and Cheriet, M. 2023. Federated Learning assisted deep q-learning for joint task offloading and fronthaul segment routing in open RAN. IEEE Transactions on Network and Service Management, 1.

[32] Chung, W.C., Chang, Y.C., Hsu, C.H., Chang, C.H. and Hung, C.L. 2023. Federated feature concatenate method for heterogeneous computing in Federated Learning. Computers, Materials and Continua, 75(1): 351–370.

[33] Grasso, C., Raftopoulos, R., Schembra, G. and Serrano, S. 2022. H-HOME: A learning framework of federated FANETs to provide edge computing to future delay-constrained IoT systems. Computer Networks, 219.

[34] Nguyen, D.C., Hosseinalipour, S., Love, D.J., Pathirana, P.N. and Brinton, C.G. 2022. Latency optimization for blockchain-empowered federated learning in multi-server edge computing. IEEE Journal on Selected Areas in Communications, 40(12): 3373–3390.

[35] Ullah, R., Wu, D., Harvey, P., Kilpatrick, P., Spence, I. and Varghese, B. 2022. FedFly: Toward migration in edge-based distributed Federated Learning. IEEE Communications Magazine, 60(11): 42–48.

[36] Taskou, S.K., Rasti, M. and Nardelli, P.H.J. 2022. Blockchain function virtualization: A new approach for mobile networks beyond 5G. IEEE Network, 36(6): 134–141.

[37] Li, X., Sun, C., Wen, J., Wang, X., Guizani, M. and Leung, V.C.M. 2022. Multi-user QoE enhancement: Federated multi-agent reinforcement learning for cooperative edge intelligence. IEEE Network, 36(5): 144–151.

[38] Jijin, J., Seet, B.C. and Chong, P.H.J. 2022. Smart-contract-based automation for OF-RAN Processes: A Federated Learning use-case. Journal of Sensor and Actuator Networks, 11(3).

[39] Chen, X. and Liu, G. 2022. Federated deep reinforcement learning-based task offloading and resource allocation for smart cities in a mobile edge network. Sensors, 22(13).

Chapter 9

Future of Medical Research with a Data-driven Federated Learning Approach

G Arun Sampaul Thomas,[1,*] *S Muthukaruppasamy,*[2]
S Sathish Kumar[3] *and K Saravanan*[4]

1. Introduction

Despite ongoing technological advancements, the process of creating a new medication is time-consuming and expensive (known as "Eroom's law"). The pharmaceutical industry is now aware that partnerships with rival firms and medical facilities may grasp the secret to retrogressive this tendency, from generating new leads to enhancing clinical probationary enterprise to keeping track of persistent sequels. It is a logical progression in the ongoing use of machine learning in our day-to-day activities, driven by both present limitations and other concurrent advances [1, 6, 14].

Federated learning facilitates collaborative machine learning without the need for centralized training data. Unlike traditional machine learning

[1] Department of AI&ML, J.B. Institute of Engineering and Technology, Hyderabad, Telangana, India.
[2] Department of EEE, Velammal Institute of Technology, Panchetti, Tiruvallur Dt, Tamilnadu, India.
[3] Department of AI&ML, J.B. Institute of Engineering and Technology, Hyderabad, Telangana, India.
[4] Department of CSE, College of Engineering Guindy, Anna University Chennai, Tamilnadu, India.
Emails: mksamy14@yahoo.com; mailsathishcse@gmail.com; Saravanan.krishnann@gmail.com
* Corresponding author: arunsam.infotech@gmail.com

setups where data is stored in the cloud, federated learning develops a shared machine learning model by retaining all the training data on a single device. This approach provides a high level of privacy and security, making it particularly beneficial in industries like healthcare, banking, and government that have stringent regulations [3].

With federated learning, users can take advantage of personalized machine learning models without compromising their private information. This article focuses on federated learning and its various applications, with an emphasis on how it can benefit the healthcare industry [7].

1.1 AI and Healthcare

Federated learning has the potential to provide ground-breaking AI-driven innovations more swiftly to the market and patients. Additionally, the healthcare industry may experience significant advancements and discoveries as a result. FL has exhibited that it can change the game while it derives from the general use of AI in healthcare. Collaborative learning expresses in what way privacy-preserving FL approaches may aid the development of robust AI prototypes that perform admirably across industries, uniform in those with restricted access to data or secret knowledge [8, 9, 14, 16].

Additionally, this method aids in capturing a wider range of patient characteristics, such as age, gender, and ethnic variances that may be highly regional in nature. Such a wide range of data sets will probably result in less biased and more accurate machine learning models. The accuracy of the different AI models may then be further enhanced with the expert input of qualified medical practitioners [5].

Federated learning consequently has the potential to deliver breakthrough AI-driven technologies to the market and patients more quickly. It might also bring about enormous improvements and discoveries in the healthcare sector [9]. While it emanates from the widespread deployment of AI in healthcare, FL has demonstrated its ability to shift the game. It's now time for:

- Through significant budgetary, regulatory, and public-private sector partnership measures, governments should support and accelerate the digital transformation of healthcare.

- Healthcare organizations should invest in data and IT foundations, sustenance close alliances with academia, and AI researchers with startup companies.

- Endeavor investment assets to champion and capitalize oon public-private sector partnerships (PPPs) that will drive the digital transformation of healthcare.

- Collaborative learning shows how privacy-preserving federated learning approaches might make it possible to build strong AI models that function effectively across businesses, even in sectors with limited access to data or industries with secret information [16].

The study's primary author, Dr. Ittai Dayan, who oversaw AI enlargement and started the healthcare start-up Rhino Health this year, stated that, "often in AI research, when you design an algorithm on one hospital's data, it doesn't perform well at any other hospital." "But we were able to design a generalizable model that can support frontline clinicians globally by leveraging federated learning and objective, multimodal data from multiple continents," he added. The extensive FL initiatives are currently ongoing in the healthcare business, including a Team research project for evaluation, mammography and medicinal company Bayer's exertion drill an AI exemplary for pique dissection. Yonder Healthcare, FL can assist energy firms in the analysis of seismic and wellbore data, financial institutions in the improvement of fraud detection models, and researchers working on autonomous vehicles in the development of AI that generalizes to driving habits in various countries.

1.2 AI in Pharma Industries

However, issues with IP protection and privacy are significant roadblocks for the sector. Due to the sensitivity and significance of these clinical datasets, the data generated from pharma-sponsored clinical trials—which account for around two-thirds of all studies (3)—are often unavailable to anybody but the sponsor until the experiment is over.

Like this, 'chemical libraries' controlled by the pharmaceutical industry—priceless collections of compounds used for in-house drug development—are not designed for sharing. Additionally, because developing in-house expertise takes time, many pharmaceutical companies have not yet implemented machine learning technology at a scale. Instead, they rely on exclusive agreements with tech firms to safeguard their ideas [2, 3]. The promise of data and insight exchange across pharma businesses is still unrealized, despite the growing interest in AI.

A new model of cooperative research in a setting of low trust is denoted by the word "coopetition" [4]. Competing pharmaceutical businesses use federated learning to aggregate insights from many datasets into a machine learning model without revealing their raw data, how they distribute it across partners in the competition, or how they gain access to other servers. Each partner will continue to be responsible for maintaining data ownership and ensuring compliance with legal requirements (such as GDPR or HIPAA). To stop data leaks, further privacy-preserving mechanisms have been implemented (safe aggregation, differentially private model training) [5, 6].

MELLODDY, a public-private cooperation involving 10 pharmaceutical companies in Europe that Owkin enabled, is an illustration of "coopetition in action." More than 10 million compounds and 1 billion tests comprise the biggest chemical compound library currently in existence, enabling each partner to choose the most promising medication candidates more effectively for development a scenario that seemed unimaginable until recently.

Contrary to popular belief, MELLODDY uses distributed ledger technology to make every activity and trade completely traceable. It also works on an open-source framework called Substrata, which is accessible to the scientific community. The first "federated run" was completed in 2020 following a successful security assessment of the platform (outsourced to a separate company), proving that a multi-task (i.e., target-agnostic) machine learning model can be trained at scale across universities. We will utilize the platform moving forward to evaluate a range of research hypotheses aimed at boosting drug development effectiveness.

These are illustrations of how linking people and organizations may revolutionize medical research through federated learning in healthcare and collaborative machine learning. In addition to the previously mentioned uses, the technology presents a wide range of other options, such as enhancing research on rare diseases by training a model on sparse datasets that are dispersed over the globe. The breadth of such partnerships can cover several study domains and generate unique solutions since federated learning frameworks are adaptable, which frequently allows for much quicker innovation than traditional machine learning techniques [8, 13].

2. Federated Learning and its Technology-Related Importance

Federated learning is a cutting-edge practice for instructing the routines we employ every day. However, what does federated learning truly entail, and is it additional secluded? You might not be familiar with the term "federated learning" unless you're committed to staying current on all things AI. While it might not be as well-known as other technological developments like Drones and 3D printing, its ramifications for machine learning and privacy may make it much more popular in the years to come. To put it plainly and simply, FL is a marginal strategy for refining the procedures that now govern many aspects of our lives, whether they are Google Maps or Facebook's News Feed [2].

In the more traditional system, our data is collected and analyzed by a central server, which then modifies the algorithm based on the relevant information. Federated learning provides a resolution that augments user discretion because most of the information is stored on an individual's device. Procedures succession themselves straightforwardly on user devices and only provide the important data summaries, so the data is not transmitted back in its full form. As a result, companies may improve their procedures without collecting all a user's data and provide a more privacy focused service.

Let's not sugar-coat it: the pits of FL may be confusing besides complex for most individuals. The study of AI includes far more arithmetic and reasoning than most of us are comfortable with, and it is outside the expertise of many individuals. Despite these challenges, federated learning is a fascinating and significant technological advancement, so it's worthwhile to try to understand it. So that you can see the larger picture, we will simplify and deconstruct the concepts to make things easier for you.

2.1 Machine Learning and Algorithms

Our life is made up of algorithms, except you employ all your time defining our data as someone from the 1930s. Algorithms are basically sets of commands that are cast off in this context to solve a tricky problem or process an anticipated outcome. They help Instagram, Twitter, and Facebook increase revenue by delivering tailored gratified that are most prospective to be of interest. The search engine Google transforms your search phrases into pages of what it believes you are seeking using complex algorithms. Algorithms in your email filter out junk, although Waze uses procedures to determine the fastest route from point A to point B.

Numerous additional algorithms support us in completing jobs, keep us entertained, or hide in the background of routine operations. If it supports the company's own goals, which are often to make money, companies are continually working to enhance these algorithms to provide you with supreme accurate, and well-organized outcomes. Since their first deployment, several of our most-used algorithms have advanced significantly. When you searched on Google in the late 1990s or the early 2000s, you had to be very particular, and the fallouts were subpar in comparison to today. Machine learning, a branch of artificial intelligence, plays a significant role in how algorithms improve at their duties.

They may advance without requiring outside forces, like a human developer, to implant these changes into them. In the past several decades, machine learning has exploded, enhancing our algorithms, assisting us in obtaining better outcomes, and expanding into new sectors. Due to its usefulness, it has also been a big source of revenue for businesses like Facebook, Google, and numerous others. The more high-quality data points there are in the data set, the bigger the data pool is, and the further precise these machine learning algorithms may be. Because an algorithm may earn further wealth the more efficient it is, data has practically become a commodity [2].

Companies have started to pay attention in the wake of a slew of data concealment crises, including Facebook's Cambridge Analytica debacle and Google+'s large information leak. They appear to be seeking ways to continue achieving their objectives without infuriating their consumers or lawmakers, since they don't want to be usurped. The statement "The future is private" made by Mark Zuckerberg at this year's F8 conference may have been the turning point. Although it's probably better to approach this movement with caution, federated learning is one of the more encouraging advances in terms of user privacy.

3. The Federated Learning Training Process

By making the most recent version of an algorithm accessible to eligible devices, federated learning helps algorithms. The model of this algorithm then gathers information from a subset of users' phones' private data. The new data is processed and then briefly transferred back to the company's computer system, never leaving the phone with the real data. This data is often encrypted on the way back to the server for security reasons. Google created the Secure Aggregation Protocol, which is seen in Figure 1, to stop the server from extrapolating specific data from the summary it has received.

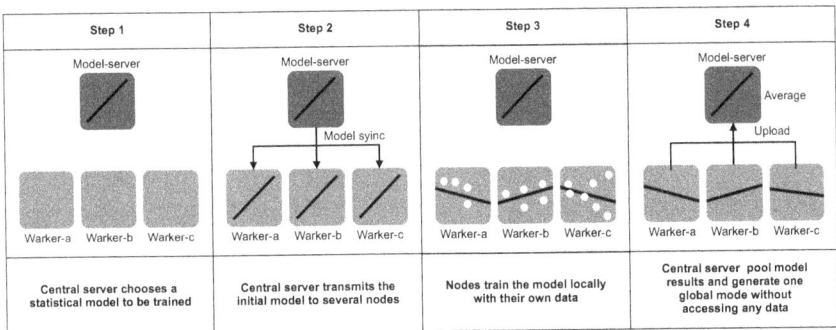

Step 1	Step 2	Step 3	Step 4
Model-server	Model-server	Model-server	Model-server Average
	Model syinc		Upload
Warker-a Warker-b Warker-c	Warker-a Warker-b Warker-c	Warker-a Warker-b Warker-c	Warker-a Warker-b Warker-c
Central server chooses a statistical model to be trained	Central server transmits the initial model to several nodes	Nodes train the model locally with their own data	Central server pool model results and generate one global mode without accessing any data

Figure 1. Federated Learning training process.

Alternatively, disparity privacy can be employed to conceal an individual's summary through the addition of random data noise, obscuring the results. The random data is added before sending the summary to the server, resulting in an accurate outcome suitable for algorithmic processing without the server having access to the actual summary data. This method ensures the privacy of the individual's data. The system uses encryption to prevent the server from receiving specific data summaries. Only after the summary has been improved with input from hundreds or thousands of other users can the server view it according to this protocol.

Disparity privacy and the Secure Aggregation Protocol are two essential methods for shielding user data from both the organization and hackers. They are necessary for federated learning to protect users' privacy. The information summaries are utilized to update the algorithm after being securely transmitted to the server. Thousands of iterations of the procedure are performed, and assessment varieties of the algorithm are also directed to innumerable consumer expedients. This enables businesses to test new iterations of algorithms using actual consumer data. Algorithms can be tested without combining user data on a central server since scrutiny is done within the precincts of user devices. The revised algorithm exemplary is distributed to user expedients to interchange the old one when the testing is finished. Then, the improved algorithm is put to work for its typical duties. It will be more precise and successful in achieving its aims if everything goes as planned.

The entire process is then repeated and again:

- The new algorithm examines data from certain consumer devices.
- It delivers reviews of this consumer's data securely to the server.
- The results from other users are then averaged using this data.
- The algorithm creates updates and evaluates them after learning from the new data.
- Users are given access to a more sophisticated version of the algorithm.

Without ever needing to keep user data on business servers, the algorithm continuously gets better over time by learning from user data. If you're still having trouble understanding what FL is and how it operates, Google has created a caricature that clarifies what makes the federated learning technique easier for you to understand.

In addition to data concealment, the federated learning exemplar provides consumers with several other benefits. The cramming process may be carried out when an expedient is incriminating, associated with WiFi, and not in practice, limiting the hassles experienced by consumers, as opposed to continuously exchanging data with the server. Users won't waste their limited data or battery life as a result when they are out and about. Overall, federated learning sends less data than conventional learning models since it just transmits a summary of pertinent data, not the actual data. Additionally, federated learning may offer both universal and unique algorithmic models. To give a more effective service, it can gather perceptions from a larger clutch of users and integrate them with data from the user.

4. Applications of Federated Learning and its Benefits

There are many possible applications for federated learning, particularly where privacy concerns and the requirement for algorithm improvement collide. The most well-known federated learning experiments to date have been carried out on cell phones, but the same methods may be used with computers and IoT (Internet of Things) devices like self-governing cars.

Some of the current and future applications include:

Google Gboard

As a fragment of Google's keyboard claim, Gboard, federated learning was first widely implemented in the real world. The business wanted to enhance word recommendations using the method without jeopardizing user privacy. Better keyboard predictions could have been created using the old machine learning method but doing so would have been extremely intrusive. Altogether our typing, self-contained communications, and odd Google explorations would have obligated to be sent to a dominant server for analysis, and who distinguishes what else the data could have been cast-off for? Providentially, Google went with its FL strategy.

Autonomous Vehicles

Federated learning may enhance user privacy by renewing the algorithms to utilize only summaries of this data rather than the entire collection of user data. For self-driving automobiles, federated learning has two key uses. The first is that it could protect the privacy of user data. The awareness of having their trip histories and other driving data transferred to and assessed on a centralized server is opposed by many people. A federated learning technique should be used since it could reduce latency. Federated learning, as opposed to standard cloud learning, which necessitates massive data transfers and a slower learning rate, may enable autonomous automobiles to function more rapidly and properly, avoiding accidents and improving safety.

Federated learning may also serve as the catalyst for upcoming developments in the sector [6].

Cloud Computing

In a market dominated by the digital behemoths Google, Amazon, and Microsoft, cloud computing has emerged as the predominant computing paradigm for machine learning.

Community Economy

Google developed its GBoard next-word predictor using federated learning. The development of additional services that rely on data obtained from mobile devices and other IoT devices as a type of sharing economy should be encouraged by the capacity to train a model without jeopardizing users' privacy.

B2B Alliance

Federated learning makes it possible for various data owners to cooperate and exchange their data at the organizational level because data never leaves its original premises.

5. Federated Learning Centered Use Cases in Healthcare

Numerous domains, including healthcare, digital health tracking, portable applications, and others, offer vast potential for federated learning. It has already demonstrated its effectiveness in healthcare applications such as COVID-19 detection, medical imaging, remote health monitoring, and health data management. For instance, Google utilized this approach to enhance Smart Text Selection on Android mobile phones. This use case enables users to quickly select, copy, and utilize text by anticipating the desired word or sequence of words based on user input. The global model receives precise feedback each time a user taps to select a piece of text and corrects the model's suggestion, improving the model's accuracy [3].

Federated learning has huge potential in several fields, including healthcare, digital health tracking, portable apps, and healthcare. Healthcare applications including COVID19 detection, medical imaging, remote health monitoring, and health data management have already been effectively implemented using it [3].

Google utilized this method to enhance Android mobile phones' Smart Text Selection as an illustration of how it may be used for mobile applications. This use case involves anticipating the required phrase or string of words based on user input, allowing users to swiftly choose, copy, and utilize content. The global model receives exact input every time a user taps to pick a text and overrides the model's recommendation, which is then utilized to enhance the model.

To improve instantaneous supervisory and concurrent data collection concerning traffic and roads, federated learning is also pertinent for autonomous cars. Self-driving cars need real-time updates, and federated learning allows for the efficient real-time pooling of the sorts of data from several vehicles.

5.1 Healthcare

In the healthcare segment, data confidentiality, and safety are exceedingly complicated. Many companies keep a lot of delicate and valuable patient data on hand, and hackers are always on the lookout for it [2, 10, 14].

Nobody wants a humiliating diagnosis to be made public. These repositories' enormous amounts of data are incredibly helpful for fraud schemes like identity theft and insurance fraud. The team's efforts showed that a deep learning prototype could be created that was 99 percent as precise as one created using more traditional methods. Most nations have put strong rules into place concerning how health data should be maintained because of the massive volumes of data and the significant hazards faced by the healthcare sector, such as the US's HIPAA requirements. To demonstrate the impending of federated learning in medical imaging, Intel worked with the Center for Biomedical Image Computing and Analytics at the University of Pennsylvania in 2018.

5.2 Medical AI's Data Dependency

Datasets that depict the primary data dispersal of the issue or populace in the tangible world are frequently used by AI systems. But this requirement is much more pressing in the field of medical AI [9].

1. Diagnostics

Machine learning models have been developed to help doctors recognize patterns more quickly and accurately. Diseases are identified by doctors based on patterns they notice in a variety of data, including symptoms, physiological signals, medical imaging, etc.

2. Drug Discovery

In the early phases of drug discovery, it is crucial to comprehend numerous aspects of substances, such as their distribution, metabolism, absorption, and excretion. Compound complexity makes QSAR models complex and high-dimensional, necessitating a significant amount of training data [22].

3. Quality of Care

The hospital mortality ratio is frequently regarded as a crucial indicator of how well patient safety and care quality are being provided. To enable clinicians to provide more individualized care and therapies, it is vital to identify the factors that affect patient mortality. Large and diverse patient populations are needed to identify the factors that influence mortality, merely supreme hospitals barely have admittance to their patient information, which confines their knack to make precise forecasts [21].

For a model to accurately represent the variances and distributions seen in the actual world, it must be visible to an extensive range of circumstances, including complex and unique circumstances as well as the intricate relationships between sickness patterns, socioeconomic factors, and genetic factors. This would only be achievable with access to several distinct datasets. Since certain dataset characteristics, such as demographics (gender, age), or technical imbalances (such as acquisition protocol or kit maker), can slant the model's predictions, this can have a negative impact on the tool's accuracy (for instance, an algorithm developed using data from one hospital may not exertion healthy when smeared to data from another hospital).

This may have a negative impact on the tool's accuracy (for example, an algorithm developed using data from one hospital may not work fine once pragmatic to data from an alternative clinic) and create a prejudice issue since certain dataset characteristics, such as details (gender, age), or nominal inequities (such as procurement etiquette, equipment firm), may distort the model's forecasts. A model must be subjected to a variety of examples to represent the variations and distributions seen in the actual world, such as the complicated and unusual cases, as well as the nuances in the interactions between socioeconomic, genetic, and illness patterns. To do this, access to numerous and varied datasets is necessary.

However, because it is strictly regulated and managed by the legal and data governance standards of healthcare organizations, this sort of quality data is difficult and expensive to access. A multi-hospital program that included 20 institutions from five different countries made a significant radical in the application of FL to medical technology. This effort, which was glimmered by the COVID-19 incident, has demonstrated how privacy-preserving FL approaches may enable the building of strong AI models that function effectively transversely to organizations, flat in sectors with restrictions due to sensitive or scarce data [8, 9].

A neural network was trained to predict how much additional oxygen a patient with COVID-19 symptoms might require 24 and 72 hours after presenting to the emergency room as part of the "EXAM" (or EMR CXR AI Model) partnership, which was led by Mass General, NVIDIA, and health start-up Rhino Health. In one of the largest and broadest clinical FL experiments ever conducted The EXAM partners created an AI prototype that learned from the chest X-ray images, patient vitals, demographic data, and lab values from each participating hospital without ever accessing the sensitive information kept on each location's private server.

On nearby NVIDIA GPUs, each hospice proficient a duplicate of the identical neural network. Only occasionally were updated model weights sent to a central server during training, where they were combined by a global neural network to create a novel comprehensive exemplary. What's intriguing and extremely encouraging for the field is that the final model not only preserved discretion but also showed an enactment boost when associated with similar facsimiles trained at any single site, with an average increase of 38% in generalizability and a 16% improvement in the AI model's standard enactment.

This is a significant advancement not just in the healthcare sector, where other broad FL efforts are now under way, including research for the evaluation of mammograms and work by the pharmaceutical company Bayer to train an AI model for spleen segmentation. This cutting-edge technology may now be used by several other businesses with comparable data difficulties (energy, banking, autonomous vehicles, insurance, etc.) to spur more innovation and enhance things like fraud detection, self-driving technologies, and risk projections.

5.3 *Machine Learning vs Federated Learning*

A machine learning method called FL (Federated learning) produces the prototypical data more reasonably than the other way around. Instead of explicitly requesting data samples, Federated Learning trains a pattern athwart plentiful distributed verge devices or servers allotting insular data models. To properly train patterns established on client interfaces with portable devices and enrich consumer proficiency for Android mobile users, FL was originally released by Google in 2017. Google praised FL for its capacity to produce cannier prototypes with lower latency, less supremacy use, and privacy protection, as seen in Figure 2.

The global model is then created by the central server without gaining access to any data by pooling the local nodes' model findings. Figure 3 demonstrates the use of homomorphic encryption techniques to perform calculations directly on encrypted material without first decrypting it. To increase privacy, the shared parameters can be encrypted before being shared across learning rounds [9–11].

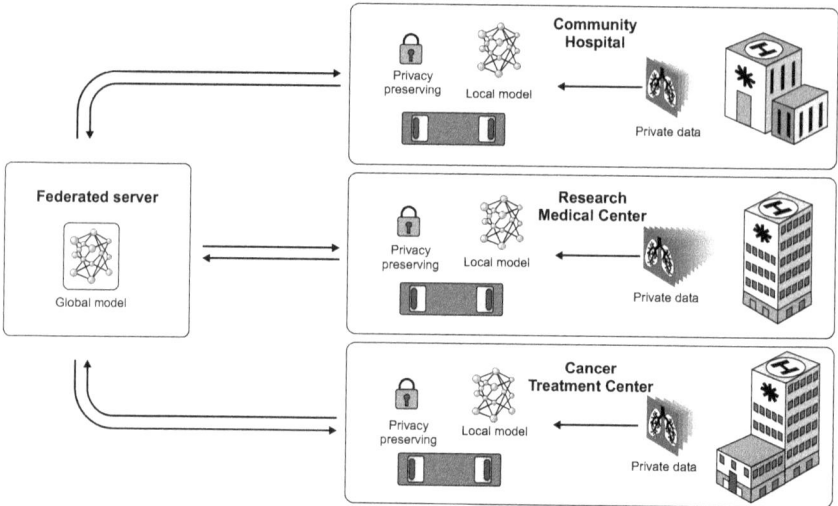

Figure 2. Federated Learning based centralized server approach.

To properly train prototypes based on user collaborations with portable devices and enhance user proficiency for Android consumers, FL was originally released by Google in 2017. The capacity to do machine learning was decoupled from the requirement for data storage on the cloud, as Google stated at the time, thanks to breakthrough technology that allowed mobile phones to jointly develop a distributed extrapolation prototypical while retaining all the training data locally on the device. As seen in Fig. 2, Google praised FL for its capacity to produce better models with reduced latency, less power consumption, and privacy protection.

Importantly, FL allows several players to create a reliable and accurate ML model in the healthcare environment without revealing the essential data, implanting privacy-by-design into the prototypical and solving significant concerns like data admittance rights, data confidentiality, and data security. Every data controller establishes their specific ascendancy procedures and solitude guidelines, manages data access, and has the authority to withdraw permissions. This covers mutually the validation phase and the training phase. Considering low occurrence rates and insufficient datasets at individual institutions, rare illness research on FL might open new avenues for investigation.

The Imminent of Digital Health with FL is an intriguing article that was published in September 2020 in Nature Digital Medicine. It reconnoiters how FL might offer a resolution for the impending of digital health and focuses on the confronts and deliberations that are still essential to be concentrated [11].

Federated learning workflows

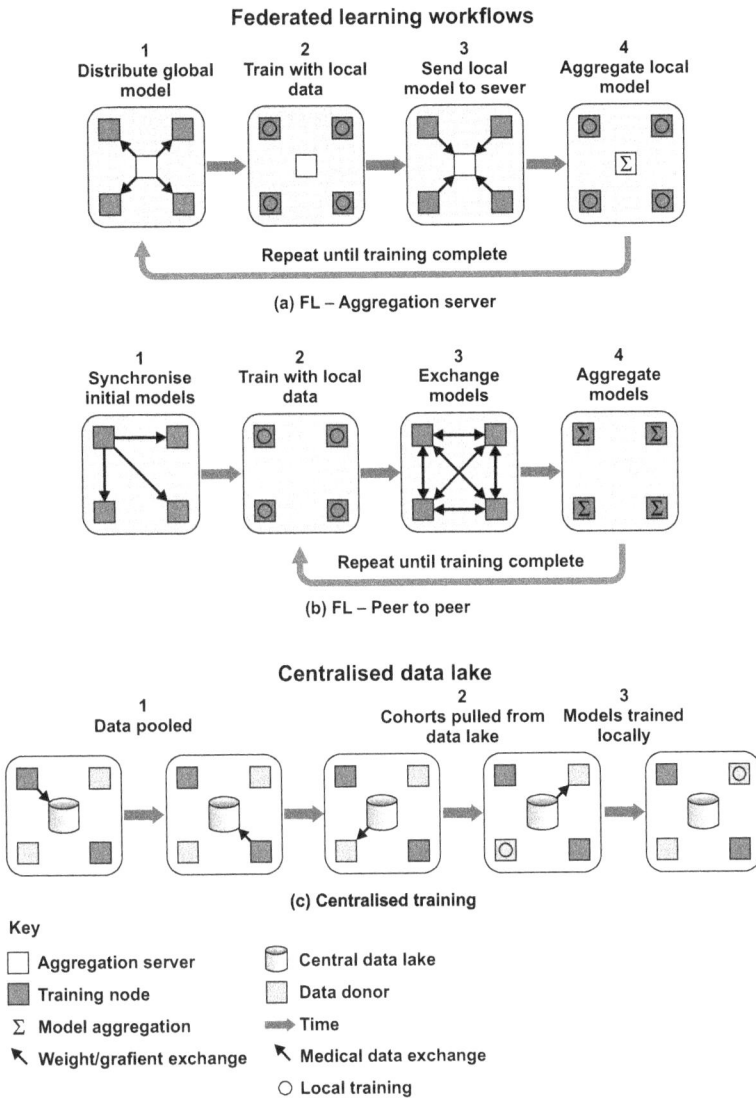

1	2	3	4
Distribute global model	Train with local data	Send local model to sever	Aggregate local model

Repeat until training complete

(a) FL – Aggregation server

1	2	3	4
Synchronise initial models	Train with local data	Exchange models	Aggregate models

Repeat until training complete

(b) FL – Peer to peer

Centralised data lake

1	2	3
Data pooled	Cohorts pulled from data lake	Models trained locally

(c) Centralised training

Key

☐ Aggregation server　　▢ Central data lake
▨ Training node　　　　　☐ Data donor
Σ Model aggregation　　　➡ Time
↖ Weight/grafient exchange　↖ Medical data exchange
　　　　　　　　　　　　　○ Local training

Figure 3. Federated Learning workflows and data lake.

The document cites FL has enormous charm for the healthcare industry since it makes it possible to get insights collectively, i.e., in the ritual of a compromise exemplary, deprived of transferring patient data outside of the institutions' firewalls. Instead, the machine learning (ML) process takes place nearby at each partaking institution, and only the model's constraints and inclines are sent to a centralized server (also known as an "aggregate server"). The global model is then created by the central server without gaining access to any data by pooling the local nodes' model findings. To increase privacy, the bartered issues can be encrypted before being shared between cramming sequences. Additionally, homomorphic encryption algorithms as shown in Fig. 3 can be utilized to perform calculations directly on the encrypted data without first decrypting it [9, 10].

How FL functions in contrast to conventional, centrally managed machine learning (used with permission from the impending of Cardinal Health with federated learning [9, 10]. This differs from the conventional centralized ML training method and roadmap, in which local, autonomous training takes place within the central data lake after data is collected from data collection stations.

FL has a great deal of potential as a large-scale enabler of exactitude medicine because it can create AI models that produce objective judgments, accurately reflect the physiology of an individual, and are penetrating rare diseases while also taking data authority and discretion apprehensions into account. This can assist in overcoming the drawbacks of conventional ML techniques, which call for a single pool of centralized data [12].

A record amount and pace of healthcare data is being gathered (36% composite yearly growth proportion!). This unseals the door for intriguing applications like accurate illness diagnosis and quick drug development to make use of cutting-edge data-driven modeling tools. However, because of how complex the healthcare system is, patient statistics are dispersed and frequently spread across several institutions. To develop high-performing models, clinical associations, communal well-being organizations, and pharmaceutical corporations lack access to the proper data (any size or variety). Data centralization across corporations is too dangerous. This is a severe privacy concern since healthcare data is sensitive and there is always a possibility of data leaking.

The popularity of FL, a two-way learning prototype, has grown because of its potential to address the problem of learning sophisticated prototypes through sliced data. We'll familiarize a few federated learning applications in healthcare in this post. Below, we will concentrate on three specific use cases and talk about how FL is applied to complete the respective assignments.

5.4 Diagnostics

To assist clinicians in more rapidly and precisely identifying trends, machine learning models have been developed. Based on patterns they see from several sources, including symptoms, physiological signals, medical imaging, etc., doctors diagnose illnesses. However, each institution typically lacks access to the vast amounts of training data that these models frequently require. A rising number of individuals are using federated learning as a solution to the data limitations in diagnostics. Li et al., researched the identification of ASD (Autism Spectrum Disorders) in 2020 using scattered time series of rs-fMRI brain imaging data. 52 to 167 patients were included in the data, which came from four different places [20].

In this model, around several parameters to learn. Due to the small number of patients available, data from any one location was not enough to generate a reliable result. The study contrasted independent models trained at a single site with a federated model developed using data from four sites collaboratively. Results indicated that the preeminent single-site model had a precision of 0.695, while the federated model recorded 0.849 on the identical assessment data, which was a considerable increase (you can see the specifics in Table 3 of the original study).

The authors of this research raised the intriguing topic of data heterogeneity between sites, which they refer to as "domain shift," which creates impediments to collaboration. The authors suggested methods for preventing domain shift, allowing data sources to be less uniform, and creating new chances for collaboration. The effectiveness and efficiency of illness diagnostics will be significantly increased by applications of federated learning in this field.

5.5 Drug Discovery

In the early phases of drug development, it is crucial to comprehend numerous aspects of substances, such as their fascination, dispersal, absorption, and evacuation, Pharmaceutical companies use a method called QSAR (Quantitative Structure Activity Relationship) investigation, which shapes extrapolative models to forecast belongings of compounds with inputs like hypothetical molecular signifiers because it is expensive and inefficient to obtain this information through chemical or biological experiments.

Compound complexity makes QSAR models complicated and high-dimensional, necessitating a significant amount of training data. However, gathering training data requires a lot of time and money. Any cross-institute collaboration has been constrained by worries about sharing compound intellectual property (e.g., proprietary structures) and other trust difficulties. The use of FL as vital to enabling alliances across organizations to collaboratively create QSAR prototypes was examined by Chen et al. in 2020. The experiment's findings (see Figure 4 below) demonstrated that the federated model outperformed any single-client

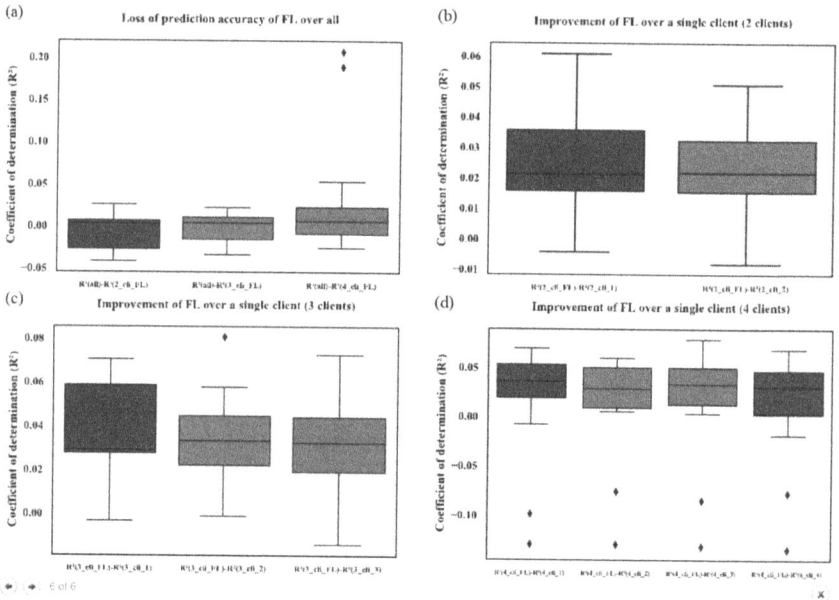

Figure 4. The forecasting performance of horizontal Federated Learning.

model proficient with secretive data in terms of prediction performance. Federated learning makes it possible for researchers to work together more effectively in the pharmaceutical sector and to find novel medications [22].

5.6 *Quality of Care*

The dispensary transience proportion is frequently seen as a crucial indicator of how well patient safety and care quality are being provided. To enable clinicians to provide more individualized care and therapies, it is vital to identify the variables that affect patient mortality. Large and diverse patient populations are needed to identify the factors that influence mortality, but record infirmaries only have admittance to their individual patient data, which restricts their capacity to make precise forecasts.

Using data from electronic health records (EHRs), Vaid et al. (2020) constructed FL models to forecast death in COVID-19 patients analyzed within seven days of hospital admittance. The study considered patient demographics, prior admission obligations, medical records, and labs data from COVID-19 positive patients obtained from Epic EHR systems of five hospitals within the Mount Sinai Health System (MSHS) in New York City (NYC) (e.g., heart rate, respiration rate, glucose). A multilayer perceptron (MLP) and logistic reversion with L1-regularization were both employed to predict mortality. AUC-ROC was used by the authors to assess the enactment of limited and coalesced models, and they found that at most locations, federated models performed better [21].

Federated learning gives healthcare professionals greater knowledge of risk variables that affect treatment choices, enabling them to provide safer, more efficient, and more individualized care.

6. Challenges of Federated Learning

Federated learning has several restrictions that thwart it from existing as a panacea for all our data privacy problems, in addition to the possible security concerns. One thing to keep in mind is that federated learning entails considerably supplementary confined way power and reminiscence to sequence the prototype than conventional machine learning techniques do [11, 12, 15].

Federated learning has several constraints that prevent it from being a panacea for all our data privacy problems, in addition to the possible security concerns. One thing to keep in mind is that federated learning takes substantially more resident device power and reminiscence to sequence the model than conventional machine learning techniques do. However, many modern devices are powerful enough to do these tasks, and this strategy also results in considerably less data being transmitted to centralized servers, lowering data use. This trade-off may be advantageous for many users if their apparatus has sufficient power [11].

A different technological problem is bandwidth. Unlike standard machine learning, which takes place in data centers, federated learning takes place across WiFi or 4G. The bandwidth rates used amid the occupied nodes and attendants in these midpoints are orders of magnitude lower than those of WiFi or 4G. Over time, device bandwidth hasn't increased as quickly as device processing power; therefore, a bandwidth constraint might potentially result in increased latency and a longer learning process than with a more conventional method. The performance of a gadget is decreased if algorithm training is done while it is in operation. By only teaching devices, while they are turned on, wrought into an aperture, and idle, Google has found a solution to this issue. This fixes the issue, but because training can only be done outside of peak hours, it slows down the learning process [12].

Devices dropping out during training is another issue; they may be used by their owners, switched off, or experience some other disturbance. A less precise algorithmic model can result from the inability to use the data from dropped devices. The transition from federated learning's idea to implementation is not without difficulties. Researchers, including those working independently of proponents of federated learning, have helped to improve knowledge of the relevant challenges [8, 11].

The effectiveness and accuracy of federated learning have been studied, but the security related difficulties listed below are, in my opinion, more pressing.

Inference Attacks: The protection of the confidentiality of the customers' data serves as the driving force behind federated learning.

Model Poisoning: Some academics have looked at the idea of unruly clients installing backdoors or launching Sybil attacks to poison the global model.

Poisoning assault with a specific target in stochastic gradient descent. The red dotted vectors represent Sybil's contributions that steer the model in the direction of a poisoner goal. As seen in Figure 5, the solid green vectors are provided by loyal customers who work to achieve the real goal.

Federated learning is a relatively new training approach that has promise for a variety of uses. The technique needs a lot more investigation because it is still in its early phases of study, making it impossible to evaluate all its potential applications or all the confidence and concealment threats it may be subject to. It's arduous to predict with precision how widely the strategy will be used in the impending until that point. The good news is that Google's Gboard has already been efficaciously used in the real world [15].

Nevertheless, it's doubtful that federated learning would completely replace conventional learning models in all circumstances because of

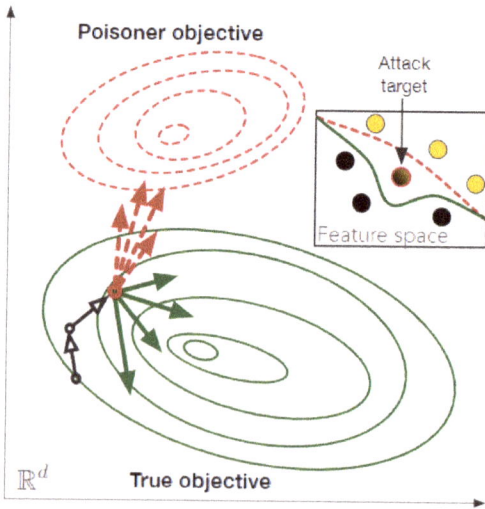

Figure 5. Targeted positioning based on objectives.

some of the drawbacks that have already been mentioned. The future will also depend on how serious our main internet corporations are about protecting privacy. We have excellent cause to be dubious at this point.

Federated Learning Challenges with Reference to Healthcare

Although there are numerous benefits to FL, it's crucial to remember that, like traditional ML techniques, FL still faces performance issues including bias, standardization, and data quality. Additionally, FL encounters the following challenges in the context of medical data [9, 11, 15]:

- **Data assortment** — medical data is by nature complicated and varied. Both organized and unstructured forms are available. The variety of the data is also impacted by factors including acquisition disparities, the brand of the medical equipment, and area demographics. It might be challenging to combine data and information from several sources. Data may not be disseminated equally among the collaborating universities.

- **Secrecy and security** — FL still bears some privacy and security risks, even if it provides far stronger privacy protection than traditional ML models. Information leaking to rival parties is a possibility. When utilizing FL, trade-offs could be necessary since, in some circumstances, performance may be somewhat worse than when using regular ML models. Researchers are investigating differential privacy to provide FL with additional degree of privacy.

- **Traceability and liability** — It's essential that medical AI solutions can be replicated and explained. In FL, however, we are apportioning with multi-party calculations carried out in settings with a wide range of networks, hardware, and software. As a result, it is difficult for researchers to examine the actual data to explain any weird or unexpected outcomes or uncover the causes by forecasting the prototype.

- **Procedure design** — The FL structure that enables admittance to the data housed by various establishments and knobs would need to consider the requirements for data veracity, encoding, and varying computational capabilities.

The Future of Federated Learning in Healthcare

We have all become familiar with the significance of and effects of developing technologies on numerous facets of our lives in recent years. Technology plays an important role in our lives, whether it's applications on smartphones cramming our penchants or using AI to succor clinicians in diagnosing. In Machine Learning (ML) and Deep Learning (DL), to get revelations from medical data [20] resultant from pathology, genetics radiology, cancer, etc., we have leapfrogged artificial intelligence innovation in the healthcare sector.

To get insights from medical data collected from radiology, pathology, etc., we have leapfrogged artificial intelligence innovation in the healthcare sector, especially in machine learning (ML) and deep learning (DL). We have all become familiar with the significance and effects of developing technology on numerous facets of our lives in recent years. Technology plays a significant role in our lives, whether it's programs on our smartphones discovering our preferences or employing artificial intelligence to help physicians diagnose [17].

However, data in the healthcare industry is not just poorly organized and extremely sensitive. Furthermore, full respect for country-specific regulatory issues like NMPA (China), FDA (USA), CDSCO (India), GDPR (Europe), etc., is required to maintain data security, data access, and privacy rights. In addition to its enormous volume, data also contains a significant amount of unstructured material and sensitive patient information, is widely dispersed geographically, and lacks compatibility across various devices and manufacturers.

AI, for instance, has shown the ability to help radiologists with computer-aided analysis and diagnosis. However, specified the difficulties in gathering, curating, and preserving high-quality data with a varied population, it is still difficult to create effective models free of bias from tiny datasets. It also costs a lot of time, money, and effort. Professionals in healthcare and the life arts may use the advantages of federated learning (FL) to handle the issues of privacy, data integrity and governance by jointly developing algorithms, with the data remaining behind the hospital's firewalls and just the prototypes being shared [18].

Recent studies have demonstrated that FL-trained models can outperform prototypes trained on centrally hosted data sets and models that only witness isolated single-institutional data. Federated Learning examines patients across multiple demographics and collects a wider range of data variability. Given access to electronic health information, FL, for example, can assist in identifying clinically comparable affected roles and forecasting hospitalizations owing to cardiac proceedings, death, and ICU stay duration. FL was initially created for several provinces, including ambulatory and edge expedient use cases, but it has lately acquired popularity for healthcare applications [19].

FL may assist in the development of replicas for organ dissection in X-ray, ultrasound, computed tomography, charming reverberation imaging, and positron emission imaging, as well as tumor characterization for disease-specific imaging techniques. FL can allow for revolutionary improvements in the future by giving a chance to acquire more variability in data and assess patients across various demographics.

The study demonstrated that the FL-generated prototypes performed better and were more generalizable than those trained on the data from a particular institute. An FL framework will be created and implemented across four institutions in France as part of the Health Chain project. With this technique, patients with melanoma and breast cancer may estimate how well their treatments will work. This can also help oncologists determine the best course of action for each patient based on the images from their dermo copy or microscopic anatomy. FL may also be applied to enhance scholarly investigation. For instance, in 2020, Federated Learning was utilized by the American College of Radiology, Diagnostics da America, Partners HealthCare, Ohio State University, and Stanford Medicine to create more accurate prediction models for determining breast tissue density for mammograms.

Additionally, compiling data rendering to the "FAIR" ideologies—Findable, Manageable, Interoperable, and Recyclable—makes the data, which might come from complicated autonomous algorithms, legible and understood to a large extent. For example, displaying data as widely recognized vocabularies like the Uniform Resource Identifier (URI) and publicly accessible ontologies might assist many in this arena of study in overcoming language hurdles in clinical data retrieved from international medical locations. Globally applicable clinical insights from large data sets can be obtained by combining FL with FAIR principles.

Numerous stakeholders, including doctors, patients, hospitals, AI researchers, and healthcare providers, may be significantly impacted by federated learning. Despite FL's benefits, researchers and AI developers must carefully consider study design, clinical procedure selection, data assortment, and data value to reduce model prejudice. A potential idea for securing reliable, secure, and impartial data models is federated learning. FL overcomes challenges associated with delicate medical data by consenting multiple parties to train collaboratively without the need to give and take or integrate data sets. Precision medicine will be greatly impacted by federated learning, which also has the latent to enhance persistent care on a worldwide scale.

7. Conclusion

Federated learning has the transformative competence to convey AI modernization to the medical roadmap, as stated in the Machine Learning Expect statement. Amending regional privacy laws, setting up sufficient standardizing code-sharing procedures, and IT infrastructure, or developing fair remuneration plans for associates contribute few of the interdisciplinary problems that remain in federated learning that can only be solved through collaboration. Federated learning collaborations must continually advance to be successful in the complicated healthcare environment of the future. With the use of this method, information sharing has seen major advancements, and hospital collaboration on machine learning has become more effective. By making use of cutting-edge machine learning and deep learning techniques, it eludes and overwhelms the challenges of functioning with enormously sensitive medical data.

References

[1] https://owkin.com/publications-and-news/blogs/federated-learning-in-healthcare-the-future-of-collaborative-clinical-and-biomedical-research.

[2] https://www.comparitech.com/blog/information-security/federated-learning/.

[3] https://www.sundeepteki.org/blog/federated-machine-learning-for-healthcare.

[4] https://bluesteens.medium.com/federated-learning-in-healthcare-7a380ffc01f9.

[5] Thomas, G.A.S. and Robinson, Y.H. 2020. IoT, Big Data, Blockchain and Machine Learning Besides Its Transmutation with Modern Technological Applications. Springer Book Chapter-Internet of Things and Big Data Applications—Part of the Intelligent Systems Reference Library book series (ISRL, volume 180), 25 February 2020, pp. 47–63, ISSN: 978-3-030-39118-8.

[6] https://medium.datadriveninvestor.com/an-overview-of-federated-learning-8a1a62b0600d.

[7] Thomas, G.A.S. and Robinson, Y.H. 2020. Real-Time Health System (RTHS) Centered Internet of Things (IoT) in Healthcare Industry: Benefits, Use Cases and Advancements in 2020. Springer's Multimedia Technologies in the Internet of Things Environment (Scopus Indexed), 29 September 2020, ISSN: 978-981-15-7965-3.

[8] https://medium.datadriveninvestor.com/an-overview-of-federated-learning-8a1a62b0600d.

[9] https://medium.com/mlearning-ai/whats-federated-learning-and-why-it-s-key-to-the-future-of-medical-ai-e53c6869a849.

[10] Rieke, N. and Hancox, J. 2020. The future of digital health with federated learning. npj Digital Medicine, volume 3, Article number: 119.

[11] Li, T., Sahu, A.K., Talwalkar, A. and Smith, V. 2020. Federated learning: Challenges, methods, and future directions. IEEE 2020: Signal Processing Magazine, 37: 50–60.

[12] Yang, Q., Liu, Y., Chen, T. and Tong, Y. 2019. Federated machine learning: Concept and applications. ACM Trans. Intell. Syst. Technol. (TIST), 10: 12.

[13] Kairouz, P., Sen Zhao and Zheng Xu. 2019. Advances and open problems in federated learning. arXiv preprint arXiv:1912.04977.

[14] Lee, J., Li Xiong and Dawn Song. 2018. Privacy-preserving patient similarity learning in a federated environment: Development and analysis. JMIR Med. Inform., 6: e20.

[15] Brisimi, T.S. and Chaoyang He. 2018. Federated learning of predictive models from federated electronic health records. Int. J. Med. Inform., 112: 59–67.

[16] https://blogs.nvidia.com/blog/2021/09/15/federated-learning-nature-medicine/.

[17] https://www.expresshealthcare.in/blogs/guest-blogs-healthcare/the-future-of-federated-learning-in- healthcare/430940/.

[18] Rieke, N., Hang Qi and Ziteng Sun. 2020. The future of digital health with federated learning. Nature Partner Journal Digital Medicine, 3: 119. https://doi.org/10.1038/s41746-020-00323-1.

[19] Kalendralis, P., Hang Qi and Zheng Xu. 2020. FAIR-compliant clinical, radiomics and DICOM metadata of RIDER, interobserver, Lung1 and head-Neck1 TCIA collections, 47(11): November 2020, Medical Physics.

[20] Li, X. and Weikang Song. 2020. Multi-site fMRI analysis using privacy-preserving federated learning and domain adaptation: ABIDE results. Medical Image Analysis, 101765.

[21] Vaid, A. and Sanmi Koyejo. 2020. Federated learning of electronic health records improves mortality prediction in patients hospitalized with COVID-19. medRxiv.

[22] Chen, S. and Daniel Ramage. 2020. FL-QSAR: A Federated Learning-based QSAR prototype for collaborative drug discovery. Bioinformatics, 5492–5498.

Chapter **10**

Collaborative Federated Learning in Healthcare Systems

Bini M Issac[1] and *SN Kumar*[2,*]

1. Introduction

Machine learning algorithms have been used for the past few years for doing complex tasks such as disease prediction or detection. However, the accuracy of the machine learning algorithms depends on the data that is used for creating the model. In addition, data privacy is a concern when using these algorithms because the data may get leaked if the server storing these data is attacked. In a standard machine-learning scenario, clients send their data to a central server, which combines the data from various clients and uses it to train a model. But this poses a serious security risk particularly in the healthcare sector as the data may be leaked while transferring between the client and the server. Research in the healthcare sector involves collection, storage, and transmission of large volumes of health information and if security is compromised, there may be a chance that this health information may be improperly accessed. If wrong information related to a patient leak, it may cause several problems for

[1] Department of CSE, Amal Jyothi College of Engineering, APJ Abdul Kalam Technological University, Kerala.
[2] Department of EEE, Amal Jyothi College of Engineering, APJ Abdul Kalam Technological University, Kerala.
Email: binimissac@amaljyothi.ac.in
* Corresponding author: appu123kumar@gmail.com

the patient such as social or psychological problems, loss of job, health insurance, and so on.

Google introduced Federated Learning (FL) as a cutting-edge Machine Learning (ML) strategy in 2016 [1] to solve the problem of training a centralized machine-learning model when the data is distributed among many clients. This brute force strategy will solve the problem of sending large amounts of data with the clients to Google. The solution proposed was to send a local model to each device, compute the optimal parameters for that client and send the computed weights to the server. The central node will aggregate all the parameters, average them, and create a global model. A small model distributed among clients paves the way toward the generation of an optimal model for huge data transfer. Federated Learning is an ML approach in which the machine-learning model is created from different data sets located at different sites without sharing the training data. Therefore, the personal data will remain with the clients (hospitals, imaging centres, etc.) reducing the possibility of personal data breaches. This approach is useful when the data generated is too large such that it cannot be transmitted and processed centrally. Since in federated learning, the multiple participants such as hospitals and other healthcare organizations train a model without sharing the data, this can be useful in the healthcare industry where data privacy is a major concern and where the organizations may be hesitant to share sensitive information.

Federated learning may be classified into three types- Horizontal, Vertical, and federated transfer learning [2]. Horizontal FL (sample-based federated learning) is introduced in scenarios where data sets share the same feature space but differ in sample space. This is also called homogeneous federated learning. In the case of the health sector, the data from various hospitals are different in their samples as data are of different patients but the features are the same, i.e., the patients' information [3]. Therefore, this corresponds to horizontal FL. In Vertical FL (feature-based federated learning) two or more data sets share the same sample space but they differ in feature space. This type is also called heterogeneous federated learning. For example, if we collect data related to different imaging modalities of the same group of users, it corresponds to vertical federated learning. In federated transfer learning [4], there is no similarity between the feature space instead the tasks are related, and it builds a model for a domain based on the knowledge from other domains. Figure 1 shows the different types of FL approaches.

The data for FL is taken from multiple hospitals/imaging centres, the data will be of a wide variety- the gender ratios, age distributions and ethnicity of patients will be different for various hospitals [5]. Therefore, generating a model from the data derived from multiple sources will be more reliable than the data derived from a single hospital. This is a major

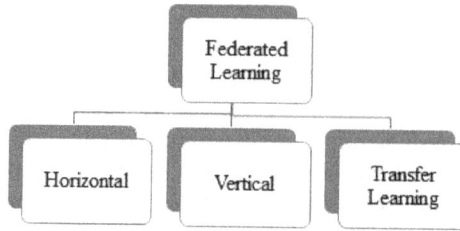

Figure 1. Types of FL approaches.

benefit of federated learning. Another benefit is that we know that periodic updating of the model is necessary for the deployment of AI models. So periodic labelling of the data by radiologists for retraining the model is necessary. However, with FL, even though some radiologists were busy with patient cases, the model can be made from the data received from other hospitals; this reduces the burden associated with training using data from individual sites. Also, auto-scaling of the model occurs as new hospitals collaborate which is an added advantage as there is no need to manually scale the system.

Collaborative Federated learning in the health sector can be viewed as a distributed privacy-aware model in which data is not centralized and it enables N clients (in this case, hospitals/imaging centres) to build a Deep Learning (DL) model or a Machine Learning (ML) model using decentralized data. Sensitive medical data resides in the clients themselves. Mathematically FL can be modelled as:

$$\text{Min}_x F(x), \text{ where } F(x): = \sum_{k=1}^{n} p_k F_k(x)$$

where n is the total number of clients participating, F_k is the objective function for the kth client and p_k is the impact of each client with $p_k \geq 0$ and $\sum_{k=1}^{n} p_k = 1$.

In collaborative federated learning, each participant trains a model by himself or herself and then the trained model is sent to a server. The server sends the updated model to all participants and the participants can then use the shared models to train a global model and can be used to make predictions. This is particularly useful when identifying patterns in healthcare data. One major benefit of collaborative federated learning is that it allows organizations to train models on large data sets than they could on their own. This increases the accuracy of the model, and it will lead to better outcomes. In addition, the data is not directly shared between the participants so sensitive information is not exposed. How machine learning is used in the healthcare industry can be increased better by improving patient care and outcomes by using collaborative federated learning. Now let us look at the benefits of FL for the different stakeholders. Clinicians will be most benefited from the FL approach, as the model will be having less bias as heterogeneous data is used for training. For the case of patients residing in remote locations, a globally trained model will give a way for getting better treatment [6]. In the case of hospitals, even small hospitals can get a chance to collaborate in the model creation and they will get an updated model.

Figure 2 shows the architecture of a collaborative federated healthcare system. Here four hospitals are participating in the learning process. They use ML/DL algorithms to train their local models and share their models with the centralized federated server. The server aggregates the models and sends the updated model back to the hospitals, which are again used for their local training, and this process continues until an accurate model is created which is used by hospitals.

The chapter organization is as follows; Section 1 introduces federated learning & highlights its features, Section 2 describes the role of federated learning in cancer diagnosis, Section 3 deals with its role in Covid-19 diagnosis, Section 4 is about its significance for steganographic applications in healthcare and Section 5 highlights the other applications in smart health care. Various challenges of FL and future applications are presented in Section 6. Finally, we conclude in Section 7.

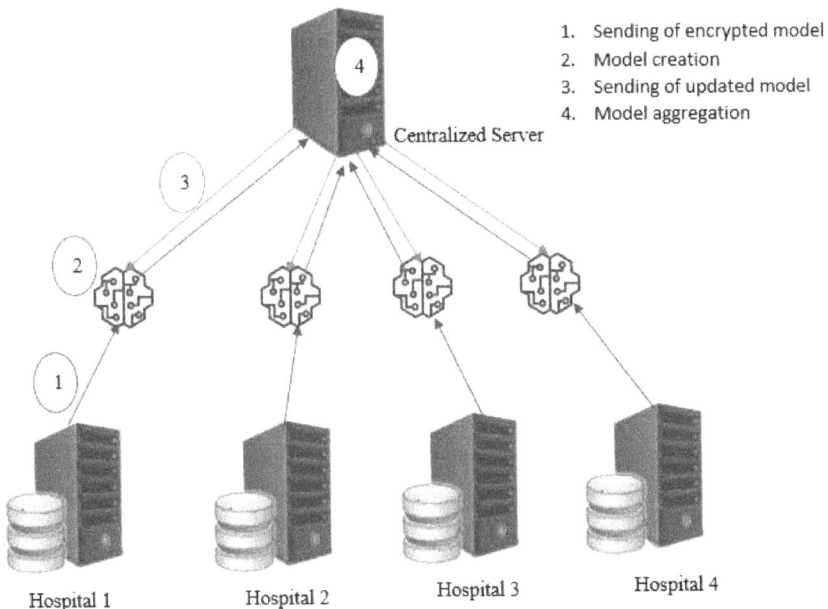

Figure 2. Architecture of a Collaborative federated health care system.

2. Role of Federated Learning in Cancer Diagnosis

Cancer is one of the main reasons of death worldwide and the parts that are affected by cancer include the breast, lungs, colon and rectum, prostate, skin, etc. When normal cells transform into tumour cells, cancer arises and normally cancer progresses in stages, i.e., from a precancerous lesion to a malignant tumour. This disease can be treated and cured effectively if detected earlier. X-ray, CT scan, MRI scan, and ultrasound is some of the imaging tests that are used to diagnose tumours. Many ML and DL techniques have been proposed for the detection of tumors and segmentation is used to segment the tumor from the MRI images. But training ML/DL model for this purpose needs a wide database covering various anatomies, pathology reports, and input data.

Some researchers have applied federated learning in studies related to brain tumours. In [7], the authors for the segmentation of brain tumours propose a federated learning architecture called SU-Net and the 'Brain MRI Segmentation' dataset from Kaggle is the dataset used for the study. They implemented the network in both a non-federated scenario and a federated scenario consisting of five clients and compared the architecture with several other deep learning architectures. Results show that AUC (Area Under Curve) and DSC (Dice Similarity Coefficient) are 99.8% and 78.7% respectively in the federated scenario which is higher than the non-federated scenario. In [8], federated learning is implemented using a BraTS data set containing MRI brain scan images of patients diagnosed with gliomas. Glioma is a kind of tumour that affects the brain and the spinal cord. Their simulation results show that 99% model accuracy is achieved in the federated environment even with imbalanced data sets. In [9], a deep learning framework 3D-Unet model is proposed and evaluated the performance in both scenarios by collecting data from 10 different institutions and the researchers found that the model reaches 99% accuracy by sharing data between different institutions. Breast cancer is one of the main types of cancer that affects women. But if it can be detected early, then the mortality rate can be reduced. Doctors recommend X-ray mammography for detection and follow-up. By using FL, the size of the dataset is increased and heterogeneous images from multiple institutions are used for the study. Federated learning is used for breast density classification using mammography data. They evaluated the performance using Cohen's linear weighted kappa to evaluate the performance locally and with federated learning and the federated learning model achieved better performance [10]. A variation of FL called the memory-aware curriculum FL approach is proposed for classifying breast cancer, which can have a great impact on the consistency of local models by penalizing for the incorrect predictions made by using a data scheduler [11]. In this approach, they combined unsupervised domain adaptation for dealing with the shifting of domains. Three data sets from various clients were utilized for their study.

Prostate cancer is one of the main types of cancers that affect men. Prostate cancer occurs when some glands grow quickly and will spread from the prostate to other body parts and detecting this cancer based on the symptoms from the early stage is difficult. For detection in the early stages, we can use FL. In [45], Yan et al., used a cycleGAN network for model creation and they used medical images from eight different institutions for the study and achieved an accuracy of 98% and AUC of 99% for prostate cancer classification. Researchers investigated the effect of adaptive federated learning on skin tumour detection scenarios [12] and they used the ISIC 2019 dataset for validating the accuracy of the model. Here, the authors use the data collected by various edge devices such as dermoscopy, laser microscope, and dermalite, and these data are uploaded to the cloud for training and classification. In [13], tumour detection in thyroid nodules is proposed using federated learning by collecting images from five different institutions. Five different deep learning architectures are implemented, and a comparison is made before and after using federated learning. They showed that despite the data heterogeneity, federated learning performed better by holding up data privacy. In [14], a Cox model for larynx cancer patients is proposed and data from patients from three different countries is used as input.

An encrypted federated learning approach is proposed for cancer image analysis by employing homomorphic encryption [15]. Here the weights will be encrypted before sending to the server. The server will perform model creation in these encrypted values and the updated model parameters are sent to the clients for decryption. Only the corresponding client knows the decryption key and so the server cannot infer any data from the clients. So, one more layer of security is added to the sensitive patient information. Since everything is happening in the encrypted environment, when working with personal data information homomorphic encryption is the best choice. For ensuring the accountability and fairness of FL systems for tumour detection, blockchain technology can be utilized. Medical image segmentation is an important function in cancer diagnosis. It is the process of identifying the Region of Interest (ROI) from the medical image. Medical images may be in a wide variety of forms and some of the literature has focused on segmentation tasks. For brain tumour segmentation, U-Net architecture is implemented in MRI images by [16] and they used Convolutional Neural Network based architecture for this task. They reported a dice score of 85% in the implementation and found that as the collaboration increases, the score considerably increases. For prostate cancer detection, a 3D Anisotropic Hybrid Network is used for segmenting MRI images [17] using FL. They obtained a dice of a score of 88.9% when data from three institutions were taken for the study.

3. Role of Federated Learning in Covid-19 Diagnosis

The covid-19 pandemic has affected the health and functioning of people everywhere in the past few years. There was a wide variety of symptoms for this disease ranging from fever or chills, cough, headache, sore throat, loss of taste or smell, and runny nose to shortness of breath or difficulty breathing. But patients who were affected with Covid-19 pneumonia have developed Acute Respiratory Distress Syndrome (ARDS), a state in which respiratory failure can occur if the air sacs in the lungs fill up with fluid and such patients need ventilator support to breathe. The main screening method for the diagnosis was the use of chest X-rays, ML and DL approaches, Computer Vision has contributed significantly to the classification of chest X-rays in the healthcare sector [18]. But we know that in Machine Learning or Deep Learning, the collection of the training data, and ensuring the privacy and security of the data is a big challenge. In that context, we can use FL for ensuring the privacy and security of medical data including medical images. FL was applied for a variety of use cases related to Covid-19 such as CT image-based Covid detection, prediction of mortality, prediction of clinical outcomes, X-ray-based covid detection, and detection of face masks in dense crowds [3]. Figure 3 shows

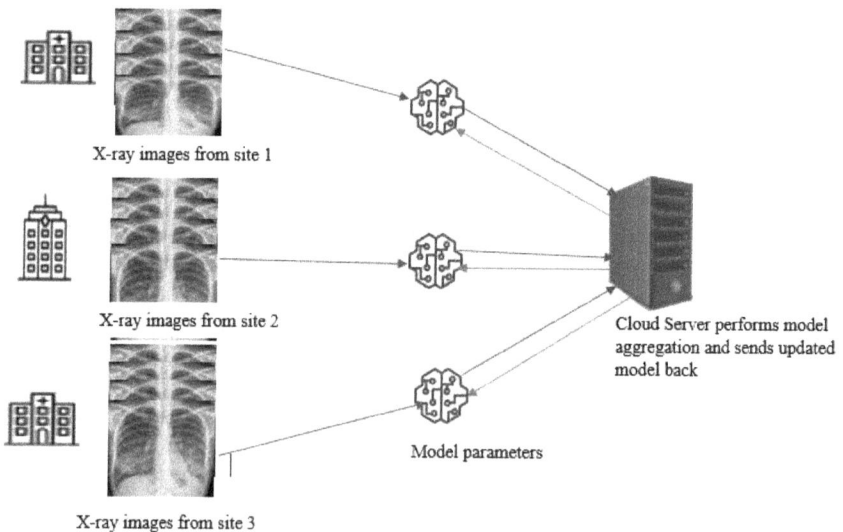

X-ray images from site 1

X-ray images from site 2

Cloud Server performs model aggregation and sends updated model back

Model parameters

X-ray images from site 3

Figure 3. Federated Learning in Covid-19 diagnosis.

a specific scenario of the application of FL in covid-19 diagnosis using X-rays from various hospitals and radiology centres.

Yan et al. proposed an FL-based approach [18] using four different networks [19–22] which are summarised in Fig. 4. The dataset they used is the COVIDx dataset, which is publicly available on Kaggle, and 13,703 images were used to train the model and 1,579 images were used to test the model. Three labels were there in the data set- normal, covid pneumonia, and non-covid pneumonia. The experiments were carried out with and without using FL and it was observed that ResNet18 architecture showed the best performance with and without FL. It was observed that ResNeXt performed well and MobileNet_v2 had a smaller number of parameters. The study concluded that models discussed in [20, 22] were best suited for Covid-19 diagnosis.

A fusion-based federated learning approach [4] was proposed which improves the efficiency and performance of the global model created by the server by selectively choosing the clients for participation based on their training time. They used 3326 images from the data set they collected and divided them into 2800 images for training and 526 images for testing. Three clients and one server are used for the experimental setup and three models such as GhostNet, ResNet50, and ResNet101 are used by the clients for local training and the models are trained using six different data sets. They also used different sizes of data sets for each client such as 600 images, 900 images, and 1300 images respectively. The accuracy, training time, and communication efficiency of these models are analysed, and dynamic fusion is applied for reducing the communication overhead through fewer model uploads. A particular time will be allotted for each client for training and model creation and if a participating client does not finish within the allotted time, then the server continues to model updating without input from this client. Also, if any client's performance is worse than its previous round, then the client request to the server to skip the

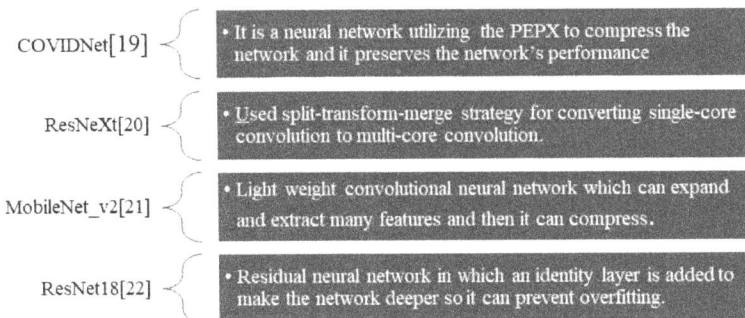

Figure 4. Different architectures implemented in FL for Covid 19 diagnosis.

aggregation of that round. A federated learning approach called EXAM (Electronic Medical Record (EMR) chest X-ray AI Model) for predicting the clinical outcomes in patients with Covid-19 was proposed in [2]. They collected data from 20 institutions around the world for training a model which predicted oxygen requirements for symptomatic patients affected with covid-19 in the future using the various parameters laboratory, vital signs, and chest X-rays.

Blockchain technology was coupled with federated learning to develop a model for detecting Covid-19 [23]. This framework uses a data normalisation technique to normalise the data values collected from various hospitals and a capsule network-based segmentation and classification used to detect patients affected with Covid-19. For data normalisation, spatial and signal normalisation is done. Spatial normalisation relates to the CT scan's size and resolution. The CT scanners' signal normalisation process deals with the intensity of each voxel, which is based on the lung window. The dataset containing 34,006 CT scan slices (images) belonging to 89 subjects of Covid-19 affected patients is used for the study. This blockchain-based FL network learns collaboratively from various radiology centres and multiple hospitals having different kinds of CT scanners. They used SegCaps [44] algorithm for segmenting images and a Capsule network is used for generalization. For ensuring privacy, the hospitals will only share the gradients with the blockchain network and the blockchain network aggregated the gradients, and the updated model is distributed to the participating institutes after generating a global model by the server. In essence, in a federated environment, the weights aggregation from various local models occurs while keeping the privacy of the data given.

For ensuring accountability and fairness, blockchain technology was used [24]. In the case of covid-19 detection, the details of X-ray images provided by hospitals will not be disclosed. Hence, the model is unable to check who is providing trustworthy data, unable to determine the clients which is poisoning the global model too. By utilizing blockchain technology coupled with FL, the model was made accountable in the above scenario. If the training data or the training method is biased, then a fairness problem occurs. If the number of images sent for model creation for various diseases differs, then there is a possibility for bias. In the case of covid-19, the normal X-ray images will be greater in number than the x-ray with positive cases. So, there is a chance for bias which may lead to the model being unfair. In such cases, blockchain technology will provide a better solution. In [25], deep reinforcement learning is utilised in a federated approach for looking into the resource availability and the trust scores of the IoT devices which can be used for the selection of clients which can be used for the Covid-19 study.

In clustered federated learning, the devices are organised into clusters. Each cluster trains a local model based on its data and the server aggregates them to get a global model. The global model is then broadcasted back to the devices in each cluster which will be used to update their local models. This process is then repeated and greater or equal performance than conventional FL can be achieved by allowing clients to arrive at more specialized models. The role of clustered FL for automatic covid-19 diagnosis is evaluated by Junaid et al. [26] in an edge computing network. Edge computing helps to process, analyse and store data closer to the location where data is generated such as a mobile device at a patient's premise or an on-premises server in a hospital. The key motivation to implement clustered federated learning with edge computing is to learn a single model from data of multiple modalities for, e.g., two modalities may be CT image and ultrasound image. 2109 images containing both X-rays and ultrasound images were used for the study and 80% of the data was used for the training and 20% of the data for testing and modified VGG-16 architecture is used for the study.

Kandati et al. [27] suggested a genetically clustered federated learning grouping the edge devices based on the hyper-tuned parameters and modifying genetically the parameters cluster-wise. A genetic algorithm is an optimisation algorithm that can be used to improve the performance and efficiency of the model in clustered federated learning. This paper studied the differential behaviour of the algorithm with various hyperparameters that includes client ratio, number of iterations (n), batch size, minimum samples, and learning rate (η). Hyperparameters are those parameters that are not directly learnt from the model but can control the learning process. Instead of image data, they used ten attributes that represent the symptoms of the disease with the final attribute storing the result of whether the person has covid-19 or not. The various parameters such as Accuracy, Precision, Recall, and F1-score of the traditional FL and genetically clustered federated learning for different numbers of clients is evaluated. It is found that only one epoch is needed for all the rounds for reducing resource consumption and training time.

A new approach for detecting Covid-19 with the FL approach using Generative Adversarial Networks (GAN) and edge computing was proposed in [28]. GAN was used to generate synthetic images of covid-19 database where there are not enough medical images are there to train the machine learning models. In GAN, there are two neural networks- a generator and a discriminator. The generator produces the synthetic image whereas the discriminator distinguishes between the real image and the synthetic image. The generator is then updated, and this process will continue until indistinguishable images are generated. In this way, each edge device trains its model and updates its trained model to a cloud server without revealing the actual parameters. So, the local GANs collaborate with the central cloud server to improve the global GAN model for generating realistic images of covid-19. A blockchain-based framework is also proposed for decentralised covid analytics. For evaluating the performance, they used the DarkCOVID dataset with 620 X-ray images and the chest COVID data set with 950 X-ray images. 80% of the data is used for training and 20% for testing.

4. Federated Learning Coupled with Steganography for Healthcare Applications

Federated learning is a distributed machine learning technique that allows multiple parties to collaborate on building a model without sharing their data directly. In healthcare, this approach is especially important due to privacy concerns around patient data. Instead of sharing data, federated learning involves training local models on each organization's data and then combining these models to create a global model. In healthcare, federated learning can be used to build predictive models for patient outcomes or to improve diagnostic accuracy. It allows organizations to collaborate on building models that are more accurate and generalizable, without sharing sensitive patient data. It also provides a way for smaller organizations to benefit from the insights gained from larger organizations without having to share their data. For example, a hospital system could train a model on patient data from multiple hospitals without sharing the data. Instead, each hospital would train the model on its data and then send the updated model parameters to a central server. The central server would then aggregate the updated parameters and use them to update the global model, which would be sent back to each hospital for further training. One of the benefits of federated learning in healthcare is that it allows for the creation of more robust and accurate models by leveraging data from multiple sources. It also allows for more privacy and security for patient data, as the data never leaves the local hospital or organization. The federated learning process starts with each organization training a local model on their data. The local model is then sent to a central server, which combines the local models to create a global model. The central server then sends the updated global model back to each organization, which uses it to train its local model further. This process repeats over multiple iterations until the global model reaches the desired level of accuracy.

However, there are also challenges associated with federated learning in healthcare. One of the main challenges is ensuring that patient data privacy is maintained throughout the process. This requires careful consideration of how data is collected, stored, and processed, as well as the implementation of appropriate security measures. Another challenge is ensuring that the local models are trained effectively and that the global model is updated frequently enough to reflect the latest insights from each organization's data. This requires coordination between the organizations involved in the federated learning process and careful monitoring of the training process. Overall, federated learning has the potential to improve healthcare outcomes by enabling collaboration and knowledge sharing without compromising patient data privacy. However, it requires careful implementation and monitoring to ensure that it is done effectively and responsibly.

Image steganography is the process in which one image called the secret image is hidden inside the cover image. Medical image steganography deals with hiding the secret medical image within the public or private cover image which is significant in the healthcare sector as everyone was relying on teleradiology applications during the covid scenario. For giving a double stage of security cryptographic techniques can be combined with steganography. Image steganalysis is the reverse process of retrieving the hidden secret image that is embedded using steganographic algorithms. Only very few works are done in creating personalized steganalysis approaches using federated learning. The work done till now in this area is summarized here. In [27], the possibility of using federated learning in training image steganalysis algorithms has been researched. They have developed a personalized distributed model called Fedsteg for doing secure image steganalysis using federated transfer learning and evaluated the model using famous state-of-the-art steganographic methods such as WOW, S-UNIWARD, and HILL. Fedsteg can collect data from isolated institutions without compromising on privacy and security and transfer learning is used to create a personalized model. We know that transfer learning uses the knowledge of creating a model in doing one task in another application domain. They used TLU-CNN for training the user and cloud models and homomorphic encryption is used for additional security. BOSS base1.01 containing 10,000 grayscale images is utilized for training the model. Results showed that Fedsteg coupled with homomorphic encryption has better results.

In [28, 31] a covert communication framework called FL-talk which enables secret information sharing between participating edge nodes is proposed. Here, the sender can encode the message using spectral steganography [30] and the receiver can correctly decode the message from the global model without interfering with FL training. In this method, they first generated covert communication parameters, then they transformed the weights into the spectral domain and then encoded the message in the weight spectrum, and inverse spectral transformation is performed. They compared this scheme with state-of-the-art methods and found that successful covert communication is possible while preserving FL convergence. In [29], another framework for covert communication between participating sites is implemented using the MNIST dataset. In cryptography, federated learning can be used to improve the security of encrypted communication systems by enabling multiple parties to collaboratively train a machine learning model on their encrypted data. This approach can help to detect potential attacks or vulnerabilities in the system by identifying patterns or features in the data that may be indicative of a security breach. The model can then be used to improve the encryption algorithms used in the system, making it more secure. In steganography, federated learning can be used to improve the efficiency and effectiveness of steganographic algorithms by allowing multiple parties to collaboratively train a machine learning model on a diverse set of data. This approach can help to identify patterns or features in the data that can be used to optimize steganographic algorithms and make them more effective at hiding information. In both cases, federated learning allows multiple parties to train a machine learning model collaboratively without sharing their raw data. This helps preserve the privacy and security of the data while still allowing for improvements in the cryptography or steganography system.

5. Other Application Areas of Federated Learning in Healthcare

Smart healthcare systems [32] will benefit from federated learning. The bottleneck in the current scenario is the lack of medical data for training machine learning and deep learning models as the data are with several private hospitals or imaging centres and they are hesitant to share these data. If the different organizations owning these data unite, then a large data set will be available for training an accurate model and better feature extraction and classification can be done on the medical data which will surely open a new door for disease diagnosis.

For example, the lack of large data sets causes a hindrance to the classification of EEG (electroencephalographic) signals in Brain-Computer Interfaces (BCI). In [33], a method is proposed using federated transfer learning for this purpose and achieved higher accuracy than other DL approaches. In [34], a CNN-based autoencoder system is proposed and federated transfer learning is applied for ensuring data availability and privacy to ECG signal classification. In [35], the use of FL in speech emotion recognition is evaluated and they used C NN with LSTM for the classification purpose. Depression is a mental illness that affects persons and as the symptoms vary from person to person and as it depends on various factors, it is very difficult to clinically confirm this disease. In [36], mood detection of a person using mobile health care data is proposed as this information can be used to detect depression in individuals. FL can also be used for autism disorder classification, the study of sepsis, and many other disease diagnosis and classification [37] which needs further study.

For patient mortality prediction, some studies have been carried out by some researchers [38–40]. They grouped Non-Independent and Identically Distributed Data (Non-IID data) of patients into different communities and they trained models for each community and the algorithm obtained an AUC score of 69.13%. In [39], Brisimi et al., implemented a Support Vector Machine (SVM) classifier in a federated environment to study the future hospital predictions of patients who are suffering from heart diseases. In [40], a Multilayer Perceptron based architecture is used for the study of patients admitted to Intensive Care Units. They reported an AUC of 97.76%. For human activity and emotion recognition, some studies have been carried out in the FL environment. In [41], a model is proposed for studying the various activities such as standing, sitting, walking, etc. using federated transfer learning and CNN-based architecture and obtained 99.4% accuracy in classifying human activities. For in-home monitoring applications [42], an edge-cloud federated architecture is proposed and used an autoencoder for model creation. And for studying speech signal, and emotion recognition activity [43], an emotion index is created using the patient's mental health data and it showed an AUC of 88%.

6. Future Directions and Challenges of Federated Learning

Deep learning solutions for Covid-19 diagnosis or cancer diagnosis depend on large amounts of data for training and if the medical images that are used for training contain unlabeled data, then they should be manually labelled by the radiologists to get an accurate model. But in the case of federated learning, data is taken from multiple institutions, and the privacy of data is also preserved as the actual data is not shared by the hospitals or imaging centres, rather local models are shared to a global server which aggregates the data from all participating sites. The possibility of application of federation learning in the steganalysis field is still not fully utilized. For secure parameter sharing between various sites, various encryption algorithms such as secure multi-party computation, garbled circuit protocol, etc., can be used instead of homomorphic encryption [27].

The use of federated learning involving multi-institution collaboration will surely have a good impact on precision medicine, which is an emerging area of research. Precision medicine is a form of medicine that uses knowledge about a person's genes or proteins and his lifestyle to prevent, diagnose or treat disease. It will help doctors to use a person's genetic information for giving personalized care. The impact of federated learning in precision medicine needs to be explored. Much research is also going on related to communication speed, convergence time of the model, working with non-IID data [3], different poisoning attacks on data and model, secure aggregation, secure sharing of data, selection of clients for participation, security, and privacy of data and so on. Some of the challenges that may come while implementing the FL approach are discussed here. The main challenge is the updating of weights. The central cloud server is responsible for the updating of weights when multiple hospitals collaborate and the weights that should be given for each hospital should not be the same as some hospitals will be superior to others [5]. Federated averaging is proposed by some researchers, but it is not a good solution. Another challenge is that when coming to the research funding for training and implementing a federated model, contributions by different hospitals will be different in the sense that some may be labelling and uploading large amounts of data sets, but the diversity of images will be less. So how to calculate the share for each hospital is still a challenge. Also, if the participating sites operate on diverse hardware and software, the process of optimization and debugging will be difficult. The fourth challenge lies in the protocol differences in image acquisition and labelling tools used by various participating hospitals.

The future of federated learning looks very promising, and it is expected to play an increasingly important role in the development of machine learning models, especially in fields such as healthcare, finance, and the Internet of Things (IoT). Here are some key trends that are likely to shape the future of federated learning:

1. Increased privacy protection: Federated learning allows data to be kept on the device or server where it is generated, instead of being sent to a central location. This approach provides increased privacy protection, which is critical in industries such as healthcare and finance where data confidentiality is paramount.

2. Improved model accuracy: Federated learning has the potential to improve model accuracy by allowing data from a wider range of sources to be incorporated into the training process. This can lead to more accurate predictions and better decision-making.

3. Greater scalability: Federated learning can be scaled to accommodate many devices or servers, which makes it ideal for applications in the IoT space.

4. Continued research and development: Federated learning is still a relatively new approach, and there is a lot of ongoing research and development happening in this area. As more researchers and developers work on federated learning, we can expect to see new and innovative applications of this technology.

One of the key advantages of federated learning is its ability to improve data privacy since data is kept on the device or server where it is generated, rather than being sent to a central location. Federated learning also has the potential to improve model accuracy by allowing a wider range of data sources to be used in the training process, and it can be scaled to accommodate many devices or servers. In the future, we can expect federated learning to become increasingly important in industries such as healthcare, finance, and the Internet of Things (IoT), and continued research and development in this area will likely result in new and innovative applications of the technology.

7. Conclusion

We have seen that for disease diagnosis or tumour detection, the creation of a model from the data generated from a single hospital will not be reliable. But with the use of federated learning, this problem is solved as data from various collaborating hospitals or imaging centres can be used for the generation of the model. Also, the fear of data breach is avoided as the data is not directly shared, instead a local model is created, and this model is shared with the global server which in turn creates a global model from the local models received. In that way, data privacy is also guaranteed. This chapter discussed the concept of federated learning, its benefits, types, and its various applications in the healthcare sector such as covid-19 diagnosis, tumour detection, and steganalysis. In the future, in many more areas, federated learning can be applied to generating accurate models while ensuring data privacy and security.

References

[1] McMahan, B., Moore, E., Ramage, D., Hampson, S. and y Arcas, B.A. 2017, April. Communication-efficient learning of deep networks from decentralized data. In Artificial Intelligence and Statistics (pp. 1273–1282). PMLR.

[2] Chowdhury, A., Kassem, H., Padoy, N., Umeton, R. and Karargyris, A. 2022. A review of medical Federated Learning: Applications in oncology and cancer research. In International MICCAI Brainlesion Workshop (pp. 3–24). Springer, Cham.

[3] Majeed, A., Zhang, X. and Hwang, S.O. 2022. Applications and challenges of Federated Learning paradigm in the big data era with special emphasis on COVID-19. Big Data and Cognitive Computing, 6(4): 127.

[4] Zhang, W., Zhou, T., Lu, Q., Wang, X., Zhu, C., Sun, H., Wang, Z., Lo, S.K. and Wang, F.Y. 2021. Dynamic-fusion-based federated learning for COVID-19 detection. IEEE Internet of Things Journal, 8(21): 15884–15891.

[5] Ng, D., Lan, X., Yao, M.M.S., Chan, W.P. and Feng, M. 2021. Federated Learning: A collaborative effort to achieve better medical imaging models for individual sites that have small labelled datasets. Quant. Imaging Med. Surg., 11(2): 852857. doi: 10.21037/qims-20-595

[6] Rieke, N., Hancox, J, Li, W. et al. 2020. The future of digital health with federated learning. npj Digit. Med., 3: 119. https://doi.org/10.1038/s41746-020-00323-1.

[7] Yi, L., Zhang, J., Zhang, R., Shi, J., Wang, G. and Liu, X. 2020, September. SU-Net: An efficient encoder-decoder model of federated learning for brain tumour segmentation. In International Conference on Artificial Neural Networks (pp. 761–773). Springer, Cham.

[8] Sheller, M.J., Reina, G.A., Edwards, B., Martin, J. and Bakas, S. 2019. Multi-institutional deep learning modelling without sharing patient data: A feasibility study on brain tumour segmentation. In International MICCAI Brainlesion Workshop (pp. 92–104). Springer, Cham.

[9] Sheller, M., Edwards, B., Reina, G.A., Martin, J. and Bakas, S. 2019. NIMG-68. Federated Learning in neuro-oncology for multi-institutional collaborations without sharing patient data. Neuro. Oncol., 21(Suppl 6): vi176–7. doi: 10.1093/neuonc/noz175.737. Epub 2019 Nov 11. PMCID: PMC6847543.

[10] Roth, H.R. et al. 2020. Federated Learning for breast density classification: A real-world implementation. *In*: et al. Domain Adaptation and Representation Transfer, and Distributed and Collaborative Learning. DART DCL 2020 2020. Lecture Notes in Computer Science, vol 12444. Springer, Cham. https://doi.org/10.1007/978-3-030-60548-3_18.

[11] Jiménez-Sánchez, A., Tardy, M., Ballester, M.A.G., Mateus, D. and Piella, G. 2022. Memory-aware curriculum federated learning for breast cancer classification. Computer Methods and Programs in Biomedicine, 107318.

[12] Hashmani, M.A., Jameel, S.M., Rizvi, S.S.H. and Shukla, S. 2021. An adaptive federated machine learning-based intelligent system for skin disease detection: A step toward an intelligent dermoscopy device. Appl. Sci., 11: 2145. https://doi.org/10.3390/app11052145.

[13] Lee, H., Chai, Y.J., Joo, H., Lee, K., Hwang, J.Y., Kim, S.M., Kim, K., Nam, I.C., Choi, J.Y., Yu, H.W., Lee, M.C., Masuoka, H., Miyauchi, A., Lee, K.E., Kim, S. and Kong, H.J. 2021. Federated Learning for thyroid ultrasound image analysis to protect personal information: Validation study in a real health care environment. JMIR Med. Inform., 9(5): e25869. doi 10.2196/25869. PMID: 33858817; PMCID: PMC8170555.

[14] Hansen, C.R., Price, G., Field, M., Sarup, N., Zukauskaite, R., Johansen, J., Eriksen, J.G., Aly, F., McPartlin, A., Holloway, L., Thwaites, D. and Brink, C. 2022. Larynx cancer survival model developed through open-source federated learning. Radiotherapy and Oncology, 176: 179–186.

[15] Truhn, D., Arasteh, S.T., Saldanha, O.L., Müller-Franzes, G., Khader, F., Quirke, P., West, N.P., Gray, R., Hutchins, G.G., James, J.A., Loughrey, M.B. and Kather, J.N. 2022. Encrypted Federated Learning for secure decentralized collaboration in cancer image analysis. medRxiv.

[16] Sheller, M.J., Edwards, B., Reina, G.A., Martin, J., Pati, S., Kotrotsou, A., Milchenko, M., Xu, W., Marcus, D., Colen, R.R. et al. 2020. Federated Learning in medicine: Facilitating multi-institutional collaborations without sharing patient data. Sci. Rep., 10: 12598.

[17] Sarma, K.V., Harmon, S., Sanford, T., Roth, H.R., Xu, Z., Tetreault, J., Xu, D., Flores, M.G., Raman, A.G., Kulkarni, R. et al. 2021. Federated Learning improves site

performance in multicenter deep learning without data sharing. J. Am. Med. Inform. Assoc., 28: 1259–1264.

[18] Yan, B., Wang, J., Cheng, J., Zhou, Y., Zhang, Y., Yang, Y., Liu, L., Zhao, H., Wang, C. and Liu, B. 2021, July. Experiments of federated learning for COVID-19 chest X-ray images. In International Conference on Artificial Intelligence and Security (pp. 41–53). Springer, Cham.

[19] Wang, L. and Wong, A. 2020. COVID-Net: A tailored deep convolutional neural network design for detection of COVID-19 cases from chest x-ray images. Sci. Rep., 10(1): 19549.

[20] Sharma, A. and Muttoo, S.K. 2018. Spatial image steganalysis based on resnext. *In*: IEEE 18th International Conference on Communication Technology (ICCT) 2018, Chongqing, People's Republic of China, pp. 1213–1216.

[21] Sandler, M., Howard, A., Zhu, M., Zhmoginov, A. and Chen, L.C. 2018. Mobile net V2: inverted residuals and linear bottlenecks. *In*: Proceedings of the IEEE Conference on Computer Vision and Pattern Recognition (CVPR) 2018, pp. 4510–4520, IEEE Conference on Computer Vision and Pattern Recognition, Salt Lake City.

[22] Ayyachamy, S., Alex, V., Khened, M. and Krishnamurthi, G. 2019. Medical image retrieval using Resnet-18. *In*: Medical Imaging 2019, Imaging Informatics for Healthcare, Research, and Applications, 10954: 1095410. Proceedings of SPIE.

[23] Kumar, R., Khan, A.A., Kumar, J., Golilarz, N.A., Zhang, S., Ting, Y., Zheng, C. and Wang, W. 2021. Blockchain-federated-learning and deep learning models for covid-19 detection using ct imaging. IEEE Sensors Journal, 21(14): 16301–16314.

[24] Lo, S.K., Liu, Y., Lu, Q., Wang, C., Xu, X., Paik, H.Y. and Zhu, L. 2022. Towards trustworthy AI: Blockchain-based architecture design for accountability and fairness of federated learning systems. IEEE Internet of Things Journal.

[25] Rjoub, G., Wahab, O.A., Bentahar, J., Cohen, R. and Bataineh, A.S. 2022. Trust-augmented deep reinforcement learning for federated learning client selection. Information Systems Frontiers, 1–18.

[26] Qayyum, A., Ahmad, K., Ahsan, M.A., Al-Fuqaha, A. and Qadir, J. 2022. Collaborative federated learning for healthcare: Multi-modal covid-19 diagnosis at the edge. IEEE Open Journal of the Computer Society, 3: 172–184.

[27] Kandati, D.R. and Gadekallu, T.R. 2022. Genetic clustered federated learning for COVID-19 detection. Electronics, 11(17): 2714.

[28] Nguyen, D.C., Ding, M., Pathirana, P.N., Seneviratne, A. and Zomaya, A.Y. 2021. Federated learning for COVID-19 detection with generative adversarial networks in edge cloud computing. IEEE Internet of Things Journal.

[29] Yang, H., He, H., Zhang, W. and Cao, X. 2020. FedSteg: A federated transfer learning framework for secure image steganalysis. IEEE Transactions on Network Science and Engineering, 8(2): 1084–1094.

[30] Chen, H. and Koushanfar, F. 2022. FL-Talk: Covert communication in Federated Learning via spectral steganography. In Workshop on Trustworthy and Socially Responsible Machine Learning, NeurIPS 2022.

[31] Costa, G., Pinelli, F., Soderi, S. and Tolomei, G. 2021. Covert channel attack to federated learning systems. arXiv preprint arXiv:2104.10561.

[32] Yang, Q., Liu, Y., Chen, T. and Tong, Y. 2019. Federated machine learning: Concept and applications. ACM Transactions on Intelligent Systems and Technology (TIST), 10(2): 1–19.

[33] Ju, C., Gao, D., Mane, R., Tan, B., Liu, Y. and Guan, C. 2020, July. Federated transfer learning for EEG signal classification. In 2020 42nd Annual International Conference of the IEEE Engineering in Medicine & Biology Society (EMBC) (pp. 3040–3045). IEEE.

[34] Raza, A., Tran, K.P., Koehl, L. and Li, S. 2022. Designing ECG monitoring healthcare system with federated transfer learning and explainable AI. Knowledge-Based Systems, 236: 107763.

[35] Latif, S., Khalifa, S., Rana, R. and Jurdak, R. 2020, April. Federated learning for speech emotion recognition applications. In 2020 19th ACM/IEEE International Conference on Information Processing in Sensor Networks (IPSN) (pp. 341–342). IEEE.

[36] Xu, X., Peng, H., Sun, L., Bhuiyan, M.Z.A., Liu, L. and He, L. 2021. Fedmood: Federated Learning on mobile health data for mood detection. arXiv preprint arXiv:2102.09342.

[37] Prayitno, Shyu, C.R., Putra, K.T., Chen, H.C., Tsai, Y.Y., Hossain, K.S.M.T., Jiang, W. and Shae, Z.Y. 2021. A systematic review of Federated Learning in the healthcare area: From the perspective of data properties and applications. Appl. Sci., 11: 11191.

[38] Huang, L., Shea, A.L., Qian, H., Masurkar, A., Deng, H. and Liu, D. 2019. Patient clustering improves the efficiency of federated machine learning to predict mortality and hospital stay time using distributed electronic medical records. J. Biomed. Inform., 99: 103291.

[39] Brisimi, T.S., Chen, R., Mela, T., Olshevsky, A., Paschalidis, I.C. and Shi, W. 2018. Federated Learning of predictive models from federated electronic health records. Int. J. Med. Inform., 112: 59–67.

[40] Shao, R., He, H., Chen, Z., Liu, H. and Liu, D. 2020. Stochastic channel-based Federated Learning with neural network pruning for medical data privacy preservation: Model development and experimental validation. JMIR Form. Res., 4: e17265.

[41] Chen, Y., Qin, X., Wang, J., Yu, C. and Gao, W. 2020. FedHealth: A federated transfer learning framework for wearable healthcare. IEEE Intell. Syst., 35: 83–93.

[42] Wu, Q., Chen, X., Zhou, Z. and Zhang, J. 2020. FedHome: Cloud-edge based personalized Federated Learning for in-home health monitoring. IEEE Trans. Mobile Comput.

[43] Chhikara, P., Singh, P., Tekchandani, R., Kumar, N. and Guizani, M. 2021. Federated Learning meets human emotions: A decentralized framework for human–computer interaction for IoT applications. IEEE Internet Things J., 8: 6949–6962.

[44] LaLonde, R. and Bagci, U. 2018. Capsules for object segmentation. arXiv preprint arXiv:1804.04241.

[45] Yan, Z., Wicaksana, J., Wang, Z., Yang, X. and Cheng, K.T. 2021. Variation-aware federated learning with multi-source decentralized medical image data. IEEE J. Biomed. Health Inform., 25: 2615–2628.

Chapter **11**

Federated Learning for Efficient Cardiac Disease Prediction based on Hyper Spectral Feature Selection using Deep Spectral Convolution Neural Network

B Dhiyanesh,[1,*] *G Kiruthiga,*[2] *P Saraswathi,*[3]
Gomathi S[1] *and R Radha*[4]

1. Introduction

Cardiovascular infection is one of the main causes of death around the world today. As a result, visualization and early analysis are vital to the success of treatment planning. In the clinical field, it decreases death

[1] Associate Professor/CSE, Dr. N.G.P. Institute of Technology, Coimbatore.
[2] Associate Professor/CSE, IES College of Engineering, Thrissur.
[3] Assistant Professor/IT, Velammal College of Engineering and Technology, Madurai.
[4] Assistant Professor/EEE, Study World College of Engineering, Coimbatore.
Emails: kirthikacsehod@gmail.com; psw@vcet.ac.in; mail2mathi86@gmail.com; radharaja100@gmail.com
* Corresponding author: dhiyanu87@gmail.com

rates and brings down clinical expenses. Despite substantial advances in determination and treatment, CVD illness remains a main source of grimness and mortality worldwide. Brainpower innovation provides another tool for deciphering fundamental information for independent clinical guidance; it has altered coronary illness practice.

Information mining and deep learning strategies provide techniques and procedures for extracting critical information required for direction. Cardiovascular infection (HD) is a type of cardiovascular illness, a sickness of the heart and veins. A broad exploration of all parts of cardiovascular disease (conclusion, treatment, electrocardiography, echocardiography, etc.) has created much information. The point of this study is to deliberately survey the utilization of AI and deep learning techniques and apparatuses in cardiovascular illness research as they connect with cardiovascular entanglements, expectations, and analysis.

Generally, 60% of the utilization is AI procedures and backing vector machines, and 30% is profound learning methods. Most information is utilized in clinical datasets. This study provides experience in selecting appropriate techniques and pathways to work on HD forecast precision. Articles chosen in this study provide valuable information extraction, a more profound comprehension of cardiovascular sickness, and new speculations for additional examination.

The features are gathered from the PHR cardiac dataset to filter the significance of marginal constructed values deemed essential attributes. The feature selection is important for dimension reduction in a cardiac dataset. The classifier, the LSTM-Convolutional Neural Network (LSTM-CNN), creates an activation function logical prediction based on the large-scale prediction to categorize the disease level based on the class by reference.

2. Related Work

This review explores the differential techniques and implementation presented by various authors and discusses their limitations.

A framework to work on persistent cleanliness and careful interaction is presented [1]. In a coordinated framework, the Video See-Through (VST) headset, shows total data about the patient's health status progressively (it should be worn by attendants and anesthesiologists and different individuals from the careful group for a medical procedure). Operating Room (OR) administrators can get imperative signs and access electronic clinical records continuously through the device. Coronary illness is the main source of death in numerous nations, and many methodologies center on exact and careful medications to treat this sickness. By and large, her 5-year endurance rate for coronary illness patients is anticipated by the example size and the limited quantity of information in conventional relapse models. Based on an extensive database, he aims to predict 5-year survival rates and introduces feature selection algorithms and machine learning tools such as random forest classifiers (RFCs) [2].

Heart disease is a major problem, a significant burden that severely impacts on the quality of life. Many technical solutions have been proposed to predict the evaluation or reduce the risk; most are based on accelerometers and gyroscopes. It is not, however, used to be almost an identification to identify the first fall. A model to predict a decrease in baseline use has proposed an analysis of short-term Heart Rate Variability (HRV) [3]. An increase in the demand for rising healthcare costs and services has asked us to use medical resources more effectively. With the random nature of the resources necessary, the medical service process will need to be more efficient. What is available is meant to reduce the uncertainty of patient demand for resources. To achieve this goal, the patients are divided into the same resource and user groups [4].

Short electrical heartbeats are intruded onto the growth tissue when irreversible high-voltage electroporation (IRE) is used. Effective treatment relies on direct openness, everything being equal to deadly electric fields; however, with such openings hard to anticipate ahead of time, it depends on clinicians to decide the ideal treatment boundaries required. The clinician depends on how the beat is halted and how the cells inside the tissue change during electroporation to screen for tissue opposition [5]. Cardiovascular disease prediction algorithm using POMDP model. The patient will calculate the alert through the mist to the doctor in an emergency. An ambulance is where the patient is transported in emergencies. The physician receives data through fog computing logic, a new field of medical computing that is gaining traction in the research community [6]. The main problem is caused by heart disease because it is being used to identify the disease; this will need a lot of experience, and knowledge is also very difficult. Depending on the condition, death will be the main cause of it worldwide. Additionally, judging and understanding patients takes a long time for doctors since they must observe their health and diet. It is a very long process, and it needs to get tested and seek medical attention because the process is so long [7].

Cardiovascular illness is the most well-known reason for death on the planet, so early discovery and routine checking of heart infections can decrease mortality. Wellbeing, gushing about various frameworks, observing the dramatic development of information from various sources, and wearable sensor gadgets communicated over the Internet deliver a lot of continuous data. Massive data analysis and machine learning, particularly streaming early confirmation, as well as creative advances can recognize the enormous impact of innovation in the clinical field of coronary disease. It will be more impressive and reasonable [8]. Heart disease is a modern world concern, and life is short. Large populations depend on healthcare systems to deliver accurate results quickly. Every day, healthcare organizations generate and obtain a large amount of data. For interesting discoveries, Data Innovation can extract data through spraying processes. Weighted association rules apply even if mining techniques were used to eliminate manual data that the data type helped remove directly from the electronic records [9].

Hypoxia is caused by a loss of blood flow due to imbalance and specific parts of the death of the cells in that region, in the form of an ischemic stroke in the heart itself. These amounts represent the definition of infarct volume, for which some neurosurgeons are extremely sensitive to treatment [10]. In most cases of death from heart disease, the diagnosis by the doctor is laborious and, in most fields, is carried out by professionals. Heart disease risk can be identified using a fuzzy expert system. There are some factors in analyzing the patient's heart disease, and it takes work to process the doctor's work. However, experts want an accurate tool to determine the risk factors based on the information provided with references [11]. The meaning of data mining techniques hard disk, including not only a brief comprehensive literature review but also classification. A review of the scientific literature shows that several data mining methods have been used on HDDs. Through mining innovation, it provides an insightful data set of critical writing on proposed orders [12]. The ensemble is based on the method used for the KNN distance applied to diagnose heart disease and displays the results. Orchestra has been implemented in the following two configurations: One used five of the three distances, and others added the average accuracy of their reference when used in the KNN method, given each distance's weighting format [13].

With the rise in heart attack rates among adolescents, organizations have had to develop systems to detect and prevent heart attacks based on early warning signs. It is normal people who, in many cases, undergo expensive tests such as electrocardiograms. Because it is not realistic is reliable and requires a convenient system to predict the probability of heart disease [14]. Heart disease is a serious chronic condition that often results in death. Medical data shows that this disease will soon wipe out the population with more knowledge than the rich and poor. As a result, an accurate diagnosis of the patient's time is based on the urgency of the medical insurance functions. A shady hospital leads to invalid diagnoses [15]. Described the use of big data for real-time ML heart disease detection. In recent years, heart disease has become the most common cause of death worldwide. Early detection and continuous monitoring of heart disease using multiple sources of information such as the Internet of Things and wearable sensors mean to reduce mortality data through exponential growth [16]. Described computational prediction models (CPM) based on feature conservation in the context of high-dimensional data processing. Using Mass Spectrometry Data High dimensional features are non-related features that reduce classification accuracy [17]. Describe the Medical Suggestions of ML in Forecasting Interpretation can presume the threat of early heart disease (HD) and chest pain in patients with cardiovascular events, and providing adequate health is essential for its positive diagnosis. It extracts the control function of a simple decision theoretical model prediction and disease classification between the sample and effective quality early detection [18]. In diagnosing cardiovascular illnesses and creating precise prediction models, sensors and remote sensor organizations, applications of artificial intelligence (AI), and LSBD are described as crucial elements [19].

Describe the Wearing of a Sensor Monitoring System Based on Real-Time Monitoring of Patient Vital Signs and Their Doctor's Reports to Improve Health Care and the Quality of HD Findings [20]. The mode of the Internet of Things is discussed with smart healthcare systems can be an advantage in remote health monitoring [21]. Contiki-NG and Open WSN, data from the most advanced open sensor condition dust, dust test two of the most widely used means available, Caozuoxitong detection of heart disease in [22], described how, with wearable sensor devices, smart healthcare systems, which have led to population aging, have paid special attention to home care and e-health change over the last few decades. Its goal is to allow them to remain in their home environment and hospital rather than providing a medical service in the patient's home to improve the patients' quality of life.

Described the Internet of Things (IoT) Topology's Congestion-Free Routing Mechanism, which collects physical, physiological, and consumer-centric electronic health or vital signs of patients in health services used by consumers [23]. Wireless sensor networks are offered during the data collection process to prevent complications. He described HD as a complex infection, and many people around the world suffer from this disease. Using machine learning to classify time and heart disease, especially in the cardiovascular field, for effective health care identification plays an important role. The Feature Selection Problem is solved by an IFS Algorithm [24]. The global incidence of heart disease is increasing due to the increasing use of heart disease grouping and extraction mining options. Cardiovascular disease classification Time and an efficient framework for cardiovascular disease play an essential role in effectively identifying health, especially in the cardiovascular field [25]. Data mining techniques to investigate how coronary artery disease risks prognostic scores can be obtained from large amounts of data [26]. These multidisciplinary fields include data mining (DM), machine learning (ML), statistics, pattern recognition (PR), artificial intelligence (AI), and data visualization. Highlight the importance of sequential BSD models for early heart disease prediction in HD. Detecting the disease at an early stage can help treat and recover from HD. It has been developed for doctors to use symbolic machine learning techniques using HD symbols [27]. "Describing a productive cardiovascular illness forecast utilizing blended models," cardiovascular sickness is one of the main sources of death in this present reality. Cardiovascular sickness is a significant test in the field of forecasting and examination of clinical information [28]. Based on an analysis of random forest algorithms [29]. Machine learning may improve the accuracy of the diagnosis of heart disease. One of the leading causes of heart failure is narrowing of the coronary arteries. The coronary artery supplies blood to the heart. As described by [30], the stacking of noninvasive and noninvasive procedures for diagnosing CHD is accurate for detecting CHD. A physical examination is not equipped to detect coronary artery disease with this surgical procedure.

DM and ML algorithms were used to reduce feature sizes as in [31]. By optimizing the algorithms for selecting features and recommending updated features, rating processes that combine subgroups will be reduced. Feature selection algorithms are evaluated according to this metric, which has a profound influence on the selection of the algorithm. A hypertensive patients' prognosis is crucial for the development of cardiovascular disease prevention is proposed [32].

To aid in the detection of cardiovascular support during model optimization, cardiovascular researchers developed a heart failure (HF) model and a learning expert system. Research shows that the prediction of early failure can be improved using HF machines and descriptive vector machines from both SVMs [33]. A new monitoring and prediction system using DL-MNN is described [34], which provides various solutions for IoT devices, with smart wearable gadgets becoming more prevalent. Unfortunately sudden cardiac arrests are associated with a lower survival rate. Describe the Random Forest algorithm (RFA), a medical explanation for heart disease. Forecast data are detected on the Internet [35]. Because classification accuracy could be improved, unevenly feeling clinical data analysis and affect-dependent weighting function without classification were used. Deep learning (DL) has been identified, and the health industry has proven effective with a large amount of forecast data.

3. Proposed Method

A deep learning-based Cardiovascular Diagnosis Detection (CADD) for disease probability proposes a precise AI model that can dependably differentiate cardiovascular infection given the clinical boundaries of patients. The suggested method is based on the following steps: (1) Outlier removal, missing value replacement, and inequality class handling are all included. (2) Feature selection using feature importance based on weights; (3) classification. A model is then trained, and the trained model is used for prediction. Our main goal is to achieve better results with less feature/ computational complexity. This research aims to reduce cardiovascular complications through a multilevel approach. Figure 1 shows the structure of the HSBCFS-DCNN to build efficient data analysis and fitness data analysis models.

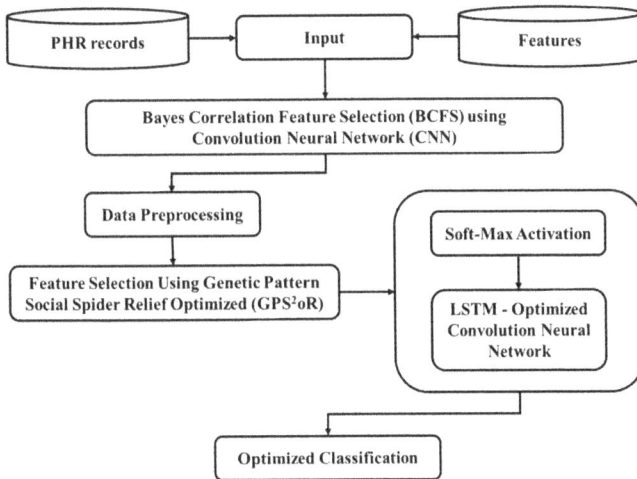

Figure 1. Proposed architecture for LSTM-Convolution Neural Network (CNN).

3.1 Cardiac Data Preprocessing

The various recordings are collected, and cardiac data is preprocessed. This dataset contains 303 patient records, with 6 missing values. Preprocessing was performed on the 297 remaining patient records after removing these six records from the dataset. Specific dataset attributes can be classified using multiclass variables and binary classifications. Heart disease can be assessed using a multiclass variable. There has been a determination of the value for patients with myocardial infarction.

Alternatively, the patient's heart disease score would be set to zero, which indicates he is healthy. Maps are converted into detection values during data preprocessing. 297 patient records were preprocessed, and 137 of them exhibited a value of 1, which indicates there is heart disease present. Compared to the 160 records without heart disease, the remaining 160 had a value of 0. The dataset was preprocessed by removing missing values and standard deviation (SD) and scaling it by min-max.

Algorithm

Initialization

Evaluate $\mu\&\beta$ from the values of the heart disease dataset

If a_{uv} is missing values the

$$S(tu, v) = \sqrt{\sum_{a=1}^{n}(\mu_u + \beta_v)^2}$$

$$\hat{X} = \frac{X - Minimum}{Maxi - Mini}$$

Set $a_{uv} = \mu u + \beta v$

Start with non-singular evaluates $\widehat{\Sigma\&\Delta}$

Calculate $X^H\widehat{X^{-1}} + \hat{X} + S(\widehat{\Sigma^{-1}})$

Update the evaluation of $\mu\&\beta$

Optimize (Q) in relation to Δ^{-1} to get $\hat{\Delta}$

Step (Δ): Calculate $\widehat{(\Delta X)}^H\ (X)^{-1} + \widehat{F(X^{-1})}$

Step (Σ)

Obtain updated estimates $\mu\&\beta$

Optimize (Q) in relation to Σ^{-1} to get $\hat{\Sigma}$

Repeat steps

Reduce the missing data

This leads to simple maximization with respect to $\Sigma^{\wedge}(-1)$ and others with respect to $\Delta^{\wedge}$. $X^{\wedge}$ is the new value selected after normalization, and X is the value selected from the numerical features

The above calculation depicts a preprocessing stage that diminishes the aspect considering traits by sifting values. Each record contains several attributes that represent information about a patient.

3.2 *Genetic Pattern Social Spider Relief Optimized (GPS²oR)*

The engineering problems that GPS²oR solves are based on sharks' ability to hunt with their keen sense of smell. As a result, it depends on the speed of accumulation and the rate of gathering. It was proposed that GPS²oR be used to overcome this problem. Overall, self-optimizing capabilities are demonstrated in existing optimization models. GPS²oR consists of four phases: initialization, advance, circular motion, and position update.

Initialization: For GPS²oR modelling, the population of initial solutions is arbitrarily generated in the search space. Each solution spider optimizes for a possible weighting level at the beginning of the search process.

3.2.1 Centroid and Fitness Evaluation

In the social spider, the female population dominates over the male population. The calculation of the number of females, Mf, is shown below, and the algorithm starts with a zero-initialized weight vector W. At each iteration, the algorithm obtains the eigenvectors that belong to random instances and the eigenvectors of the model closest to each class. The most direct and homogeneous example is the immediate hit, and most cases of different km are comparable at close range.

$$Mf = level\ [(0.9\text{-rand} \times (0.25)) \times P] \tag{1}$$

where P = total population rand = random value within [0, 1]

$$Low-risk\ Centroid\ (C),\ C_{Hd} = \frac{\Sigma_n^0 C_{hd}}{N} = C_{Hd2} = \frac{\Sigma_n^0 C_{hd^2}}{N} \tag{2}$$

Here, the female population is chosen randomly, 65–90% of the whole population "P".

Calculation of male population Mm is given as fitness

$$fitness(f) = f_h = \Sigma \sqrt{(C_{hd}1 - C_{hd}1)^2} + \sqrt{(C_{hd}2 - C_{hd}2)^2} \tag{3}$$

$$Mm = P - Mf \tag{4}$$

3.2.2 Fitness

Everyone's fitness in the total population PP is calculated by receiving a weight week. In high-risk situations, display the outcome value of each spider "n" unbiased of gender analysis.

$$w\ risk\ Centroid\ (C),\ x_{Hd} = \frac{\Sigma_n^0 x_{hd}}{N} = x_{Hd2} = \frac{\Sigma_n^0 x_{hd^2}}{N} \tag{5}$$

As a result, everyone's weight is determined by their high fitness risk.

$$fitness(f) = f_h = \Sigma \sqrt{(x_{hd}1 - x_{hd}1)^2} + \sqrt{(x_{hd}2 - x_{hd}2)^2} \tag{6}$$

$$Mm = P - Mf$$

$$wn = J\ PPn - worst\ PP;\ best\ PP - worst\ PP$$

where *worst PP* = worst individual
best PP = best individual
J PPn = the value of fitness, which is given by the evaluation of spider position PPn as per the objective function.

3.2.3 Vibration in their Webs

During optimization, usually, the spiders interact through the vibration of their strings in their given webs. The spider gets the vibration, which is dependent on the distance within them and the spider size that sends the vibration.

Vnan, $t = wt\ .\ e - dn$, 2(4) dk, = Vnan = vibration formed by spider "n" that is given by spider "t" within the nth and t^{th} spiders.

In this method, the spiders feel only three vibrations from other spiders like-(i) nearby spider with higher fitness (*Vn an, c*), (ii) the spider that is better in the swarm (*Vn an, best*) and (iii) the female spider that is nearby and accessible only to male spiders (*Vn an, f*).

3.3 Bayes Correlation Feature Selection based on Weights (BBCFS)

Selecting appropriate, reliable, and accurate features is necessary to improve the algorithm's performance. Feature selection selects the subset of features (attributes, predictors) most useful to structure a model.

$$\Sigma_n^0 f(n) = e + u_1 v_1 + u_2 v_2 + \cdots + u_n v_n \tag{7}$$

$$\Sigma_0^n f(0) = Weights + \Sigma_0^n w_x x_y \tag{8}$$

An attribute score is calculated based on its correlation with the predicted parameter. To determine the selection, the feature score is used. Relationships between attributes and other details show how attributes are related to one another. When the target attribute's value increases, there are two classes of correlation: (1) positive correlations and (2) negative correlations.

$$C(x_{f1}, x_{f2}, \ldots x_{fn} \mid a) = \prod_{x=1}^{n} C(x_{f1} \mid a) \tag{9}$$

$$C(x_{f1} \mid a) = \frac{C(x_{f1} \mid a) * f}{c(x_u)} \quad a \in \{values, attributes\} \tag{10}$$

Finally, classify the test data according to their associated probabilities.

$$C_{xy} = \arg\max c(C(x_{f1})\prod_{x=1}^{n} C(x_{f1} \mid a)) \qquad (11)$$

Independent features characterize Bayes' rule. Each instance of data is categorized according to its probability and weighed accordingly.

3.4 Soft-Max Activation (SMA)

At this stage, the classifier settles on its clinical proposals given the patient's gamble and the sigmoidal enactment capability per that classifier. All elements determined in the hidden layer by the Softmax enactment capability were instated by the predisposition weight control. The initiation capability teaches neurons to convert the following groups into normal profundity values, resulting in the neuron's rationale rules.

The feed-forward network has been upgraded to support continuous brain organizations. In this organization, versatile scrounging calculation rules are determined for preparing decisions that adjust neurons to the nearest weight forecast. As of now, the brain's organization can change loads of associations to match the actual result.

The activation function trains remain the f(x) = $\begin{cases} y = 1 & if \ \Sigma_{i-1}^{n} w_i x_i \geq b \\ y = 0 & otherwise \end{cases}$ (12)

Infraclass' logistic transformation is used to estimate f(x) from the logistic activation of a neuron trained with it, $w_{(t+1)} = w_t - \mathbb{N}\ \Delta\ w_t$ and $b_{(t+1)} = b_t - \mathbb{N}\ \Delta\ b_t$. The training function is iterated until the neuron gets the closer mean value from the weight w(t) at trains' t' at $\mathbb{N}\rightarrow$ number of neurons activation links are checked.

This makes a coherent portrayal, given the weightage of elements prepared into the neurons. The Weight 'w' at 'I' and 'j' stays at the maximum furthest reaches of the elements weight to foresee the class, but the capability 'y' stays at the preparation level from the component of 'x'. A completely coordinated feed-forward of the neuron is delineated in the development of the organization.

$$net_{i(t)} = \Sigma_{j=1}^{j} w_{ij} y_{j(t)} + x_{i(t)}, \ i = 1...j \qquad (13)$$

and

$$Ti\frac{dy(t)}{dt} = -yi(t) + \varphi(net_i) + x_{i(t)}, \ i = 1...j \qquad (14)$$

The weights are optimized to get each neural layers *net i(t)* from function 'x' be the features trained at logical 'y' functions, the frequent neuron weights are constant at Ti be trained from x(i) and y(i) from the average mean weight w(i).

Algorithm
Input: Its elements as Current Example scs, Adjusted feature $rule\ net_{i(t)} -\!\!\to$ Apt.

Output: enhanced class design

Stage 1: Process the illness impact rate and element Max weight term information values
      and Scs.

      For each fed layer class Pc

Stage 2: Process the secret layer neurons weight to c as

$$set = \int_{i=1}^{size(Apt)} \sum Apt(x).class = a]$$

Stage 3: Nearest design Pps = nearest design (Cset).

      For each nearest design on a relative connection from the Impact rate, each example

Stage 4: Register for every likeness highlight is delegated class in light of chance of Max disease
      impacted rate.

      Disease Proficient feature selection DPfs = $\dfrac{\int_{i=1}^{size(p)} \sum P(i)==Scs(i)}{size(p)}$

      End

Stage 5: Register the combined pace of cardiovascular impact rate.

      Compute cumulative rate DPFS = $\dfrac{\sum_{i=1}^{size(PPS)} Pfs}{size(PPS)}$

      End

      Optimized Cardiac personal recommendation (CPR) = PFs return set maximum values

      Stop

A neural weight that represents a category is labeled to each input feature trained on the classification structure. The input factors help segment the regions so that we can learn about the features associated with hidden neurons.

3.5 LSTM Optimized Convolution Neural Network (LSTM-CNN)

To solve the binary classification problem, the LSTM-CNN Neural Algorithm includes a hidden layer that is a function of artificial neuron activation. Perceptron uses activation functions for each neuron. Thus, iterative neurons transfer weights close to the clustering group properties to find the optimal class according to the neural architecture. In the activation function, the weights of the perceptron that determine each neuron's weighted inputs are changed. This reduces the layer count of both layers. The cognitive category is the bias weight used to classify cardiac disease categories based on the weights of the feature subscales.

Algorithm

Input: Cardiovascular component Test features.

Output: Test features

Begin

    Cardiovascular weightage and infection rate

    For each example class Pc

        The weight of the neurons in the secret layer is set to

$$c = \int_{i=1}^{size(Apt)} \sum Apt(i).\,class = c$$

        Closest pattern Pps = Closest pattern (Cset).

        For each nearest design on the relative connection, each example p

            By every comparability, highlights are characterized, ass classes.

$$Pfs = \frac{\int_{i=1}^{size(p)} \sum P(i)==Scs(i)}{size(p)}$$

    End

        Compute cumulative PFS $= \frac{\sum_{i=1}^{size(PPS)} Pfs}{size(PPS)}$

    End

    Ps = PFs return set maximum value with optimised behavioural pattern

    Stop

The preceding calculation orders the coronary illness class, which is prepared with a prepared set in which the coherent highlights are evaluated. A brain classifier predicts illness prevalence considering pretreatment proposals because of a class addressing the sickness prevalence rate.

3.5.1 Evaluation of LSTM Layers

In addition to a cell, input and output gates, and forget gates are all included in LSTM cells for short-term and long-term memory. In addition to controlling information flow and storing values for arbitrary periods, three gates control how information enters and leaves cells.

$$F_1 = \sigma(w_f \mid a_{f-1}, x_d \mid + y_f) \tag{15}$$

$$X_1 = \sigma(w_f \mid a_{f-1}, x_d \mid + y_f) \tag{16}$$

An LSTM cell with gate manipulation is shown in Fig. 2. Each LSTM gate has a sigmoidal activation function and a point-wise multiplication operation, and has three gates (input, forget, and output).

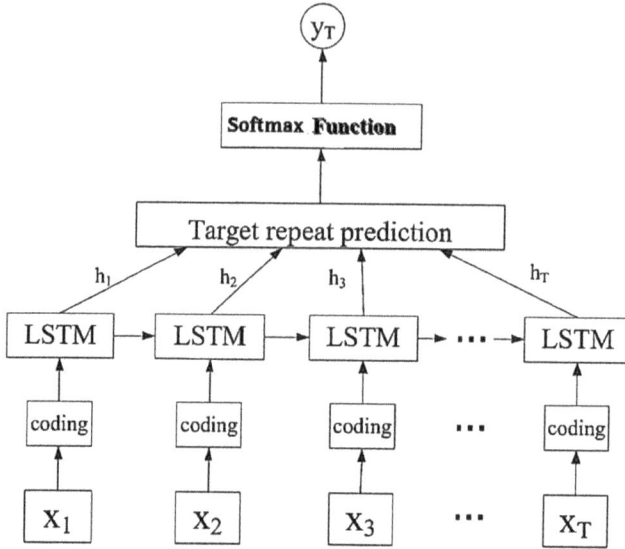

Figure 2. Cardiovascular disease prediction using LSTM.

4. Results and Discussion

The proposed technique is carried out and assessed under various boundaries in the UCI store heart dataset. These techniques assess sickness forecast viability based on a variety of elements and their characteristics. The assessment results are examined with the help of different strategies. The outcomes are introduced in this part.

Details used to assess the efficiency obtained by the different methods are shown in Table 1. Therefore, the effectiveness of these methods for various parameters was determined. The obtained results are detailed in this section.

According to Fig. 3, each method produces a different classification performance. With LSTM-CNN, we can cluster different diseases in an accurate manner.

$$Accuracy = \frac{TP + TN}{TP + TN + FN + FP}$$

As shown in Fig. 4, different methods generate different levels of accuracy in predicting disease. There is greater accuracy in disease prediction in each category when using the proposed LSTM-CNN method than when using other methods.

$$Precision = \frac{TP}{TP + FP} X100$$

Table 1. Evaluation detail.

Simulation parameter	Value
Tool used	Python, Jupiter notebook
Dataset used	Cardiac Dataset
No. of features	30
No. of data	500

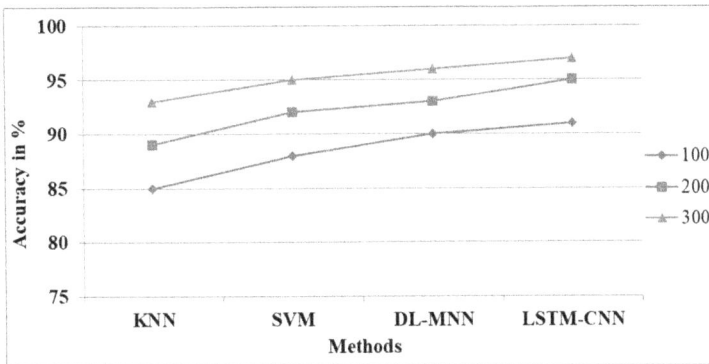

Figure 3. Accuracy in classification.

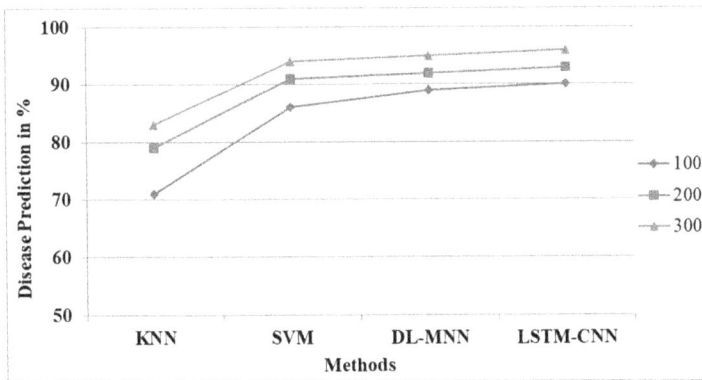

Figure 4. Accuracy of disease prediction analysis.

Figure 5 shows that the precision (also known as positive predictive value) is the number of prediction errors divided by the total number of feature inputs. In the proposed method, LSTM-CNN produces 92% of correct values, which is better than the existing methods.

$$Recall = \frac{TP}{TP + FP} \times 100$$

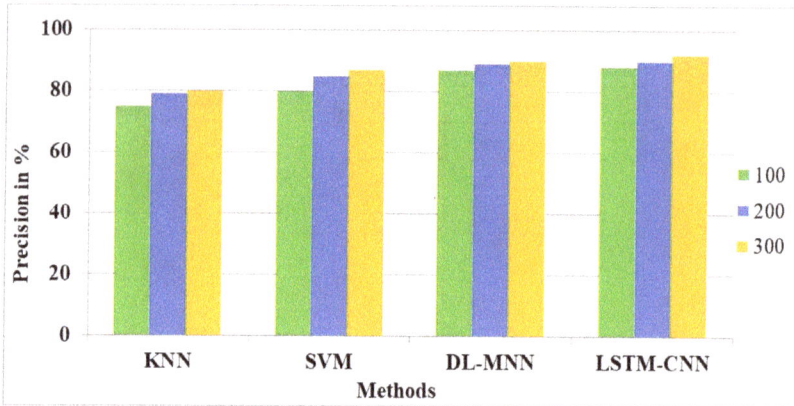

Figure 5. Performance of precision.

It is a measure of how many correct class predictions are made about the total number of input features for that class shown in Fig. 6. In the case of proposed methods that contain better results than previous algorithms, this is an indication of how complete the classifier is in the recall calculation.

A comparison of disease prediction accuracy generated by different methods is shown in Fig. 7. It has a higher accuracy than other methods in predicting diseases in each category using the LSTM-CNN algorithm.

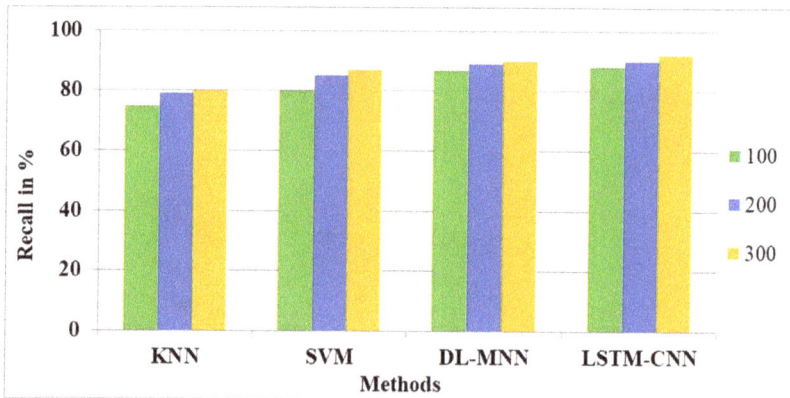

Figure 6. Performance of recall.

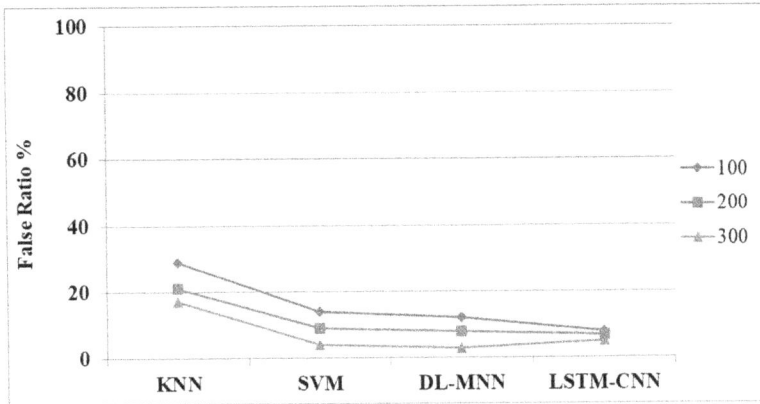

Figure 7. Analysis of false classification ratio.

5. Conclusion

To conclude, cardiac disease prediction-based federated learning using the Deep Spectral Time Variant Feature Analytic Model and Soft Max LSTM-CNN has produced the best prediction accuracy. The proposed system analyses the Cardiac Disease Influence Rate (CDIR) to select cardiac features depending on marginal accuracy. This reduces the big data dimensionality problems, which makes features trained on recurrent neural networks the best training features. The resultant factors prove that the best classification accuracy is achieved with regard to the cardiac disease influence rate, which is up to 97%. This supports a better way to make early-risk disease predictions for premature treatment to reduce cardiac attacks.

References

[1] Arpaia, P., Cicatiello, M., Benedetto, E.D., Anna Dodaro, C., Duraccio, L., Servillo, G. and Vargas, M. 2020. A health 4.0 Integrated system for monitoring and predicting patient's health during surgical procedures. 2020 IEEE International Instrumentation and Measurement Technology Conference.

[2] Mei, X. 2017. Predicting five-year overall survival in patients with non-small cell lung cancer by relief algorithm and random forests. 2017 IEEE 2nd Advanced Information Technology, Electronic and Automation Control Conference.

[3] Castaldo, R., Melillo, P., Izzo, R., De Luca, N. and Pecchia, L. 2017. Fall prediction in hypertensive patients via short-term HRV analysis. IEEE Journal of Biomedical and Health Informatics, 21(2): 399–406.

[4] Kumar, A. and Anjomshoa, H. 2018. A two-stage model to predict surgical patients' lengths of stay from an electronic patient database. IEEE Journal of Biomedical and Health Informatics, 1–1.

[5] Beitel-White, N., Martin, R.C.G. and Davalos, R.V. 2019. Post-treatment analysis of irreversible electroporation waveforms delivered to human pancreatic cancer patients. 2019 41st Annual International Conference of the IEEE Engineering in Medicine and Biology Society.

[6] Latha, R. and Vetrivelan, P. 2019. Blood viscosity based heart disease risk prediction model in edge/fog computing. 2019 11th International Conference on Communication Systems & Networks.

[7] Sudeshna, P., Bhanumathi, S. and Hamlin, M.R.A. 2017. Identifying symptoms and treatment for heart disease from biomedical literature using text data mining. 2017 International Conference on Computation of Power, Energy Information and Communication.

[8] Ed-Daoudy, A. and Maalmi, K. 2019. Real-time machine learning for early detection of heart disease using big data approach. 2019 International Conference on Wireless Technologies, Embedded and Intelligent Systems.

[9] Chauhan, A., Jain, A., Sharma, P. and Deep, V. 2018. Heart disease prediction using evolutionary rule learning. 2018 4th International Conference on Computational Intelligence & Communication Technology.

[10] Qidwai, U. 2018. Fuzzy data to crisp estimates: Helping the neurosurgeon making better treatment choices for stroke patients. 2018 IEEE-EMBS Conference on Biomedical Engineering and Sciences.

[11] Kasbe, T. and Pippal, R.S. 2017. Design of heart disease diagnosis system using fuzzy logic. 2017 International Conference on Energy, Communication, Data Analytics and Soft Computing.

[12] Meena, G., Chauhan, P.S. and Choudhary, R.R. 2017. Empirical study on classification of heart disease dataset-its prediction and mining. 2017 International Conference on Current Trends in Computer, Electrical, Electronics and Communication.

[13] Pawlovsky, A.P. 2018. An ensemble based on distances for a kNN method for heart disease diagnosis. 2018 International Conference on Electronics, Information, and Communication.

[14] Gavhane, A., Kokkula, G., Pandya, I. and Devadkar, P.K. 2018. Prediction of heart disease using machine learning. 2018 Second International Conference on Electronics, Communication and Aerospace Technology.

[15] Raju, C., Philipsy, E., Chacko, S., Padma Suresh, L. and Deepa Rajan, S. 2018. A survey on predicting heart disease using data mining techniques. 2018 Conference on Emerging Devices and Smart Systems.

[16] Ed-Daoudy, A. and Maalmi, K. 2019. Real-time machine learning for early detection of heart disease using big data approach. 2019 International Conference on Wireless Technologies, Embedded and Intelligent Systems (WITS), Fez, Morocco, pp. 1–5.

[17] Tuan D. Pham, Honghui Wang, Xiaobo Zhou, Dominik Beck, Miriam Brandl and Gerard Hoehn. 2008. Computational prediction models for early detection of risk of cardiovascular events using mass spectrometry data. IEEE Transactions on Information Technology in Biomedicine, 12(5): 636–643.

[18] Joo, G., Song, Y., Im, H. and Park, J. 2020. Clinical implication of machine learning in predicting the occurrence of cardiovascular disease using big data. IEEE Access, 8: 157643–157653.

[19] Boudra, H., Obaid, A. and Amja, A.M. 2014. An intelligent medical monitoring system based on sensors and wireless sensor network. IEEE International Conference on Advances in Computing, Communications and Informatics, pp. 1650–1656.

[20] Tomasic, I., Khosraviani, K., Rosengren, P., Jörntén-Karlsson, M. and Lindén, M. 2018. Enabling IoT based monitoring of patients' environmental parameters: Experiences from using OpenMote with OpenWSN and Contiki-NG. 2018 41st International

Convention on Information and Communication Technology, Electronics and Microelectronics (MIPRO), Opatija, pp. 0330–0334.

[21] Catarinucci, L., Donno, D., Mainetti, L., Palano, L., Patrono, L., Stefanizzi, M.L. and Tarricone, L. 2015. An IoT-aware architecture for smart healthcare systems. IEEE Internet of Things Journal, 2(6): 515–526.

[22] Varatharajan, R., Manogaran, G., Priyan, M.K. and Sundarasekar, R. 2017. Wearable sensor devices for early detection of Alzheimer disease using dynamic time warping algorithm. Cluster Computing, pp. 1–10.

[23] Chanak, P. and Banerjee, I. 2020. Congestion free routing mechanism for IoT-enabled wireless sensor networks for smart healthcare applications. IEEE Transactions on Consumer Electronics, 66(3): 223–232.

[24] Li, P., Haq, A.U., Din, S.U., Khan, J., Khan, A. and Saboor, A. 2020. Heart disease identification method using machine learning classification in E-healthcare. IEEE Access, 8: 107562–107582.

[25] Kavitha, R. and Kannan, E. 2016. An efficient framework for heart disease classification using feature extraction and feature selection technique in data mining. 2016 International Conference on Emerging Trends in Engineering, Technology and Science, pp. 1–5.

[26] Jabbar, M.A., Chandra, P. and Deekshatulu, B.L. 2012. Prediction of risk score for heart disease using associative classification and hybrid feature subset selection. 2012 12th International Conference on Intelligent Systems Design and Applications, pp. 628–634.

[27] Haq, U., Li, J., Memon, M.H., HunainMemon, M., Khan, J. and Marium, S.M. 2019. Heart disease prediction system using model of machine learning and sequential backward selection algorithm for features selection. 2019 IEEE 5th International Conference for Convergence in Technology, Bombay, India, pp. 1–4.

[28] Mohan, S., Thirumalai, C. and Srivastava, G. 2019. Effective heart disease prediction using hybrid machine learning techniques. IEEE Access, 7: 81542–81554.

[29] Javeed, A., Zhou, S., Yongjian, L., Qasim, I., Noor, A. and Nour, R. 2019. An intelligent learning system based on random search algorithm and optimized random forest model for improved heart disease detection. IEEE Access, 7: 180235–180243.

[30] Jikuo Wang, Changchun Liu, Liping Li, Wang Li, Lianke Yao, Han Li and Huan Zhang. 2020. A stacking-based model for non-invasive detection of coronary heart disease. IEEE Access, 8: 37124–37133.

[31] Pasha, S.J. and Mohamed, E.S. 2020. Novel Feature Reduction (NFR) model with machine learning and data mining algorithms for effective disease risk prediction. IEEE Access, 8: 184087–184108.

[32] Alkhodari, M., Islayem, D.K., Alskafi, F.A. and Khandoker, A.H. 2020. Predicting Hypertensive patients with higher risk of developing vascular events using heart rate variability and machine learning. IEEE Access, 8: 192727–192739.

[33] Liaqat Ali, Awais Niamat, Javed Ali Khan, Noorbakhsh Amiri Golilarz and Xiong Xingzhon. 2019. An optimized stacked support vector machines based expert system for the effective prediction of heart failure. IEEE Access, 7: 54007–54014.

[34] Sarmah, S.S. 2020. An efficient IoT-based patient monitoring and heart disease prediction system using deep learning modified neural network. IEEE Access, 8: 135784–135797.

[35] Guo, C., Zhang, J., Liu, Y., Xie, Y., Han, Z. and Yu, J. 2020. Recursion Enhanced Random Forest with an Improved Linear Model (RERF-ILM) for heart disease detection on the internet of medical things platform. IEEE Access, 8: 59247–59256.

Chapter **12**

A Federated Learning based Alzheimer's Disease Prediction

S Suchitra,[1,*] *N Senthamarai,*[2] *M Jeyaselvi*[2]
and *RJ Poovaraghan*[3]

1. Introduction

In India, the overall prevalence rate for Alzheimer's according to the 2021 statistics was 0.84% for the population age 55 or older and 1.36% for the population age 65 or older. The cost of its treatment is extremely high. As neuroimaging in India costs more than $5,000 each time it is done, this is also the reason that there are not many people in India who could afford such an expensive detection measure. The likelihood that a patient will benefit from treatment is increased if Alzheimer's disease is discovered early. With a smaller number of healthcare professionals in India as compared to the population, it is extremely important to develop a model that can detect the early stages of Alzheimer's disease. There was a 17% increase in deaths due to Alzheimer's in 2020 during COVID. There are not enough Alzheimer's disease care professionals available in India who could properly handle the number of cases presented in the

---

[1] Data Science and Business Systems, SRM Institute of Science and Technology, Kattankulathur, Chennai, Tamil Nadu, India.
[2] Network Communication Systems, SRM Institute of Science and Technology, Kattankulathur, Chennai, Tamil Nadu, India.
[3] Information Technology, Jaya Engineering College, Thiruninravur, Chennai, Tamil Nadu, India.
Emails: senthamn@srmist.edu.in; jeyaselm@srmist.edu.in; poovaraghan1976@gmail.com
* Corresponding author: suchitras_2000@yahoo.com

country. This calls for a system to be deployed to cater to the needs of the patients. Computers, cell phones, mobile phones, and the internet are readily available in developing countries, just as they are in the rest of the world.

In the existing system, there is a need to address the following issues (i) Both Alzheimer's disease and ageing can cause brain mass to shrink. This makes it difficult to distinguish Alzheimer's patients from healthy elderly people solely based on MRI scans. (ii) The next major issue is the unavailability of sufficient data. (iii) Deep learning solutions have difficulties, such as high algorithm complexity, overfitting, and underfitting, which are a bit difficult to handle. (iv) Some Alzheimer's symptoms are like those of other diseases, making differentiation difficult.

The goal of the project was to use a federated learning approach to deliver precise results for early disease identification to accelerate the treatment of Alzheimer's disease. This process can cut down the time taken in the treatment and detection processes. Help from MRI images of a patient can also free up a lot of human resources, which can be used so that doctors in hospitals can concentrate on the treatment process rather than the detection. It will also ensure that this brain disorder does as little damage as possible to a patient and slow down its progression. This is possible with an accurate, fast, and reliable system for the detection process.

Early Alzheimer's disease diagnosis gives patients a greater opportunity to benefit from the disease's treatments and qualifies them for several clinical studies. The medications provide a greater number of benefits in the early stages of the onset of the disease. It allows for time to plan for medical and financial decisions related to the treatment of the Alzheimer's patient. It also opens doors to the right services and support to control the progressive condition. It can help rule out other treatable conditions associated with dementia like symptoms. Drug and non-drug treatments can help a patient only when the disease is diagnosed at an earlier stage. This calls for an effective, reliable, and fast system to detect the disease and start the necessary treatment as soon as possible. The multifaceted nature associated with the medical system as well as its activities leads to the fragmentation of medical data in most cases. However, a healthcare facility might only be able to retrieve diagnostic information of patients who fall under a specific patient group. These documents include highly private health information about specific people. Several regulations have been created to regulate the gathering and analysis of such data. Federated learning is a model that has lately gained popularity because it shows great potential for learning from private information that is dispersed. Instead of collecting data from multiple sources at once or relying on the traditional finding then replication technique, it enables training a shared, global model using a central server while retaining all the data in the community where they originated.

This proposed method suggests the use of deep learning techniques to develop a CNN architecture model to detect Alzheimer's disease at an early stage of onset using the image classification algorithms, which could be done efficiently with the help of feature extraction offered by the CNN along with input from the neuroimaging technique in the form of MRI by federated learning and it supports prediction without disclosing private user data and enables multiple users to develop a distributed, trustworthy, and accurate ML model, The dataset that has been used in the model has been obtained from the Alzheimer's dataset, which consists of 6400 MRI images that have been classified into four various categories. These categories represent the four progressive stages of Alzheimer's disease. The images in the dataset belong to people aged 45 to 85. The dataset is thoroughly inspected through several channels, and after inspection, it was found that it does not contain any repeated MRI images of the patients, and multiple images from the same patient are also not found in the dataset. The suggested technique is thoroughly examined using metrics like precision, recall, and accuracy on the Alzheimer's Dataset, which comprises 6400 MRI pictures. The proposed method was shown to have a higher precision when compared to modern methods. The contributions of our proposed deep learning model are as follows: (1) To provide a more reliable and accurate Alzheimer's disease detection system. (2) To improve the accuracy of all the previously proposed models. (3) To propose an efficient and powerful CNN architecture using different neural layers via federated learning.

Following is the remainder of the chapter: The literature research, which is shown in the second section, provides a quick summary of a few studies that have already been conducted on this topic using current models. It creates a detailed understanding of the strategies that various previous researchers have used. It also illustrates how the suggested model offers a better solution to the difficulty in diagnosing Alzheimer's disease. The third section highlights the difficulties encountered during the model's training as well as the proposed approach for the model that has been used to address the issue of Alzheimer's disease detection systems. The system for detecting Alzheimer's disease is concluded in the fourth section, along with plans for improvements.

2. Related Works

The early phases of Alzheimer's disease in patients have been extensively studied and examined. But there were certain drawbacks to the current design. Many of them created and used algorithms that had very low success rates for early disease detection. Some methods employed image processing methods like the watershed algorithm and image gradient, which could only be used to explain instances in which the brain was not already damaged when the detection was made. On the MRI image datasets, other researchers employed ML techniques like SVM, regression, and naive bayes to achieve an accuracy of about 90%. Some of them employed transfer learning techniques to achieve high levels of accuracy, while the patients with complete brain damage experienced difficulties. Utilising the CNN architecture was a different strategy that provided good accuracy but had the drawback of having a complex algorithm.

2.1 Biomarkers and Neuroimaging

Suhad et al. [1] suggested concentrating primarily on two elements, specifically neuroimaging and biomarkers. Image analysis was its main area of study. The work does not describe how AD was first found because most of the subjects selected for investigations were already known to have AD. During this procedure, significant AD datasets were reviewed, along with diagnostic methods and detection strategies. Although it was possible to use this method for early-stage neuroimaging research, the outcome was not sufficient. The significance of CFS in Alzheimer's disease identification was also demonstrated [12]. A few markers that can be used to diagnose Alzheimer's disease include the MSE, CDR, and NWB Volume. The process's DT algorithm produced results with a very poor accuracy of less than 80%, but the XG Boost algorithm could only manage a somewhat higher accuracy of 80%. The accuracy needed to surpass 81% was not met by SVM or RF classifiers. The methods used to diagnose dementia that progresses to Alzheimer's disease have been examined [5]. In recent work for the early diagnosis of AD utilising various methodologies of machine learning, IoT, artificial intelligence, etc., various analysis and assessment strategies are also examined.

A deep learning method using prior knowledge of the model dependency in the optimization was also put forth [9]. It did both DR and DF at the same time. Additionally, it demonstrates how more CSF may be extracted using a multi-layered learning model and hyper parameters when applied to the medical histories of patients with smaller sizes. Another unsupervised ML-related technique was put forth [14]. To predict the progression of Alzheimer's disease, he used structured MRI image data, and he showed that the outcomes are on par with supervised ML techniques.

2.2 Image Segmentation

Image segmentation was used in the implementation by Shrikant et al. [2]. The degree of enlargement in it assisted in placing the individual in one of the four classes this model defines as MD, MOD, VMD, and ND. In essence, they divided the patients into these four groups. To determine whether there are cavities in brain atrophy, a watershed algorithm is employed to identify brain atrophy in accordance with the gradient of an image. This process had a simple methodology and required little time to complete. The benefit of this approach was that it avoided brain damage while resolving the issue of prior detection. JRip, the Nave Bayes method, and RF were used [13] to classify the various input images of the data. The Weka software application was used to analyse the data. Then, using an explorer, knowledge flow, and APIs, the image data was classified. It was determined after comparing the results that the JRip and RF algorithms are superior to the Naive Bayes algorithm.

2.3 Feature Extraction

In MRI image data from OASIS, [3] employed three distinct views of slices of grey matter and white matter. Each slice contained the first-order feature that was extracted. Then, using the features that were retrieved, a correlation heat map is generated. The dimensionality was subsequently decreased using the principal component analysis method. The performance and accuracy of his suggested technique were then examined using a variety of machine learning algorithms, including regression (logistic), SVM, AaaBoost, and Naive Bayes. The testing outcomes showed a maximum accuracy of 90.9%, which was better than the findings of previous ML algorithms. The mathematical statistics that are to be retrieved from the electroencephalography signal were used [6] in several ways and with various features, including mean, median, mode, range, standard deviation, etc. Measurements revealed several abnormalities. Utilising the proper feature extraction tool, various characteristics were extracted.

2.4 Transfer Learning Method

The pre-trained VGG16 model employed [4] is a component of the TensorFlow and Keras learning transfer networks. It served as a tool for him to draw out features from it. Using a performance metric, they were able to determine the model's accuracy. The algorithm that split the data into 20% testing data and 80% training data had the highest accuracy, with a testing accuracy of 95.31% and a training accuracy of 97.49%.

2.5 Neural Networks and Deep Learning Techniques

By utilising convolutional neural networks with DL architecture, [7] were able to identify Alzheimer's disease with 96.86% accuracy (LeNet). Concurrently, testing and training the data were done using an enormous number of photos. The degree of accuracy they were able to accomplish was really great, which concluded that the network design chosen to detect Alzheimer's disease was appropriate.

A method for AD identification based on neuroimaging employing DL techniques was proposed [8]. After a CNN, he employed an RNN. He was able to better understand the connections between the visual sequences for each class as a result. Based on all the input images, a final decision was then made. He lacked sufficient data to train a NN. It was shown by Chenhui et al. [10] that DL holds several beliefs about understanding the connection between MRI image data and physiological significance. Brain area correlation input provides a higher degree of representation to the DL technique auto-encoder. With significantly better prediction accuracy, it was attained. The significance of diagnostic detection in MRI images was demonstrated [11]. Despite this, the precision was not adequate. Different brain scan neuro images require statistically comprehensive datasets as input [15], demonstrating the growth of deep learning (DL) modelling techniques. Both positron emission tomography (PET) and magnetic resonance imaging (MRI) are effective visual modalities that can demonstrate changes in the brain that occur as Alzheimer's disease (AD) advances. Using structural MRIs to create 3D deep convolutional neural networks [16], were able to successfully discriminate mild Alzheimer's disease dementia from mild cognitive impairment and cognitively normal persons. The deep learning model was able to distinguish between subjects with MCI or mild Alzheimer's disease and those with cognitively normal subjects with an area-under-the-curve (AUC) of 85.12%.

2.6 Federated Learning

The creation of federations of learning information and resource repositories was the focus of several initiatives that came after [24–26]. Rehak et al.'s [26] 2005 reference model describes how to federate repositories to construct an interoperable repository infrastructure. The contributing repositories' metadata is gathered into a central registry under this architecture, providing a single point of access and discovery. Sensitive clinical data can persist inside local organisations or with specific consumers without leaving them throughout the federated learning process for both provider-based software, such as developing a model to estimate the hospital return risk using digital patient medical records [27], and patient (consumer)-based software, such as detecting the presence of atrial fibrillation with electrocardiograms collected by smartwatches [28]. With this strategy, patient privacy is well protected.

All the research that has been mentioned so far used traditional ML techniques, in which data storage and model training are managed by a single computer or server. Because of this, all the data must be gathered in one location before training. Conventional machine learning algorithms [17], demand for a model to be trained from many samples for training, which may be challenging to get due to privacy concerns. Before the obtained weights are sent to the main server for aggregation, each model is trained independently on local hardware in the FL environment. As a result, only model attributes like weights, gradients, parameters, etc. are transmitted to the central server. The hurdles that must be solved were highlighted [21] as they demonstrated how FL tackles the issue of collecting medical datasets. The models were consequently aggregated in accordance with the total amount of time used to train the data, and the given strategy was to choose the top models constructed on their performance. In their investigation, they used a little improved FL model to identify COVID-19 more accurately from medical imaging data.

3. Proposed Method

The model uses deep learning methods via federated learning to analyse and predict the result. The model makes use of machine learning methods like random forest (RF) and KNN, as well as deep learning algorithms like CNN architecture, and compares our suggested model to current models and classifiers. The training dataset and the testing/validation dataset are the two sets of images that the model will first use as input, process through various processes, and then split into two sets. After finishing this procedure, we will define our own pure CNN architecture, which will include various layer types, including a 2D convolution layer, a dropout layer, an activation layer, a flattened layer, a rectified unit layer, and a softmax layer.

The FL approach was applied in this proposed method to identify Alzheimer's disease while maintaining data privacy. First, a deep learning model that can precisely identify Alzheimer's disease was constructed using CNN architecture and transfer learning. The CNN model was then used to generate an FL framework for Alzheimer's disease detection and classification. The suggested model's contribution is to show that outcomes are not considerably worsened by applying sophisticated algorithms in a federated framework. Consequently, utilising the FL technique, we can easily create reliable classifiers in the therapeutic area even if there are privacy concerns with data acquisition.

3.1 Overview

Figure 1 shows the proposed model's general layout. The framework is made up of several single blocks that are connected in series and parallel arrangements; some of the blocks just have a single pointed forward arrow, while others have both forward and backward arrows. The MRI images are first input, and after passing through several filters, they are prepared to be processed by the model. The image data has now been divided into two new datasets: training and validation/testing, in the ratio of 80:20, respectively. The newly built neural network, which comprises various neural layers with various functions and filters, is now used to process the data. After the data has been trained, it is time to evaluate our model, which may be done by obtaining the accuracy and loss functions. Finally, it's necessary to visualise the data and contrast it with other models that have been suggested in the past and with machine learning techniques already in use.

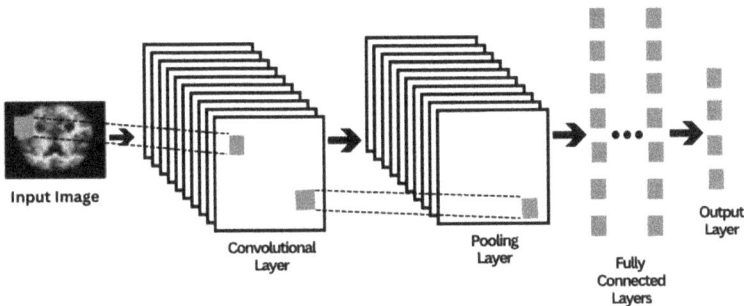

Figure 1. Overview of the proposed model—Alzheimer's disease prediction using deep learning techniques.

3.2 Dataset

The Alzheimer's disease dataset consists of 6,400 MRI images that are divided into four groups that match the four types of Alzheimer's disease that are present in the patients. The individuals in the MRI pictures range in age from 45 to 85. The pictures from each class are as follows:

- Mild Demented – This class consists of 896 MRI images of different patients.
- Moderate Demented – This class only consists of 64 MRI images of different patients.
- Non-Demented – This class has 3200 MRI images of different patients.
- Very Mild Demented – This class consists of 2240 MRI images of different patients.

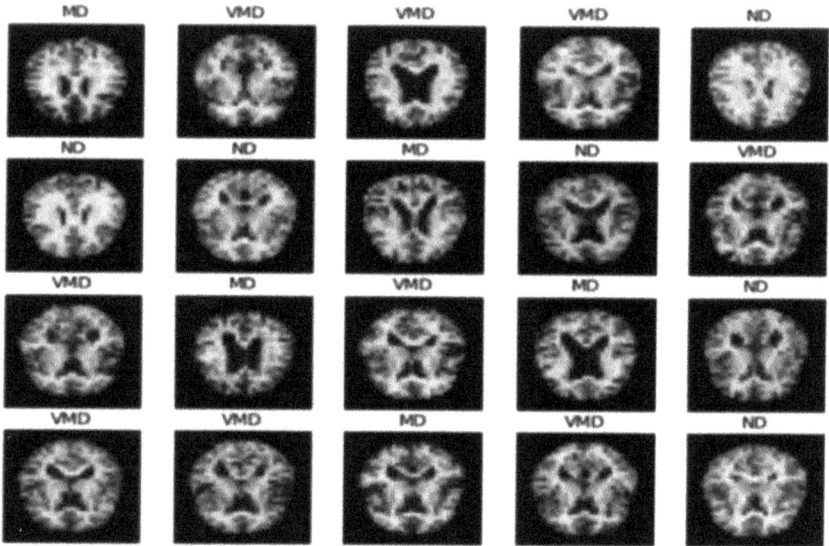

Figure 2. Different variations of MRI images of Alzheimer's disease dataset.

The dataset was thoroughly examined using several different channels, and it was discovered that neither multiple images from the same patient nor repeated MRI scans of the patients were present in the dataset. The data is divided into two categories: 80% and 20%. 80% of the images can be used to train the model. The model will be tested and validated using the final 20% of the images. Figure 2 displays the dataset's random representations of several MRI Alzheimer's disease image variations.

3.3 Pre-processing

The Geometric Transformation Technique is an algorithm that is employed for the pre-processing of the images. Using this method, the input images are scaled and resized to meet the specifications. The images have a resolution of 176 by 208 pixels. The model needs to be 32 by 32 in size. Thus, all the input images are subjected to this dimensionality reduction function.

3.4 Building Convolutional Neural Network

The convolutional neural network belongs to the deep learning subset of artificial neural networks. It consists of several neural layers, including the convolution layer, the pooling layer, the dropout layer, the denser layer, the activation layer, the flattened layer, the softmax layer, the rectified unit layer, etc. The network that is most frequently employed for image classification and recognition issues is this one. Shift-invariant, space-invariant artificial neural networks (SI-ANN), or ConvNet are other names for it. The Multi-Layer Perceptron (MPL), which is a fully connected network in which every single neuron from the previous layer is connected to every single neuron from the next layer, is an older version of the Multi-Layer Perceptron (CNN) architecture. This fully connected layer makes data more susceptible to overfitting. To avoid overfitting, CNN, an updated version of the neural network, incorporates a variety of regularisation approaches.

This model is made up of various layers of neural networks that all fit into a sequential model. Multiple inputs and outputs are not permitted in a sequential model, which means that each layer of the model only has one output and one input at a time. The layers of the CNN architecture consist of several hyper parameters, and it is necessary to know the meaning behind these filters. A filter of size *FxF* is applied to *C* channels present in the input image, so the *FxFxC* is the volume that will perform the convolutional of the input size of *IxIxC* and produce the output feature map of *OxOxK*. Here, *K* is the size of the filter.

Kernel convolution is a technique in which the small matrix of size 3 x 3 containing the numbers is passed through the input image and change it according to the small matrix which is also known as filters as well as kernels. The values of all the feature maps are now calculated in a sequence manner. It denotes the input image by *f* kernel by *h* and the indexes of rows and columns of the result matrix are marked with '*m*' and '*n*' respectively. The output feature map will be denoted as *G*[*m*, *n*], where *G*[*m*, *n*] can be denoted as given below.

$$G[m, n] = (f * h)[m, n] = \sum_j \sum_k h[j, k] f[m - j, n - k] \qquad (1)$$

The process is similar to matrix multiplication in which each value from the small matrix, kernel is multiplied in pairs with the subsequent value of the input image after they are placed over the pixels of the image. After the addition of all the result is now saved in an accessible memory location and the output is in the form of a feature map. Max-Pooling is a technique in which the maximum integer value from each of the divided multiple regions in the form of a matrix is selected and it is sent to the next layer for further processing.

This model used the Adam optimizer. The optimization done is based on adaptive estimation of first-order and second-order moments which is known as the stochastic gradient descent method. The second argument that it takes a loss function that is set to categorical cross-entropy which is a loss function that is used in the image classification model which has multi-class classification. The last argument that it takes is the metrics that are set to accuracy. The model takes several arguments such as train_x which is the input data that has been pre-processed the next argument is the train_y which is the output data, the format is the same as the input data, the batch size is set to 32, epoch or the number of iterations is set to 350, verbose is set to 1 for displaying the logs of each iteration.

3.5 Federated Learning

A recently developed machine learning technology called federated learning (FL) aims to solve the issue of stringent data privacy laws and the limitation of readily accessible data sets [18–20]. There are two ends in federated learning: a global server or central server as well as a local server or client-end. In contrast to the central server, which houses a global model, the client-server host's local data. The model is trained on a range of decentralised end devices with data maintained locally on the servers thanks to FL's ability to train models without transferring information. The revised weights are set by the neighbourhood model, which was trained on the client dataset. As a result, it maintains the anonymity, which is essential for the precise diagnosis of clinical information [21–23]. No previously trained CNN model was used in this scenario since FL aims to prevent the global model from being coupled to any training data. The scenario entails sharing system settings and training decentralised clients or nodes using local data. Using a server, a global model is created by the aggregation of system or model parameters. Using FL does not need the user to share any confidential information. Instead, users access the network locally and share the model with a centralised entity called the server. Each user must, however, be able to send their gained model through the server in FL practically and effectively. An iterative technique is used by the server in FL to combine the local data into a global network. Figure 3 shows the architecture of the suggested federated learning model.

3.6 Performance Evaluation

The proposed method obtained very good results, with an accuracy of 98.98%. The proposed method's confusion matrix is shown in Fig. 4. The graphs were plotted for the comparison of accuracy and loss function between the testing and validation data. The plotting of the graph for the model accuracy and model loss can be seen in Figs. 5 and 6, respectively.

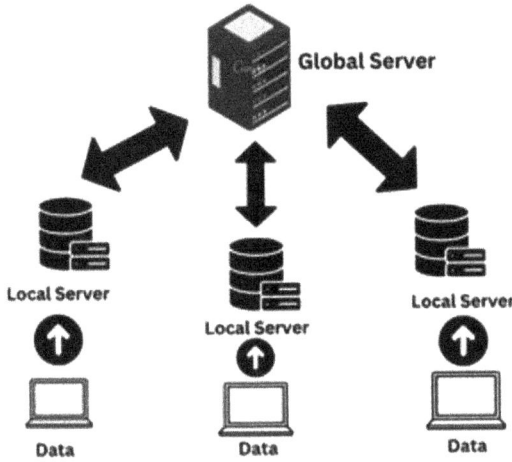

Figure 3. Federated Learning model (proposed method).

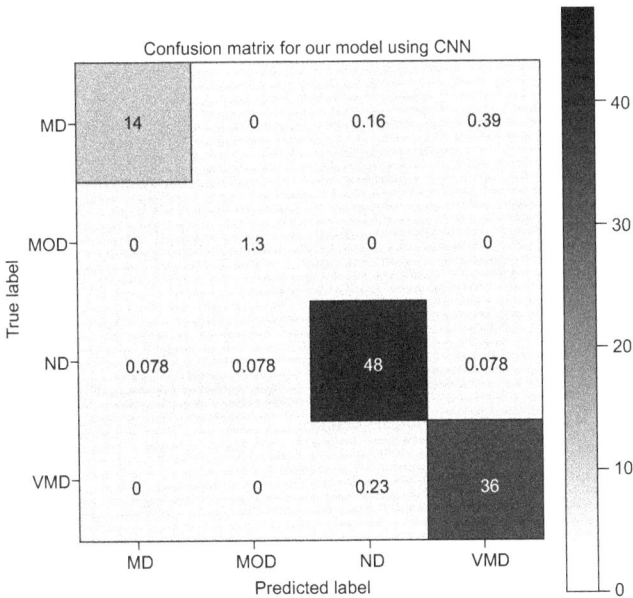

Figure 4. Confusion matrix for CNN model (proposed method).

3.7 Challenges Faced During the Training of the Model

3.7.1 Overfitting

The image data in the model is divided into two sets: one is the training data, and the other is the validation and testing data. Overfitting of the data

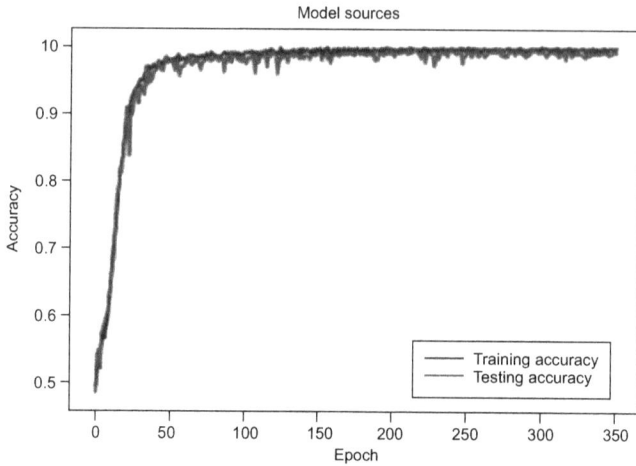

Figure 5. Model accuracy (proposed method).

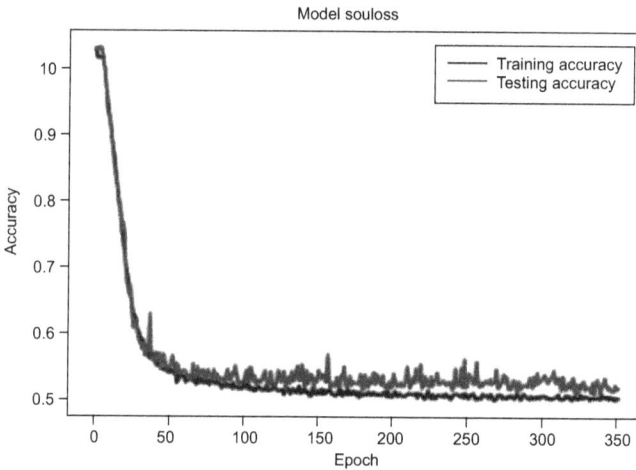

Figure 6. Model loss (proposed method).

in the CNN model means that the data training is eagerly attempting to make the data fit too well. This will result in very good and high accuracy for the training dataset, while the accuracy for the testing and validation dataset will come out to be very bad and low. In a single sentence, over-fitting means high variance and low bias. It can be removed by increasing the number of data points in the dataset as well as reducing the model complexity, or by using the dropout layer in the neural network. As illustrated in Fig. 5 the y-axis denotes the accuracy of the model, the x-axis denotes the number of epochs used to train the model, and 'x' represents the training data while 'o' represents the testing/validation data. And as a result, the model is overfitting the data. Overcome this by adding the dropout layer and increasing the number of data points.

3.7.2 *Underfitting*

The data in the model is divided into two sets: training data and validation and testing data. Underfitting of the data in the CNN model means that the model is not able to fit the data very efficiently. This will result in very low accuracy for both the training dataset and the testing and validation dataset. This may occur due to the smaller number of data points in the dataset or because the size of the dataset is too small to train the model. In a single sentence, over-fitting means low variance and high bias. It can be removed by increasing the model complexity as well as removing the noise from the dataset, or by increasing the time used for training the model as well as increasing the number of epochs. As shown in Fig. 5 the y-axis denotes the accuracy of the model, the x-axis denotes the number of epochs used to train the model, and 'x' represents the training data while 'o' represents the testing and validation data. And as a result, it can see the model is underfitting the data. Overcome this by removing the noise from the data and increasing the number of epochs.

3.8 **Comparison with Existing Models**

3.8.1 *K-Nearest Neighbor Classifier*

The KNN classifier is part of machine learning, which falls under the category of supervised machine learning. The KNN classifiers classify the new input image based on its similarity to the nearest neighbouring data point in a particularly specified class or group. This algorithm is widely used due to its low complexity. It is one of the simplest algorithms to use in machine learning for image classification. The accuracy of the KNN classifier was found to be 95.31%, which is less than the accuracy of the proposed CNN classifier. The accuracy of the KNN classifier is given below in Fig. 7.

3.8.2 *Random Forest Classifier*

Random forest is another machine learning algorithm; this also falls under the category of supervised machine learning. This algorithm is mainly used for classification and regression problems. In this case, the algorithm develops multiple decision trees from a different set of input parameters and takes the vote of the majority of the decision tree results for the classification of the data. This algorithm takes less time to run in

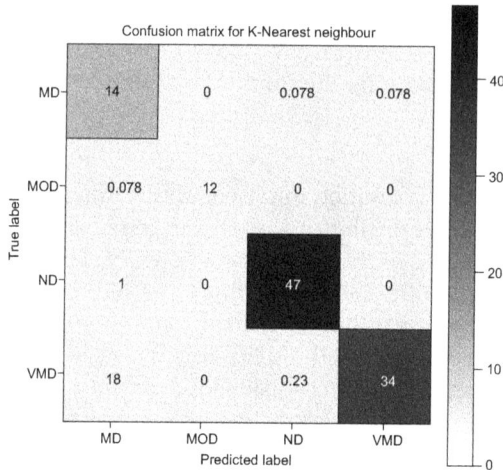

Figure 7. Confusion matrix for KNN classifier (existing method).

comparison with the other algorithms. It is one of the fastest algorithms to use in machine learning for image classification. Although the accuracy of the random forest classifier was found to be very low during the image classification for this problem of Alzheimer's disease detection when compared to the other two algorithms, namely the proposed method of the CNN classifier and the KNN classifier, The classification report and the confusion matrix for the random forest classifier are given in Fig. 8. The

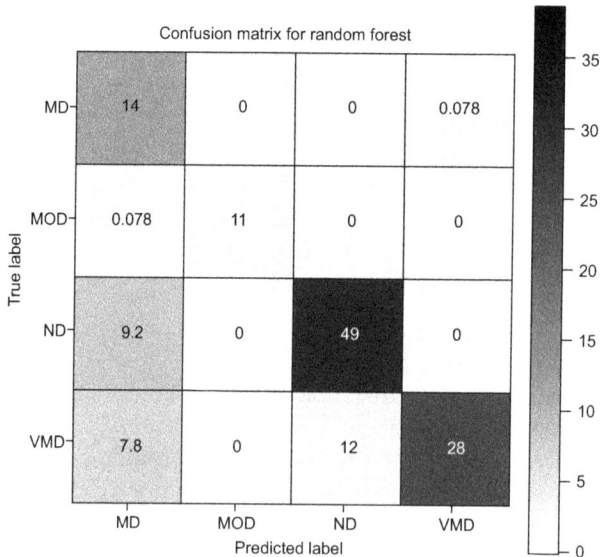

Figure 8. Confusion matrix for random forest classifier (existing method).

accuracy of the random forest classifier was discovered to be only 75.28%, which is significantly lower than the accuracy of the CNN classifier's recommended method.

Table 1 shows a comparison analysis of the proposed method with other existing methods. The proposed method obtained very good results, with an accuracy of 98.98%. The proposed CNN classifier was more accurate than the KNN classifier, which had an accuracy of 95.31%. The random forest classifier's accuracy was found to be 75.28%, which is significantly lower than the accuracy of the CNN classifier's suggested techniques. Figure 9 shows the Comparison of the proposed method's average performance metrics.

Table 1. Comparison of the proposed method's average performance metrics.

S. No.	Method name	Accuracy (%)
1.	Random Forest Classifier	75.28%
2.	KNN Classifier	95.31%
3.	CNN Classifier (proposed method)	98.98%

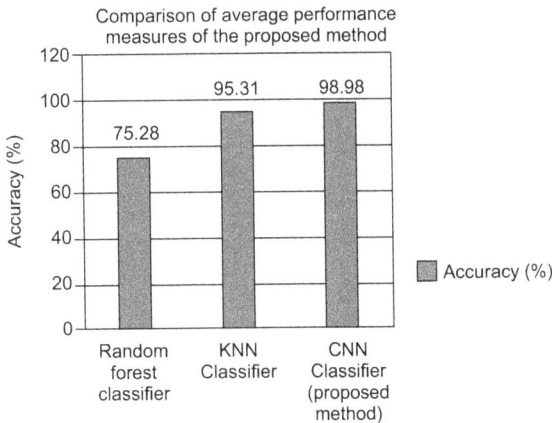

Figure 9. Comparison of the proposed method's average performance metrics.

4. Conclusion

The main objective of the suggested method is the development of a CNN architecture employing deep learning techniques, with input from MRI-based neuroimaging, to identify Alzheimer's disease at an early stage. The proposed model, which is based on the CNN architecture, compares the performance of CNN to that of other machine learning algorithms, including random forest and KNN classifiers. The dataset, which comprises 6400 MRI images, is used in the model created from the Alzheimer's dataset. With the Alzheimer's dataset, the suggested technique is thoroughly examined using metrics for precision, recall, and accuracy. It was found that the suggested method achieved greater precision when compared to contemporary methods. The best accuracy of the model after several compilations can be found to be 98.98% which is higher than all the previous classifiers such as KNN and random forest. When the proposed method, the CNN model, is compared with the KNN and random forest methods, the average accuracy is 98.98%, 75.28%, and 98.98% respectively. Future work can be extended to the growth of mobile technology, which tends to be both commercial and timesaving, with a focus on the medical community, which would greatly benefit from quick identification, the distinction of proportions of disease progression, and multi-class classification.

References

[1] Al-Shoukry, S., Rassem, T.H. and Makbol, N.M. 2020. Alzheimer's diseases detection by using deep learning algorithms. IEEE Access, 8: 77131–77141.

[2] Patro, S. and Nisha, V.M. 2019. Early detection of Alzheimer's disease using image processing. International Journal of Engineering Research & Technology (IJERT), ISSN: 2278-0181, 8(05).

[3] Aruchamy, S., Haridasan, A., Verma, A., Bhattacharjee, P., Nandy, S.N. and Vadali, S.R.K. 2020. Alzheimer's disease detection using machine learning techniques in 3D MR images. IEEE 2020 National Conference on Emerging Trends on Sustainable Technology and Engineering Applications-Durgapur, India.

[4] Ebrahim, D., Ali-Eldin, A.M.T., Moustafa, H.E. and Arafat, H. 2020. Alzheimer disease early detection using convolutional neural networks. 2020 15th International Conference on Computer Engineering and Systems (ICCES).

[5] Thankaraj, S., Khilar, R. and Sahoo, S.K. 2020. An early prediction and detection of Alzheimer's disease: A comparative analysis on various assistive technologies. IEEE International Conference on Computational Intelligence for Smart Power System and Sustainable Energy (CISPSSE-2020), July 29–31, 2020, Odisha, India.

[6] Thakare, P. and Pawar, V.R. 2016. Alzheimer disease detection and tracking of Alzheimer patient. IEEE 2016 International Conference on Inventive Computation Technologies (ICICT) - Coimbatore, India.

[7] Sarraf, S. and Tofighi, G. 2016. Deep learning-based pipeline to recognize Alzheimer's disease using fMRI data. IEEE 2016 Future Technologies Conference (FTC) - San Francisco, CA, USA.

[8] Ebrahimi-Ghahnavieh, A., Luo, S. and Chiong, R. 2019. Transfer learning for Alzheimer's disease detection on MRI images. The 2019 IEEE International Conference on Industry 4.0, Artificial Intelligence, and Communications Technology (IAICT).

[9] Liu, S. and Liu, S. 2014. Student Member, IEEE, Weidong Cai, Member, IEEE, Sonia Pujol, Ron Kikinis, Dagan Feng, Fellow, IEEE. Early Diagnosis of Alzheimer's

Disease with Deep Learning. 11th International Symposium on Biomedical Imaging (ISBI).

[10] Hu, C., Ju, R., Shen, Y., Zhou, P. and Li, Q. 2016. Clinical decision support for Alzheimer's disease based on deep learning and brain network. IEEE ICC 2016 - 2016 IEEE International Conference on Communications - Kuala Lumpur, Malaysia 2016 IEEE International Conference on Communications (ICC).

[11] Kadhim, K.A., Mohamed, F., Khudhair, Z.N. and Alkawaz, M.H. 2020. Classification and predictive diagnosis earlier alzheimer's disease using MRI Brain Images. 2020 IEEE Conference on Big Data and Analytics (ICBDA).

[12] Shah, A., Lalakiya, D., Desai, S., Shreya and Patel, V. 2020. Early detection of Alzheimer's disease using various machine learning techniques: A comparative study. Fourth International Conference on Trends in Electronics and Informatics.

[13] Sheshadri, H.S., Bhagyashree, S.R. and Krishna, M. 2015. Diagnosis of Alzheimer's disease employing Neuropsychological and classification techniques. 5th International Conference on IT Convergence and Security.

[14] Fedorov, A., Hjelm, R.D., Abrol, A., Fu, Z., Du, Y., Plis, S. and Calhoun, V.D. 2019. Prediction of Progression to Alzheimer's disease with Deep InfoMax. EMBS International Conference on Biomedical and Health Informatics.

[15] Nguyen, H. and Chu, N.N. 2021. An introduction to deep learning research for Alzheimer's disease. IEEE Consumer Electronics Magazine.

[16] Liu, S., Masurkar, A.V., Rusinek, H., Chen, J., Zhang, B., Zhu, W., FernandezGranda, C. and Razavian, N. 2022. Generalizable deep learning model for early Alzheimer's disease detection from structural MRIs. Scientifc Reports.

[17] Islam, J. and Zhang, Y. 2018. Deep convolutional neural networks for automated diagnosis of Alzheimer's disease and mild cognitive impairment using 3D brain MRI. Brain Informatics, BI 2018, Lecture Notes in Computer Science), vol 11309. Springer, Cham. https://doi.org/10.1007/978-3-030-05587-5_34.

[18] Aledhari, M., Razzak, R., Parizi, R.M. and Saeed, F. 2020. Federated Learning: A survey on enabling technologies, protocols, and applications. IEEE Access, 8: 140699–140725.

[19] Zhang, W., Lu, Q., Yu, Q., Li, Z., Liu, Y., Lo, S.K., Chen, S., Xu, X. and Zhu, L. 2020. Blockchain-based federated learning for device failure detection in industrial IoT. IEEE Internet Things J, 8(7): 5926–5937.

[20] Sarma, K.V., Harmon, S., Sanford, T., Roth, H.R., Xu, Z., Tetreault, J., Xu, D., Flores, M.G., Raman, A.G. and Kulkarni, R. 2021. Federated Learning improves site performance in multicenter deep learning without data sharing. J. Am. Med. Inform. Assoc., 28(6): 1259–1264.

[21] Zhang, W., Zhou, T., Lu, Q., Wang, X., Zhu, C., Wang, Z. and Wang, F. 2020. Dynamic fusion based federated learning for covid-19 detection. arXiv:2009.10401.

[22] Aich, S., Sinai, N.K., Kumar, S., Ali, M., Choi, Y.R., Joo, M.I. and Kim, H.C. 2021. Protecting personal healthcare record using blockchain & federated learning technologies. IEEE, pp. 109–112.

[23] Stripelis, D., Ambite, J.L., Lam, P. and Thompson, P. 2021. Scaling neuroscience research using federated learning. IEEE, pp. 1191–1195.

[24] Barcelos, C., Gluz, J. and Vicari, R. 2011. An agent-based Federated Learning object search service. Interdisciplinary Journal of E-Learning and Learning Objects, 7(1): 37–54.

[25] Mukherjee, R. and Jaffe, H. 2005. System and method for dynamic context-sensitive federated search of multiple information repositories. US Patent App. 10/743,196.

[26] Rehak, D., Dodds, P. and Lannom, L. 2005. A model and infrastructure for Federated Learning content repositories. *In*: Interoperability of Web-based Educational Systems Workshop, vol 143. Citeseer.

[27] Min, X., Yu, B. and Wang, F. 2019. Predictive modeling of the hospital readmission risk from patients' claims data using machine learning: A case study on copd. Sci. Rep., 9(1): 2362.

[28] Perez, M.V., Mahaffey, K.W., Hedlin, H., Rumsfeld, J.S., Garcia, A., Ferris, T., Balasubramanian, V., Russo, A.M., Rajmane, A. and Cheung, L. 2019. Large-scale assessment of a smartwatch to identify atrial fibrillation. N. Engl. J. Med., 381(20): 1909–1917.

Chapter 13

Detecting Device Sensors of Luxury Hotels using Blockchain-based Federated Learning to Increase Customer Satisfaction

Moyeenudin HM,[1,]* Shaik Javed Parvez,[2]
Jose Anand A,[3] Anandan R[4] and Sam Goundar[5]

1. Introduction

A kind of Artificial Intelligence branch known as Federated Learning (FL) enables to train of several models (clients) and combines their learnings to create a single final model (server). Each client model is updated with

---

[1] School of Hotel & Catering Management, Department of Hotel & Catering Management, Vels Institute of Science, Technology and Advanced Studies, (VISTAS) Deemed to be University, Pallavaram, Chennai, India.
[2] Department of Information Technology, Hindustan Institute of Technology & Science, Chennai, India.
[3] Department of Electronics and Communication Engineering, KCG College of Technology, Chennai, India.
[4] Department of Computer Science Engineering, Vels Institute of Science, Technology and Advanced Studies, (VISTAS) Deemed to be University Pallavaram, Chennai, India.
[5] Department of Information Technology, RMIT University, Ha Noi, Vietnam.
* Corresponding author: moyeenudin@gmail.com

the server model from the previous era before the start of the next one [1]. Dividing the Data segments, making machine learning models, using blockchain for security and privacy method, building communication architecture of data, and scale of the federation are the five categories of FL Partitioning of data [2]. Where the Aggregator Node (AN) of the Blockchain with federated evaluation model in connection with Sensor Node (SN) IDs are authenticated and stored in the blockchain [3]. Thus, these ANs and SNs are already authenticated on private and public blockchains, respectively. However, hostile actions are carried out by untrusted nodes when someone uses network resources. The SNs are also vulnerable to attacks from malicious nodes and have low energy, transmission range, and computing capabilities. The malicious nodes then communicate inaccurate route information and increase the number of retransmissions, which causes the SNs' energy to be quickly used up. The quick energy dissipation of the SNs shortens the lifespan of the wireless sensor network. Additionally, when malicious nodes are present in the network, throughput, and packet loss both rise. The SNs' trust values are calculated to eliminate the network's harmful nodes. To accomplish secure routing, the network considers the SNs' trust values and remaining energy. Additionally, an asymmetric key cryptosystem is employed to secure data transfer. The simulation results demonstrate the suggested model's usefulness in terms of a high packet delivery ratio. Aggregation typically uses FL. An example that is simple to grasp is a dataset that only includes breast cancer patients from a specific hospital. When clients are exposed to various feature spaces but the same or comparable sample spaces, vertical data partitioning enters the picture [4]. Using entity alignment approaches, overlapping samples are discovered in the client data and this overlapping data is then used for training. A dataset of students' GPAs obtained from institutions in several nations might be a suitable illustration. The grading scale and assessment metric are different aspects of the feature space. Horizontal and vertical data partitioning are used to create hybrid data partitioning. An example that would be simple to understand would be a collection of universities wishing to construct an FL System and evaluate student performance in all branches.

The AI Models from the dataset and issue statement are typically taken into consideration while selecting an ML model. Neural networks (NN) are among the most often used models. In addition to NNs, decision trees are used because they are highly effective and simple to understand. Homogeneous or heterogeneous models may make up an FL system. In the first scenario, all clients use the same model, and the server uses gradient aggregation. However, since each client has a unique model, there is no concern about aggregating in the latter situation. In the case of heterogeneous models, ensemble methods like max voting are used on the server in place of aggregation techniques. The datasets of many clients in horizontal data partitioning share the same properties, yet there is limited sample space intersection. FL's approach to privacy is the part that has drawn the most criticism. The main goal is to prevent client information from leaking [5].

The server can use learning gradients to decrypt client data in the absence of encryption. Consequently, it's crucial to hide the gradients. In a FL system, typically cryptography techniques are used to resolve privacy issues and differential privacy. To disguise the gradients of data or model parameters a differential privacy method is used as a technique along with the introduction of random noise. The model accuracy is considerably reduced through this method because more noise is a big disadvantage. Common privacy techniques used in FL systems include the transformation of one data set into another using homomorphic encryption and this method assists in computation by secured multi-party method. The process is applied rather to genuine clients with encrypted data to the server, and the server obtains the outcome, the encrypted input and output are finally decrypted. Although these techniques offer protection against a wide range of threats, they are computationally expensive. There are two types of FL systems centralized and decentralized. Both types of architecture continue to work similarly; the distinction is in client-server communication [6]. If the architecture is centralized, we have a different model that acts as a server, and all parameter modifications are carried out in this global model. In centralized FL systems, learning can be both synchronous and asynchronous, as we will examine in the following article. There is a difference between decentralized design and the design of the client's alternate function as servers. Every epoch, a client is chosen at random to update the global model and convey to other clients the global model. Decentralized FL systems are challenging to develop and generally fall into one of three categories: P2P, graph, or blockchain. This chapter will not address any of these topics; they will be covered in subsequent ones. Size of the federation Cross-silo and cross-device federation scales are the two main types. It is simplest to understand these distinctions by relating cross-silo to enterprises and cross-device to mobiles. Cross-silo often has a low client count but a high computational capacity. The number of customers is huge, and their computational power is modest when it comes to cross-device. We can rely on organizations (cross-silo) to always be accessible to train, but it is not the same thing as reliability. concerning mobile phones (cross-devices). A bad network may make the gadget less accessible [7].

2. Literature Review

2.1 Federated Learning with IoT

Federated learning is a new branch of AI that comes under the machine learning paradigm that minimizes bias in training the models and protects privacy. Multiple clients (such as businesses, data centres, internet of Things devices, or mobile devices) are chosen for each cycle of federated learning to train the data into models internally to give a global model as the raw data are stored in client's databases and not traded or transported [8]. Recently, IoT federated learning has used blockchain to ensure data truthfulness to draw in enough client data and computing resources for training. However, systematic and comprehensive architecture designs are lacking, making it difficult to enable methodical development and effective approaches to the HIoT's device failure detection challenge of data heterogeneity. In the proposed system, raw data is locally stored. Thus, a central server is communicating with client servers to process the global model for knowing the faulty devices in the hotel. Blockchain technology is used in the design to provide the demonstrable truthfulness of guest data and financial benefits for guest contribution in completing the survey. To lessen the effects of data similarity and monotonous responses from guests. Although this chapter has the major contribution towards the federated learning modules which are hosted to coordinate with client servers, to modify the design by shifting all with our future development, the modules from guest servers to hotel guest gadgets with more potent computational and storage capacities. Additionally, we intend to investigate in more detail how to make client devices more reliable for the aggregation process [9].

2.2 Hotel Room Reservation by Online Travel Agencies (OTA)

An application may suggest a different route to get to the destination if there is a significant amount of traffic on a certain road. Similarly, we frequently read numerous customer evaluations when looking for a hotel to book for the upcoming family vacation. Websites like TripAdvisor or Booking.com, provide almost everything, including the cleanliness of the location, the staff's customer service skills, and much more. Hotels have varied data of guests who visited the property and the data of those who are yet to visit, without sending confidential user data to servers, Federated Learning is a machine learning framework that enables researchers to train statistical models using user data. It is a distributed training technique in which training and testing are carried out locally at the edge, and only the meta-data is transmitted to a centralized server, which integrates numerous model updates (from various clients) into a new model to sort out the issues like including car accidents, traffic jams, police speed traps, cars parked on the side of the road, etc. As a result, other users can benefit from this cooperation and make wiser driving choices. As an easy illustration [10].

2.3 *Horizontal Data Partitioning*

Most FL architectures use horizontal partitioning. Aggregation at the server is made simpler by the ability to utilize a similar model for all clients. The concept is to outsource a certain process using a highly dispersed platform. This method is divided into discrete, easy activities that novices may do in a short period. These tasks could include. A significant variety of diverse clients, including mobile phones and Internet of Things devices, are used in an increasing number of application fields. The majority of the current federated learning algorithms presumptively share common global model architecture with local models. A new federated learning framework was recently created to satisfy the needs of heterogeneous clients with a variety of computing and communication capabilities. While still providing a single precise global inference model, the HeteroFL approach can enable the training of heterogeneous local models with dynamically variable computation and identically distributed data intricacy. Last but not least, utilizing crowd-sourcing platforms like Amazon Mechanical Turk to speed up machine learning development is rather prevalent [11].

The IoT Devices of hotels play a crucial role in keeping the hotel guest's satisfaction further they also coordinate with various applications of the hotel, which in turn increases the hotel room occupancy through the feedback using digital gadgets and through OTA Online Travel Agents (OTA). These are the online websites that act as a source for hotel occupancy, where the customers can leave their feedback on their visit and this assist in opting for the hotel based on requirements having a faulty mechanism in a hotel may reduce the room occupancy [12].

2.4 *Distributed Systems*

In distributed systems, pattern analysis techniques typically have a broader range of applications. Finding patterns in a database that are structurally similar is the main goal of pattern analysis. Retail businesses employ consumer purchasing patterns to update stock levels as a well-designed crucial decision mechanism to obtain an advantage over rivals. Businesses employ modelling techniques to understand client behaviour.

For instance, the client's purchasing trends for certain services or goods, preference for prices and timing, and the effects of a decision taken after analysis. This analysis aids retail businesses in providing better customer service, which boosts their revenues. Retail organizations and enterprises needed cutting-edge technologies and to constantly be looking for data to maintain the market's dynamic nature. This leads to the following query: Can data be combined in a common location by being transported around the business using several databases? Maintaining a database on the number of sales points is extremely difficult, if not impossible. Mobile wireless Point of Sales devices is also frequently utilized in stores where real-time data is used to complete sales transactions. The exchange of data sources is often hampered by obstacles. Multiple sorts of data are needed by pattern analysis techniques. For instance, in a product suggestion system, the businesses or owners have the records (product information and data about the user's purchase), but not the information that specifies the payment habits and purchasing power. Existing databases are common in most sectors in the shape of small islands. This is a result of fierce market rivalry, client privacy concerns, challenging administrative processes, and even opposition to data integration inside departments. As a result, it is either economically impractical or impossible to combine disparate data from various institutions and agencies across the nation [13].

2.5 *Data Security with Blockchain*

Companies stress data security issues due to the rise in occurrences involving data and user privacy. Data leaks harm businesses' reputations while also raising issues with the public and the government. For instance, the recent Facebook data hack sparked a variety of complaints. As a result, some American states reviewed their legislation governing data protection. The development and use of is the best illustration of data enforcement. The ability for users to have their data erased or withdrawn is a legal safeguard provided by the law. Companies found to violate the law will face severe fines. China and the United States both made the same move. Data security is mandated under the Cybersecurity Law and the General Principles of Civil Law. To prevent data leakage problems when conducting data transactions with other parties, the law also enforces the contract (which is legally enforceable and protected). They make the contracts adhere to the law's requirements for data protection. A secure data society is created as a result of these legal obligations and the adoption of new legislation. Applying the popular pattern analysis technique to the data without disclosing it, however, presents a new issue [14].

Transaction models used in pattern analysis models frequently incorporate phases like data collection, data transfer to a server, data cleansing, and party implementation of fusing requirements. Then, depending on the application, the clean data utilized by the third party to integrate, construct, and extract patterns aids in creating a suitable model for making judgments. Typically, they offer the model as a service to other businesses. The following new data regulations and legislation now face problems as a result of the legal binding of data protection laws. The traditional approaches breach laws like the GDPR because users are unaware of how the models will be used in the future [15].

To create a smart home system, appliance makers work to get consumer feedback on how to improve their goods and services. We create a federated learning (FL) system using a reputation mechanism to aid home appliance makers in training a machine learning model based on consumer data to assist manufacturers in developing a smart home system. Manufacturers can then anticipate future consumer demands and consumption patterns. The system's operation is divided into two parts. First, users train the manufacturer's initial model using both a mobile device and a mobile-edge computing (MEC) server. Customers use their smartphones to collect data from various household appliances, after which they download and train the basic model using local data. After obtaining regional customer-signed models, which they then transmit to the blockchain. We use the blockchain to swap out the standard FL system's centralised aggregator if users or manufacturers are hostile. The unaltered nature of blockchain records makes it possible to track the actions of dishonest buyers or producers. Manufacturers choose specific individuals or groups as miners in the second stage, who will calculate the averaged model using the models they have received from customers. One of the miners, chosen as the temporary leader at the end of the crowdsourcing work, uploads the model to the blockchain. We impose differential privacy (DP) on the retrieved characteristics and provide a new normalization method to safeguard customers' privacy while enhancing test accuracy. We experimentally show that our normalization method performs better than batch normalized when DP protection is applied to features. Additionally, we create an incentive system to reward participants to encourage more users to take part in the crowd-sourced FL work [16].

3. Federated Learning

Federated Learning is almost similar to distributed learning as it has a close relationship. Distributed computing and distributed storage make up a traditional distributed system. In some ways, distributed computation is like the first FL of model update for Android clients that have been presented. Although FL placed a lot of emphasis on privacy preservation, the most recent research in distributed machine learning also gives privacy-preserving distributed systems a lot of attention. Using a communication network, numerous computers in various places can be connected and managed by a central server to perform distributed processing, in which each computer completes various components of a single task. As a result, FL concentrates on developing a collaborative model without privacy leaks while distributed processing is primarily focused on expediting the processing step. To clarify the differences between distributed learning and the following FL traits more particularly Universality in situations involving many organizations FL, as proposed by Google, is essentially an encrypted distributed machine learning technique that enables users to create a training model jointly while maintaining the underlying data locally.

The basic FL concept was broadened to include any privacy-preserving decentralized collaborative machine learning methods. As a result, FL can handle both vertically partitioned data according to features and horizontally partitioned data according to samples in a collaborative learning environment. Cross-organizational enterprises could be integrated into the federal framework by expanding FL. For instance, a bank with information on customers' purchasing power could work with an online marketplace that has information on product features to suggest purchases. Therefore, create a joint model for numerous entities intelligently [17]. The different feature dimensions and various data sources make it possible for everyone to reap the value of cross-platform and local co-creation while upholding the data privacy number of equipment Clearly, FL is better suited for model improvement in this situation. Due to the diversity of device resources, FL is focused on unbalanced and non-IID data as opposed to distributed systems, which focus primarily on balanced and IID data distribution. This technology is decentralized. Technically speaking, decentralization does not imply total decentralization, but neither does it imply the existence of a fixed centre [18].

Decentralization merely serves to spread knowledge of the central node. Each customer influences the central model; there is no centre to categorize them. The network that the client creates will provide a non-linear relationship due to the effect between nodes.

The central server that dominates data distribution is primarily used by parameter servers, a typical distributed and centralized technology and computing power to create a productive collaborative paradigm. There is a double communication overhead because of this type of centralized data processing technology. Because if any datasets dispersed across various

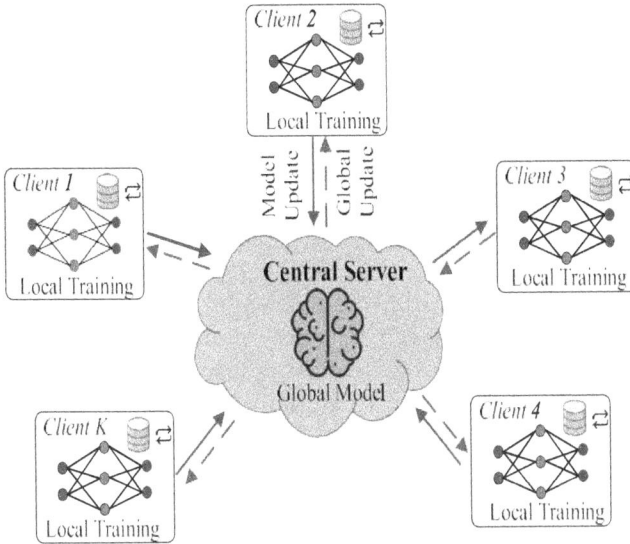

Figure 1. Federated Learning centralized model.

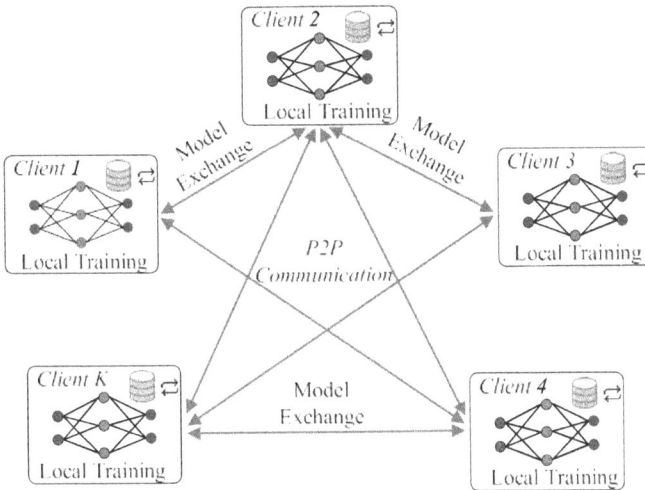

Figure 2. Federated Learning decentralized model.

databases are gathered for training, these data need first be replicated before being sent to a central server. Then, for distributed computation, a central server will distribute data to each distributed client. It intensifies the system's rigorous assessments for computing power, storage, and bandwidth. For FL situations, each client is independent, the data is not distributed by the centre, and the training is not controlled by the server. FL is a system that integrates data fusion and machine learning models through decentralized collaboration [19].

A distribution that is Hugely Non-Identically Independent Data is present in tens of thousands of edge nodes and mobile devices in FL. Each node's data capacity is limited to the total number of nodes. While in distributed systems, the fundamental goal is to enhance the degree of parallelism to reduce the workload on the central server for computation or storage. In a distributed system, there couldn't be as many nodes as there are in an FL. The world has now entered an era where wearable technology is widely employed for health monitoring. Each gadget only produces a small amount of data, making it impossible to compare it to the total. The value of data science in industrial engineering is becoming more and clearer as storage and processing capability increases. Industrial engineering has witnessed the rapid advancement of artificial intelligence, machine learning, smart production, and deep learning in recent years. However, there are two significant obstacles to the advancement of data science in this domain. The most important component is data governance, to start with. Some information is privatized due to legal reasons. Users now have complete ownership over their data thanks to the General Data Protection Regulation. Without prior consent, no institutions or organizations are permitted to use a user's data. Second, data silos are another challenging issue that restrains the growth of the contemporary industry because they require additional training. Data would enhance training effectiveness [20]. For instance, the comparison to the earliest models with Alpha Zero of chess data produced by humans and machines is capable of defeating experts. Additionally, data annotation in some industries, like the medical sector, depends on expert individuals, which may increase the rarity of correct data. The lack of labelled data is also harmful to the advancement of the industry. To tackle these difficulties in the industry, FL eventually emerged. FL is a developing machine learning strategy that seeks to address the issue of Data Island while maintaining data privacy. For decentralized machine learning settings, it refers to numerous clients coordinated with one or more central servers. Google first proposed it in 2016 to anticipate users' text input within tens of thousands of Android devices [21].

3.1 Blockchain-based Federated Learning

Smart hotel systems that are HIoT (Hotel Internet of Things) enabled have been very popular in recent years since they aim to improve quality of life. According to a Statistical analysis, there will be many smart hotels worldwide by 2022. The IoT devices with smartphones, contemporary wireless communications, cloud & edge computing, big data analytics, and artificial intelligence are primarily what make this concept of a smart hotel possible (AI). These cutting-edge technologies allow manufacturers to keep a smooth link between their smart hotel gadgets. Numerous data are generated as smart hotel technology spreads. Without having to upload data to a centralized server, federated learning (FL) enables analysts to study and use locally generated data in a decentralized manner. Blockchain is even while data are locally kept, their functionality is well maintained. We develop an FL-based system to assist hotel facilities effectively and conveniently using data generated in consumers' equipment. A cell phone is used to periodically gather data from home appliances and train the system, which treats hotel facilities of the same brand in a family as a unit. The local machine learning model is created due to the restricted processing power and battery life of mobile devices; to transfer some of the training tasks to an edge computing server [22]. Then, by averaging the total of locally trained models supplied by users, a global model is created using the blockchain smart contract. Source data are meant to ensure security and privacy in this federated manner. However, it was shown that gradient updates could provide a lot of information about the training records of clients. Data from gradients uploaded by customers

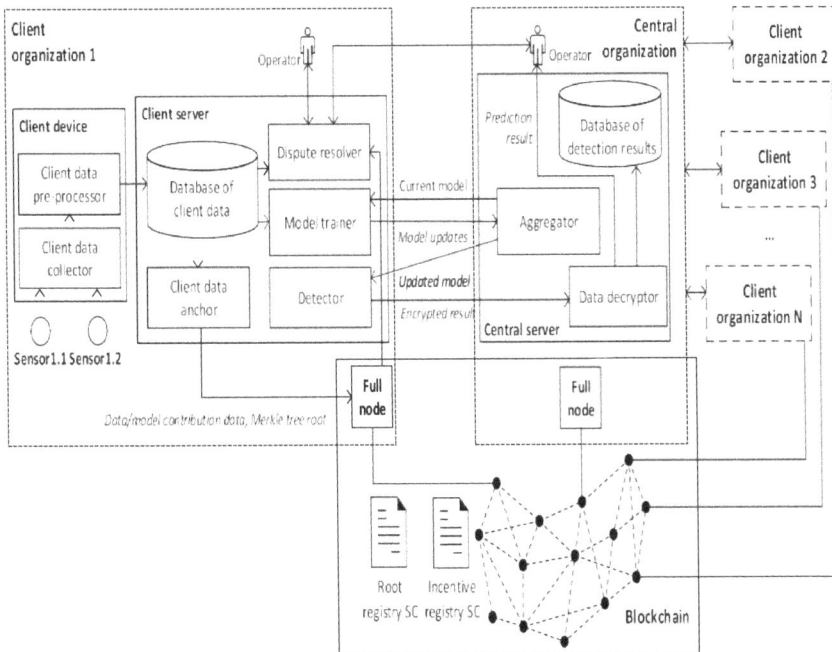

Figure 3. A blockchain-based Federated Learning model for fault detection in IoT.

can be recovered by attackers. Additionally, the federated approach to model training is vulnerable to model poisoning assaults. Additionally, the mobile-edge computing (MEC) server owned by a third party poses a danger of information leaking. We use blockchain technology and differential privacy to overcome the aforementioned security and privacy challenges. It's important to note that Apple is effectively implementing DP in FL to enhance the privacy of its well-known voice assistant service. Manufacturers specifically upload an early model with initialised parameters. Customers can take the model from the blockchain and use it to train with local data [23].

The blockchain helps the crowd-sourcing requester verify whether any fraudulent customer updates have been made. While our system uses blockchain to track crowd-sourcing activities, the old system is maintained by a third party, who charges clients exorbitant service costs. Customers and the requester can thus avoid paying expensive service costs while yet maintaining the effectiveness of the crowdsourcing method. We suggest using the InterPlanetary File System (IPFS) as the distributed storage due to the block size restriction when the model size is huge, a solution. More specifically, clients extract the features of their data on the mobile using the deployed feature extractor and disturb the recovered features in the first phase by adding noise with a formal privacy guarantee. Customers train fully connected model layers with perturbed features in the MEC server as the second stage. Additionally, we enhance the conventional batch normalization by removing the mean value and variance requirements while imposing a bound where N specifies the batch size. Customers transmit locally trained models to the blockchain after training by signing hashes of encrypted models with their private keys. To obtain the global model, selected miners download models, check the senders' identities, and then average all the model parameters [24].

3.2 IoT-Based Device Sensor of Hotels

The deployment of wireless sensors and actuators makes the choice of IoT-based automation that enables various physical devices to digitally interact with one another, enabling the Infrastructure Manager to easily monitor and oversee operations of massive infrastructures from anywhere in the globe. Through a centralized cloud platform, it enables data collection and analysis utilizing AI/ML algorithms according to the required perspective [25]. To flourish in this fast-paced climate and take the top rank, the hotel industry is in dire need of IoT-based automation. Hotels will be able to offer a high-quality service, optimize occupant comfort, and reduce costs in this valuable business with the help of comprehensive monitoring.

Device Sensor Connectivity Using IoT in Cloud Platform

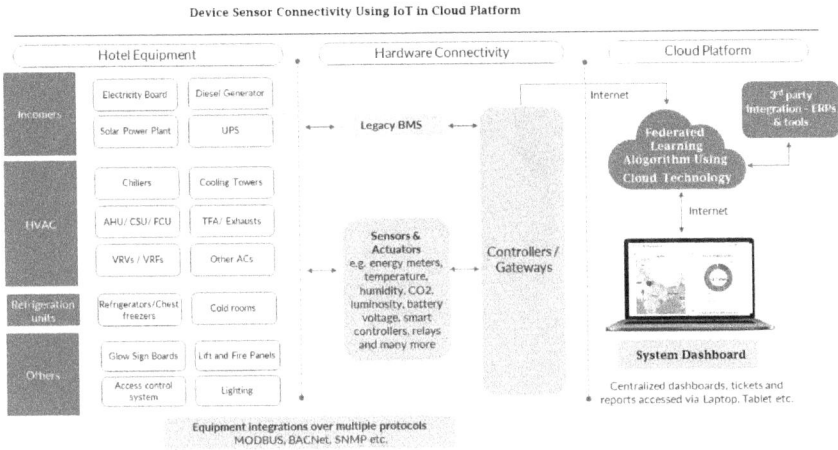

Figure 4. Connectivity of device sensor through IoT.

The equipment integrates using a variety of protocols, including MODBUS, BACnet, etc. The classic BMS or Sensors and Actuators are used to collect data about the operation and operation of all sorts of equipment. In the end, the Controller or Gateway uses the IoT to send the data collected to the cloud platform. The data and statistics are personalized and shown on the central Dashboard, which can be accessed via laptops, tablets, etc., with the use of AI/ML algorithms. The dashboard gives the Infrastructure Manager the ability to remotely monitor an entire facility or a collection of facilities and take the appropriate action. According to the requirements, the management can set up these assets' operations to be based on timetables or sensors [26]. This kind of individualized control guarantees that power is utilized effectively. The dashboard also creates maintenance tickets, alerting the management to take preventive action to fix the equipment that is prone to failure to prevent any form of interruption in service delivery. Let's anticipate a few useful advantages that IOT-based automation can offer to hotels. A great first impression is something we can never get back. Similarly, when guests enter a hotel from the front desk area, the hotel must maintain a comfortable temperature for them. At a well-lit front-desk area, a comfortable temperature and enough ventilation are required. This energizes the visitor and makes them appreciate the hotel's ambience. With appropriate HVAC and illumination system upkeep and monitoring using IOT-based technology, such Ambience can be kept as required. However, these controls can be adjusted to create a pleasant ambience in dining rooms, lobbies, and recreational spaces as well.

The occupancy rates over the term may also be considered in this customization. It has been observed that such an update increases guest

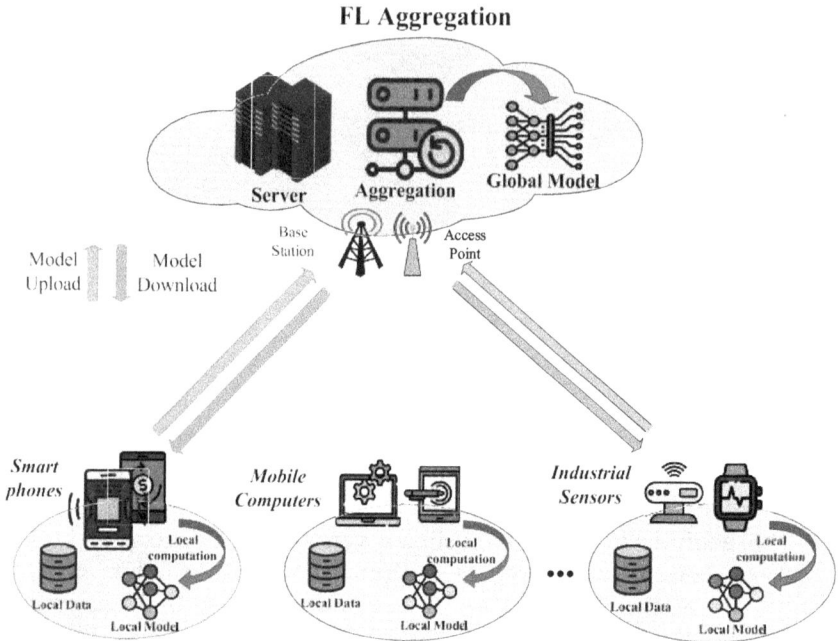

Figure 5. Design of IoT with Federated Learning.

comfort by 40–45%. Many hotels keep a supply of perishable food on hand to cater to guests around the clock. To keep the food fresh, they can use automated control to maintain the temperatures of chillers and cold rooms. A consumer will always gravitate toward your hotel if the food is good and fresh. Other amenities like dryers, refrigerators, dishwashing machines, laundry machines, and water heaters can also be easily regulated so that they are constantly available and effectively meet the demands of the residents aside, it is vital to provide these services consistently to provide the visitors with exquisite comfort [27]. Since these are the primary amenities that make one hotel stand out from others in the hospitality sector, hotels cannot afford any form of breakdown of such assets. Continuous monitoring is a solution that maintains the assets under control constantly. The facility manager is immediately notified of any kind of slight fluctuation in any of the equipment via a ticket, allowing him to take preventive action to have the equipment checked and repaired before it fully stops working. This makes it possible to reduce equipment breakdown costs by 20–25%. An understanding of the amount of energy used to provide such facilities across different various time frames.

It enables the manager to control costs and only spend what is necessary. Both the environment and the hotels benefit from these savings.

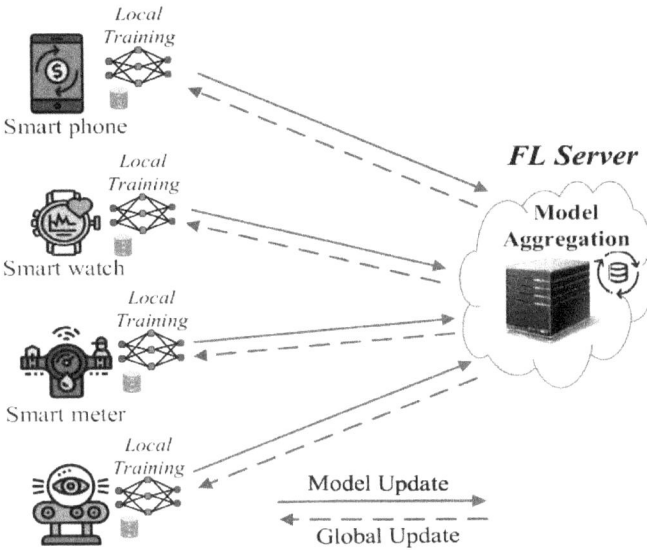

Figure 6. The Federated Learning for IoT system.

Continuous monitoring results in a profit of one cent for every cent saved. In a sector where the cost of electricity makes up such a large amount of overall spending, who would want to save 10–20%? Even this monitoring of various facilities around the nation is made possible by the Dashboard. Hotels can adopt the procedures that are the most effective when they have control over such centralized statistics of different properties. To make effective decisions regarding their power usage and spending, a manager or hotel owner must have complete information. The three main purposes of IOT automation for hotels are to increase customer comfort, reduce energy costs make facility monitoring simple, and offer uniform visibility across all locations. Technology is rapidly advancing, and we shouldn't miss an opportunity to take advantage of its advantages. We find it most amazing how simple it is to install such automated structures. To make conventional hotels more energy-efficient, Solutions offers IoT-based energy and asset management solutions for the hospitality industry. The renovation of your hotel structure is merited through the deployment of sensors connected with IoT [28].

4. Types of Sensors and Detectors used in Hotel

The hotel sector requires a variety of sensors and detectors for reliable, efficient, and safe operation due to the extremely unique nature of the business. The following is a list of some of them: Smoke detectors are

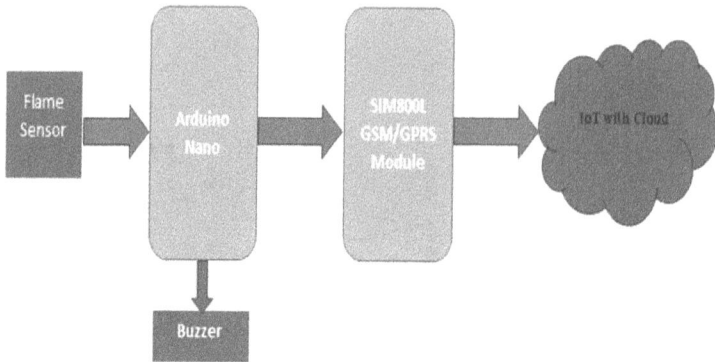

Figure 7. Sensor connectivity to Arduino.

devices that detect smoke, usually as a sign of an impending fire. As a component of a fire alarm system, commercial security devices provide a signal to a fire alarm control panel. Heat Detector: A heat detector is a gadget that alerts the user to an abrupt change in the environment's temperature [29]. The water sprinkler system receives a signal if the temperature rise exceeds a set threshold. Motion or occupancy detectors are devices that turn on or off lights and other equipment when they detect movement depending on whether there are individuals in a particular location in the sensor's field of view, turning on or off.

In several hotels, facilities such as the water treatment plant, boiler, swimming pool, and sewage treatment plant, pH detectors are employed to keep an eye on the pH level of the water. LPG leaks from gas banks are found using gas leak detectors. Gas detectors are used to find different gases that could be dangerous in a certain location. They are employed to find CO_2, poisonous, and combustible gases. Flame Detector looks for flames using optical sensors. Carbon monoxide level detector this sensor device alerts users to the presence of carbon monoxide by sounding some sort of alarm to be reduced by taking steps like increasing the area's airflow and/or evacuating the inhabitant's Photo sensors - These sensors are used to measure the amount of daylight that is present, adjust the light output of lighting tubes and bulbs as necessary, and turn on and off the lights throughout a building. It aids in energy saving and lowers the property's power usage bill Closed-circuit television CCTV is a system that employs the digital eyes of video cameras to broadcast a signal about human activity in a particular location to a storage area or a constrained number of cameras. The Internet of Objects, which refers to connecting things and people over the Internet, has established itself as a new way of doing business across a variety of industries. IoT improves the process and offers emergency services to enable quick and effective reactions in real-time. Managers with the knowledge and communication needed to utilize those resources. In this research, it is suggested that an IoT-based model be used to evaluate and examine a quick reaction to fire hazards. One of the main causes of accidental deaths worldwide is fire [30].

4.1 Arduino IoT Fire Detection

A low-cost Wi-Fi module, a gas detection sensor, a flame detection sensor, a buzzer to alarm, and temperature sensors are used to create this proposed system. With the information gathered by the sensors, the sensors identify local emergencies and notify the system and warnings from hotel organizations such as fire services, police stations, and hospitals by delivering the precise position to both the user and operator through a module that is closely connected with all of them. IoT is used to create an integrated intelligent system to manage such risks to innocent lives and property. There is the type of surveillance equipment that is most frequently employed [31]. The Flame sensor is attached to the Arduino Nano's Digital Input pin as indicated in the circuit diagram. If you're still interested, you may look at some of the other straightforward fire alarm circuits we've developed in the past. Since the SIM800L operates on 3.3v logic, it is connected to the Arduino Nano using logic shifting resistors. Due to the SIM800L module's 3.4–4.2V DC operating range and the Arduino Nano's 5V DC external supply, separate power is provided to it. Instead of using two power supplies, a 3.7–5V Boost converter can be used in this situation.

The Internet of Things-based fire alarm and monitoring system this project has suggested is most suited for commercial and residential applications. The primary factor in accidental death, where valued lives and expensive goods are lost, is fire. The main characteristic of fire is its exponential growth, which spreads quickly and consumes whatever it comes into contact with. Therefore, early fire detection is crucial to preserving many lives as well as property. It can detect smoke, an increase in temperature, a rise in flames, etc., and communicate that information to a remote controlling unit via GSM to create the necessary measures and notify the closest domestic aid. The suggested system can detect fire, various gases, and smoke [32]. The system that will be supplying Hazards is mapped to the local emergency services, such as the fire department, police, and hospitals. This IoT framework-based fire and gas sensing system focuses on the public safety and livelihood service sectors. Figure 8 displays the fire detection system using standardized IoT design techniques. The flammable gas sensor MQ-6 is used to detect gases like LPG/LNG, the spark detection sensor PT333B is used to detect sparks, and

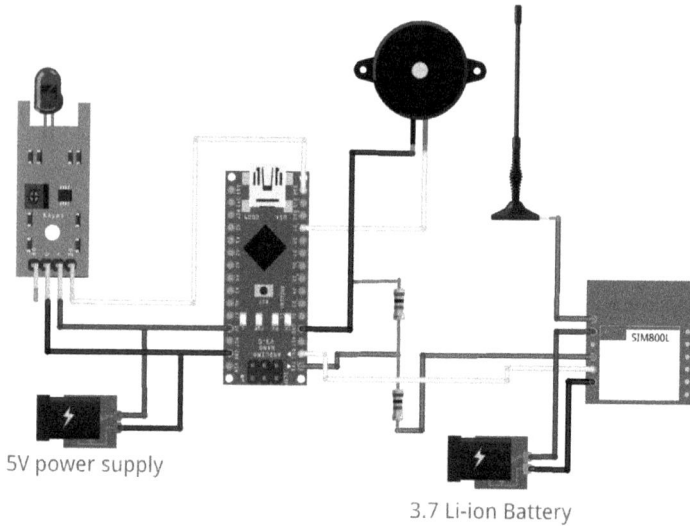

Figure 8. Diagram for Arduino IoT fire detection.

the GPS module is used to determine the location of the device. These sensors are connected to a Wi-Fi microcontroller over the Internet, which allows it to transmit the status of any hazards to the closest service centres for any type of assistance. The extension of online access to tools, gadgets, and everyday objects is known as the Internet of Things (IoT) is equipped with a wide variety of equipment, (for instance, sensors), these tools share information over a remote system, and these tools can be meticulously monitored, and modifications may be made periodically using them. Because of various advancements, including run-time analysis, framework learning (machine learning), sensors, traditional fields of inserted load-up engraved frameworks, unwired sensor systems, control frameworks, robotization including hotel and building computerization, and others, the definition of the Internet of things has changed [33].

4.2 *Revolutions of IoT*

IoT innovation is best known in the consumer market for products that relate to the concept of the "keen house," including home appliances and gadgets that support at least one typical way of life and can be controlled by telephones and other devices like speakers and phones that are connected to that framework as remotes. We are using IOT in this project to quickly assess the risks that may arise and that can be kept at a safe distance from them. IOT helps to make a few things simpler for the client. With that as an option, it provides us with the ability to expand the system whenever necessary to make it more reliable and proficient. Fire is a terrible scenario that threatens lives and property, and it may spread quickly. Every year, fire casualties result in the loss of thousands of homes and lives. A lot of money can be saved by preventing a fire from starting and taking the necessary safeguards beforehand destruction. In industrialized nations, homes are furnished with safety features. However, in developing or undeveloped regions the model size is huge, and a solution is required. More specifically, clients extract the features of their data on the mobile using the deployed feature extractor and disturb the recovered features in the first phase by adding noise with a formal privacy guarantee. Customers train fully connected model layers with perturbed features in the MEC server as the second stage. Additionally, we enhance the conventional batch normalization by removing the mean value and variance requirements while imposing a bound where N specifies the batch size. Customers transmit locally trained models to the blockchain after training by signing hashes of encrypted models with their private keys. To obtain the global model, selected miners download models, check the senders' identities, and then average all the model parameters [34]. The global model is encrypted and uploaded to the blockchain by one

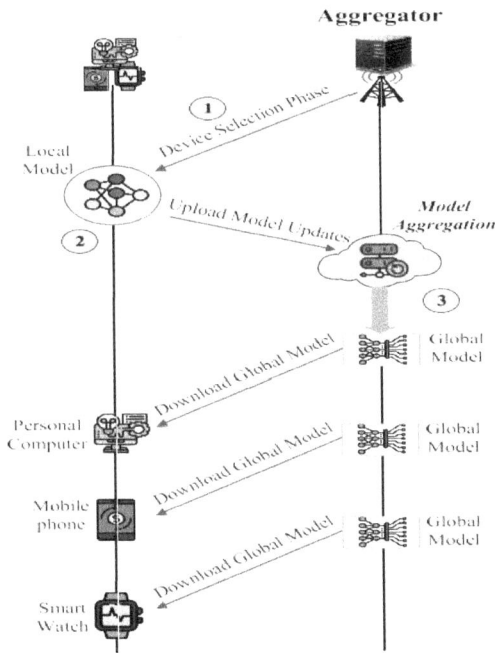

Figure 9. IoT to Federated Learning communication model.

miner chosen as the temporary leader. Additionally, we use a reputation-based crowd-sourcing incentive mechanism that promotes trustworthy customers and penalizes dishonest ones to encourage more users to engage in crowd-sourcing work and prevent harmful and poisonous updates.

The main centre of IoT administration is information sharing, which points to sending information over a shared arrangement to advantage conclusion clients in a specific application. To permit shrewd IoT systems with moo idleness and protection conservation, FL offers the sharing of learning comes about as an elective for sharing crude IoT information. The work gives a collaborative information-sharing approach for mechanical IoT applications that empowers a fast and secure information stream among decentralized different parties between information proprietors and information requestors. An FL conspire is made to empower the estimation of the sorts of information-sharing demands with questions given by a requester in arrange to return the accurately computed comes about for these inquiries for sharing typically done due to the asset restrictions of IoT clients in hotels. Additionally, protecting A blockchain is associated with the FL design to form immutable data squares that are claimed by all parties for straightforwardness and to extend information proprietorship without requiring central authority. This secures the sharing and prepares against outside dangers. The failure to alter or overhaul information once it has been put away within the blockchain diminishes the plausibility of information spillage and improves organized security as a result. The convenience of the proposed FL-based sharing framework with the highest learning exactness and improved security is bolstered by reenactment which comes about by utilizing real-time datasets. Agreeing with the discussion in this regard, the creators made a mechanical IoT unified tensor mining system based on the FL thought to coordinate multisource information and empower tensor-based mining with security affirmations. Particularly, by sharing their information, which has been scrambled employing a homomorphic encryption strategy, with a central server, industrial facilities collaborate to take part in tensor mining. In this way, crude information is held at neighbourhood production lines, ensuring information protection for tensor mining, whereas the server fair assembles the cypher content information and federates it into a tensor. Indeed, whereas aggressors can study cypher content on communication channels and busybodies can ambush the central server to take the amassed cypher content, they are incapable of getting the key for information unscrambling. The proposed combined approach may increment mining precision by 24% compared to the privacy-preserving compressive detecting strategy and give tall security by leveraging two mechanical collisions in Beijing and Shanghai with over 100 IoT hubs per factory for recreation. The assault location that's viable and exact without influencing execution [36]. Distributed information sharing in-vehicle systems has been realized utilizing FL. Agreeing to the investigation, a non-concurrent combined information sharing system is proposed, in which each vehicle capacities as an FL client to together share information with a total server at a large-scale base station.

4.3 *Multilevel Blockchain System for IoT*

Devices can ask for information sharing from the Multilevel Blockchain System (MBS) with an assortment of benefit desires, such as activity expectation or way choice. The MBS turns the sharing handle into a computer errand to address vehicle-sharing demands utilizing an actor-critic fortification learning system by running a shared worldwide show based on gathered vehicular datasets. With this procedure, you'll learn to assess the negative behaviours of taking part hubs or advantageous hubs, to help the brilliant information sharing choice by choosing the planned hubs that minimize the fetched of sharing. Blockchain innovation has been utilized to advance and fortify the security and steadfastness of information sharing [37]. This innovation is extraordinary in confirming the demonstrated parameters and storeing them in unchanging records. The outline appears in the graph for such a data-sharing framework based on FL and blockchain. Offbeat FL is additionally proposed for asset-sharing in-car systems. Here, slope plummet preparation is combined with a nearby differential protection strategy that can secure the protection of nearby upgrades to empower secure and dependable FL sharing. A decentralized FL demonstration is made that empowers the total of the vehicle to demonstrate changes at farther MBSs, particularly to maintain a strategic distance from tall communication costs and lower security dangers brought on by the centralized FL engineering. from tests with predominant precision and protection assurance than ordinary centralized FL frameworks, commonsense datasets illustrate the promising come about of the proposed decentralized FL conspire. For information sharing in-car systems with progressive blockchain, an FL plot is additionally taken into thought. Two essential chains, a beat chain, and a ground chain, make up the recommended sharing design. To be clear, ground chains incorporate various vehicles that work as FL clients and actualize nearby learning utilizing their claim equipment, as well as street-side units that act as decentralized FL aggregators and securely collect exchanges inside their scope. The computation of the FL demonstrations is done by several RSUs that are kept within the best chain. After that, the FL discoveries are included in the piece record for dispersion to RSU and vehicles for traceability and security. Besides, an HFL procedure is proposed for the shared learning between edge gadgets as members and a cloud server as the aggregator to permit the information sharing of non-free and indistinguishably dispersed (non-IID) among edge gadgets. A combined swapping demonstration is further developed based on a couple of shared information amid the HFL that can reduce the negative effect of non-IID information to address the issue of weight disparity delivered by conventional FL. Moreover, a semi-supervised learning approach is utilized by edge gadgets to foresee objects for video investigation applications [38]. The proposed FL strategy can boost picture classification precision by 3.8% and question following execution by and large by utilizing real-world video information. Comparison to the conventional approach comes about in a 2.1% enhancement within the detection work. Be that as it may, employing an inaccessible cloud solely for FL operation can cause intemperate communication idleness. The reasonable optimization strategy for this issue is to facilitate the edge of gadgets and the cloud by cutting down on communication delays. This includes collaboratively advancing the planning of shared information, confirmation control, and exactness alteration. The reenactment comes about illustrates that the proposed approach may be executed with progressed security and diminished inactivity in an assortment of arranged arrangements. Comparable to this, ponder moreover takes into consideration an FL plot on a versatile IoT organize that comprises versatile gadgets and cloud servers as learning clients. Through it utilizes a video proposal; the potential of FL in portable clouds is inspected [39]. The factorization models improve guest satisfaction on their visit. A combined suggestion framework is made to empower data combination among clouds with communication delay mindfulness by utilizing computed neighbourhood upgrades. The need for security protection architecture for cloud-based learning within the FL handle may be an imperfection in this proposed strategy. To address this issue, a blockchain ledger-enabled secure information participation system for unified information learning among different parties, counting open and private information centres, is being inquired about [40]. Sometimes recently being sent to the central server for secure accumulation, the information utilization occasion and FL changes are offloaded to the blockchain, whereas the Clients have full control over their information.

5. Conclusion

This chapter gives a federated learning strategy for understanding the IoT device failure detection in the hotel industry to provide an accurate system of service with flawless device sensors. In the proposed system, raw data is locally stored and a central server coordinates client servers to train a shared global model for detection. Blockchain technology is used in the design to provide the verifiable integrity of client data and financial incentives for client contribution. An FL algorithm is suggested to lessen the effects of data heterogeneity based on the separation of each client data set's positive class and negative class. A real-world use case is used to develop a proof-of-concept prototype utilizing the suggested architecture. To assess the strategy's viability, detection precision, and performance a federated learning algorithm is used. The evaluation's findings demonstrate that the suggested strategy succeeds in achieving all the goals. Even though most federated learning modules are proposed to alter the concept by relocating all the modules from the client servers used in this chapter to client devices with more potent computation and storage capabilities. For makers of IoT devices to better understand their customers, this study presents a design for a blockchain-based FL system. The system is built using several cutting-edge technologies, including blockchain, distributed storage, and FL. In addition, this system enforces DP to safeguard customer data privacy. A novel normalizing method is developed that, when features' privacy is protected by DP, outperforms batch normalization to increase the FL model's accuracy. Customers are more likely to participate in hotel tasks if an appropriate incentive system is offered. The benefits of IoT gadgets are providing instant information to the customers of hotels. As part of the federated training, the blockchain will audit all consumer updates so that the system can hold model upgrades accountable and thwart dishonest users or manufacturers. We plan to run further tests and evaluate our system using datasets from actual home appliance models in the future.

References

[1] Qiang Yang, Yang Liu, Yong Cheng, Yan Kang, Tianjian Chen and Han Yu. 2019. Federated Learning. Synthesis Lectures on Artificial Intelligence and Machine Learning, 13.3: 1–207.

[2] Li, L., Fan, Y., Tse, M. and Lin, K.Y. 2020. A review of applications in Federated Learning. Computers & Industrial Engineering, 149: 106854.

[3] Nguyen, D.C., Ding, M., Pham, Q.V., Pathirana, P.N., Le, L.B., Seneviratne, A., Li, J., Niyato, D. and Poor, H.V. 2021. Federated Learning meets blockchain in edge computing: Opportunities and challenges. IEEE Internet of Things Journal, 8.16: 12806–12825.

[4] Pokhrel, S.R. and Choi, J. 2020. Federated Learning with blockchain for autonomous vehicles: Analysis and design challenges. IEEE Transactions on Communications, 68.8: 4734–4746.

[5] Nguyen, D.C., Ding, M., Pathirana, P.N., Seneviratne, A., Li, J. and Poor, H.V. 2021. Federated Learning for Internet of Things: A comprehensive survey. IEEE Communications Surveys & Tutorials, 23.3: 1622–1658.

[6] Hao, M., Li, H., Luo, X., Xu, G., Yang, H. and Liu, S. 2019. Efficient and privacy-enhanced Federated Learning for industrial artificial intelligence. IEEE Transactions on Industrial Informatics, 16.10: 6532–6542.

[7] Zhang, W., Lu, Q., Yu, Q., Li, Z., Liu, Y., Lo, S.K., Chen, S., Xu, X. and Zhu, L. 2020. Blockchain-based Federated Learning for device failure detection in industrial IoT. IEEE Internet of Things Journal, 8.7: 5926–5937.

[8] Wang, R. and Tsai, W.T. 2022. Asynchronous Federated Learning system based on permissioned blockchains. Sensors, 22.4: 1672.

[9] Javed, A.R., Hassan, M.A., Shahzad, F., Ahmed, W., Singh, S., Baker, T. and Gadekallu, T.R. 2022. Integration of blockchain technology and Federated Learning in vehicular (IoT) networks: A comprehensive survey. Sensors, 22.12: 4394.

[10] Lu, Y., Huang, X., Dai, Y., Maharjan, S. and Zhang, Y. 2019. Blockchain and Federated Learning for privacy-preserved data sharing in industrial IoT. IEEE Transactions on Industrial Informatics, 16.6: 4177–4186.

[11] Cheng, X., Tian, W., Shi, F., Zhao, M., Chen, S. and Wang, H. 2022. A blockchain-empowered cluster-based Federated Learning model for blade icing estimation on IoT-enabled wind turbine. IEEE Transactions on Industrial Informatics.

[12] Moyeenudin, H.M., Bindu, G. and Anandan, R. 2021. OTA and IoT influence the room occupancy of a hotel. Data Management, Analytics and Innovation. Springer, Singapore, 265–274.

[13] Kim, Y.J. and Hong, C.S. 2019. Blockchain-based node-aware dynamic weighting methods for improving Federated Learning performance. 2019 20th Asia-Pacific Network Operations and Management Symposium (APNOMS). IEEE.

[14] Liu, H., Zhang, S., Zhang, P., Zhou, X., Shao, X., Pu, G. and Zhang, Y. 2021. Blockchain and Federated Learning for collaborative intrusion detection in vehicular edge computing. IEEE Transactions on Vehicular Technology, 70.6: 6073–6084.

[15] Liu, H., Zhang, S., Zhang, P., Zhou, X., Shao, X., Pu, G. and Zhang, Y. 2021. Blockchain and Federated Learning for collaborative intrusion detection in vehicular edge computing. IEEE Transactions on Vehicular Technology, 70.6: 6073–6084.

[16] Aïvodji, U.M., Gambs, S. and Martin, A. 2019. IOTFLA: A secured and privacy-preserving smart home architecture implementing Federated Learning. IEEE Security and Privacy Workshops (SPW). IEEE.

[17] Messaoud, S., Bouaafia, S., Bradai, A., Hajjaji, M.A., Mtibaa, A. and Atri, M. 2022. Network slicing for industrial IoT and industrial wireless sensor network: Deep federated learning approach and its implementation challenges. In Emerging Trends in Wireless Sensor Networks. IntechOpen.

[18] Cao, H., Liu, S., Zhao, R. and Xiong, X. 2020. IFed: A novel Federated Learning framework for local differential privacy in power Internet of Things. International Journal of Distributed Sensor Networks, 16.5: 1550147720919698.

[19] AbdulRahman, S., Tout, H., Ould-Slimane, H., Mourad, A., Talhi, C. and Guizani, M. 2020. A survey on Federated Learning: The journey from centralized to distributed on-site learning and beyond. IEEE Internet of Things Journal, 8: 5476–5497.

[20] Ferrag, M.A., Friha, O., Hamouda, D., Maglaras, L. and Janicke, H. 2022. Edge-IIoTset: A new comprehensive realistic cyber security dataset of IoT and IIoT applications for centralized and Federated Learning. IEEE Access, 10: 40281–40306.

[21] Lakhan, A., Mohammed, M.A., Kadry, S., AlQahtani, S.A., Maashi, M.S. and Abdulkareem, K.H. 2022. Federated Learning-aware multi-objective modeling and blockchain-enable system for IIoT applications. Computers and Electrical Engineering, 100: 107839.

[22] Zhao, Y., Zhao, J., Jiang, L., Tan, R., Niyato, D., Li, Z., Lyu, L. and Liu, Y. 2020. Privacy-preserving blockchain-based federated learning for IoT devices. IEEE Internet of Things Journal, 8.3: 1817–1829.

[23] Rahman, M.A., Hossain, M.S., Islam, M.S., Alrajeh, N.A. and Muhammad, G. 2020. Secure and provenance enhanced internet of health things framework: A blockchain managed federated learning approach. IEEE Access, 8: 205071–205087.

[24] Singh, S., Rathore, S., Alfarraj, O., Tolba, A. and Yoon, B. 2022. A framework for privacy-preservation of IoT healthcare data using Federated Learning and blockchain technology. Future Generation Computer Systems, 129: 380–388.

[25] Zhong, C.L., Zhu, Z. and Huang, R.G. 2015. Study on the IOT architecture and gateway technology. 14th International Symposium on Distributed Computing and Applications for Business Engineering and Science (DCABES). IEEE.

[26] Liu, S. and Gao, Y. 2021. Media tourism and hotel management development based on wireless sensor network and embedded system. Journal of Ambient Intelligence and Humanized Computing, 1–12.

[27] Colapinto, K.B., Linder, B., Canning, J. and Cengic, M. 2020, June. Scalability in IoT Sensor Networks for Smart Buildings. In Optical Fiber Sensors (pp. W4–65). Optica Publishing Group.

[28] Galbraith, A. and Podhorska, I. 2021. Artificial Intelligence data-driven Internet of Things systems, robotic wireless sensor networks, and sustainable organizational performance in cyber-physical smart manufacturing. Economics, Management & Financial Markets, 16.4.

[29] Salameh, H.B., Dhainat, M. and Benkhelifa, E. 2019. A survey on wireless sensor network-based IoT designs for gas leakage detection and fire-fighting applications. Jordanian Journal of Computers and Information Technology, 5.2.

[30] Yadav, C.S.B. and Sheshadri, R. 2018. Wireless Intrusion Detection and Prevention System for IEEE 802.11 Based Wireless Sensor Network.

[31] Hotel, Hard Rock. IEEE Topical Conference on Wireless Sensors and Sensor Networks (WiSNeT).

[32] Kholod, I., Yanaki, E., Fomichev, D., Shalugin, E., Novikova, E., Filippov, E. and Nordlund, M. 2020. Open-source federated learning frameworks for IoT: A comparative review and analysis. Sensors, 21.1: 167.

[33] Imteaj, A., Thakker, U., Wang, S., Li, J. and Amini, M.H. 2021. A survey on federated learning for resource-constrained IoT devices. IEEE Internet of Things Journal, 9.1: 1–24.

[34] Campos, E.M., Saura, P.F., González-Vidal, A., Hernández-Ramos, J.L., Bernabé, J.B., Baldini, G. and Skarmeta, A. 2021. Evaluating Federated Learning for intrusion detection in Internet of Things: Review and challenges. Computer Networks, 108661.

[35] Giménez, N.L., Grau, M.M., Centelles, R.P. and Freitag, F. 2022. On-device training of machine learning models on microcontrollers with Federated Learning. Electronics, 11.4: 573.

[36] Kim, H., Park, J., Bennis, M. and Kim, S.L. 2019. Blockchained on-device federated learning. IEEE Communications Letters, 24.6: 1279–1283.

[37] Majeed, U. and Hong, C.S. 2019. FLchain: Federated learning via MEC-enabled blockchain network. 2019 20th Asia-Pacific Network Operations and Management Symposium (APNOMS). IEEE.

[38] Li, Y., Chen, C., Liu, N., Huang, H., Zheng, Z. and Yan, Q. 2020. A blockchain-based decentralized Federated Learning framework with committee consensus. IEEE Network, 35.1: 234–241.

[39] Ali, M., Karimipour, H. and Tariq, M. 2021. Integration of blockchain and Federated Learning for Internet of Things: Recent advances and future challenges. Computers & Security, 108: 102355.

[40] ur Rehman, M.H., Salah, K., Damiani, E. and Svetinovic, D. (2020, July). Towards blockchain-based reputation-aware Federated Learning. IEEE INFOCOM 2020-IEEE Conference on Computer Communications Workshops (INFOCOM WKSHPS). IEEE.

Chapter 14

Navigating the Complexity of Macro-Tasks

Federated Learning as a Catalyst for Effective Crowd Coordination

S Mayakannan, N Krishnamurthy, K Vimala Devi,*
R Deepalakshmi, Sandya Rani and *Jose Anand A*

1. Introduction

Crowdsourcing, the process of completing a project by soliciting, reviewing, and implementing user suggestions, has become increasingly popular in recent years [1]. Numerous definitions exist for "crowd

---

[1] Assistant Professor, Department of Mechanical Engineering, Vidyaa Vikas College of Engineering and Technology, Tiruchengode, Namakkal, Tamilnadu.

[2] Department of Mathematics, Veltech Multitech Dr. Rangarajan Dr. Sakunthala Engineering college, Avadi, Chennai-62.

[3] Associate Professor, School of Computer Science and Engineering, Vellore Institute of Technology, Vellore.

[4] Assistant Professor (Computer Applications), Department of Inter-Disciplinary Studies, The Tamilnadu Dr. Ambedkar Law University, Chennai, Tamilnadu.

[5] Assistant Professor, Department of Information Technology, St. Martin's Engineering College Secunderabad Telangana India.

[6] Professor, Department of ECE, KCG College of Technology, Chennai, Tamil Nadu, India.

Emails: kannanarchieves@gmail.com; krishnamurthy@veltechmultitech.org; profdeepalakshmi@gmail.com; sandhya.marri@gmail.com; joseanandme@yahoo.co.in

* Corresponding author: vimaladevi.k@vit.ac.in

sourcing," but the most common centre is on "outsourcing" tasks to an online community [2]. The term "crowdsourcing" is often defined today as the collaborative effort of a big number of people engaged in some sort of online activity. CrowdFlower and Mechanical Turk are two popular crowdsourcing platforms that draw a significant number of people who are willing to complete online tasks [3]. The "crowd workers" that make up these online groups are a wealth of information and experience that can be put to use in solving difficult issues [4].

Crowdsourcing has been used, despite its promise, mostly for very easy micro-tasks. Micro-tasks, on the other hand, don't necessitate the same level of teamwork as larger projects [5]. If crowdsourcing is going to be used for addressing more complicated problems, it needs to find a way to deal with the problem of macro-tasking [6]. Sometimes complex group work, but not always, breakable down into smaller jobs is called "macro-tasking" [7]. In general, crowdsourcing larger jobs is more difficult than microtasks [8]. Work processes including ideation, integration, and group decision-making are essential for macro-tasking. Coordinated effort amongst a multitude is essential for macro-tasking, since it helps to both divide and consolidate the results of individual efforts [9].

For the most part, the requestor is responsible for crowd coordination in the HCI/CSCW areas, which leads to micro-tasking. Before the crowd gets involved, the requestors split and assign tasks, and in many cases, they never combine the results. However, the effectiveness of crowds in solving hard problems is hindered by this approach of organizing the crowd [10].

Consider the following scenario: Crowdsourcing is going to be used by one corporation to find its next big product idea. The group issues a request for fresh concepts from the public and sets a deadline for submissions. The company has asked the public to vote on the greatest new product proposal they have received. Eventually, the votes are counted, and the victor is revealed [11]. The focus of this type of crowdsourcing is on little jobs. Every member of the crowd can comprehend and follow the procedure, which is a big plus. While the result may be unpredictable, the steps taken to get there are not. There is little need for coordination between crowd workers because the jobs performed through crowdsourcing need little interaction or dependence among them.

Now consider a different scenario: To inform the company's marketing plan for the new product, the public's input is sought. The work is neither straightforward nor reasonably well-formulated, as there are several ways to do the task at hand. There is less certainty about both the procedure and the outcome than there was before [12]. The crowd is responsible for creating a single marketing strategy; thus its members will need to agree on how to allocate their time and whether their efforts should be combined. Cooperation among the crowd workers is required for the successful completion of this mission. Crowdsourcing in this context is focused on large-scale tasks, which necessitates cooperation and teamwork among the participants. If we want to maximize crowdsourcing's efficiency, need to put in more work to coordinate the crowdsourcing of large-scale projects.

The term "coordination" can mean a variety of things [13]. General definitions of coordination are provided in this chapter for the benefit of clarity.

Individuals are coordinated to focus and coordinate their actions toward a common objective.

Additionally, coordination in a crowd as:

The method of getting a bunch of individuals to all pull in the same direction to accomplish something.

Understanding macro-tasking in crowdsourcing is a focus of this chapter, and its focus on coordination challenges is intended to further our knowledge of the topic. There are three aims that hope to achieve. To begin, we look back at some of the more well-known and cutting-edge ideas of coordination from the fields of organization and computing. The concepts of transactive memory systems (TMS), stigmaria coordination, coordination theory, relational coordination, role-based coordination, and an integrative model of coordination are all introduced and discussed in this article. After reviewing the research on macro-task coordination in HCI and CSCW, organize the numerous methods into one of the accepted theories of coordination. The limited but quickly expanding body of HCI/CSCW literature on the topic of coordination in crowdsourcing reveals a shift in focus from micro-tasking to macro-tasking. We assess the state of the art in coordination concepts and current HCI and CSCW work on coordination in macro-tasking and provide a research agenda for further exploration based on our findings.

Federated learning offers a promising solution for scenarios where data is distributed across numerous devices, such as smartphones or Internet of Things (IoT) devices, and where transferring data to a central location is either impractical or undesirable due to privacy, security, or bandwidth constraints. By enabling participants to contribute to a common model without exposing their data, federated learning provides an opportunity to leverage the collective knowledge of a diverse and widespread user base, while simultaneously maintaining data privacy and ownership.

In this article, we will delve into the principles, techniques, and applications of federated learning, beginning with an overview of the core components that define this approach, such as distributed data sources, local training, model aggregation, and the global model. We will also discuss the unique challenges and requirements associated with federated learning, including the development of efficient communication protocols, privacy-preserving techniques, and federated optimization algorithms.

Furthermore, we will explore the potential of federated learning to enhance various domains, such as healthcare, finance, and smart cities, where privacy and data ownership are of paramount importance. By examining real-world examples and case studies, we will illustrate the practical applications and benefits of federated learning in these contexts.

Finally, we will identify areas for future research and development within the field of federated learning, highlighting the need for continued exploration and collaboration to fully harness the potential of this decentralized approach to machine learning.

2. Related Works

2.1 *Organization of Micro and Macro-tasking*

It's natural to wonder what sets macro-task coordination apart from micro-task coordination. For several reasons, macro-tasks call for far more collaboration between employees than do their smaller counterparts. Many projects are broken down into smaller, more manageable chunks called "micro-tasks," which are then given to specific people to complete. In contrast, individuals in a crowd don't necessarily need to communicate with one another to do discrete, self-contained tasks. Many large-scale projects, however, cannot be delegated to a single worker and instead necessitate a group effort. The interrelated nature of macro-tasks makes coordination among crowd workers important. It is more likely that the target audience will break down a macro-task into smaller tasks than the requester. Coordination among crowd members is necessary for both the breakdown of macro-tasks and the subsequent combination of microtasks.

2.2 *Coordination Theories*

2.2.1 *Transactive Memory*

Coordination of tasks in a TMS is based on individuals within the system knowing one another and sharing information. Sharing or dividing up the mental work helps get the job done [14–16]. Information systems, organizational behavior, psychology, and communications are just a few of the many areas that have studies showing that TMS improves coordination and, by extension, performance [17, 18]. In particular, TMS's advantages in terms of team coordination have resulted in emergent and adaptive patterns of behaviour, which in turn have facilitated efficient and unobtrusive forms [19]. TMS is a proven method for coordinating the efforts of multiple teams.

Five essential components make TMS work. To begin, everyone in the group needs to be an expert in their own right. Second, the group needs to create a mental diagram of how everyone on the team contributes to the whole when it comes to this niche expertise. Third, each member of the group should be responsible for a certain set of tasks according to his or her area of expertise [20–22]. Everyone in the group needs to have confidence in one another and trust that everyone else is competent in their respective roles [23, 24]. Fifth, there needs to be constant two-way communication so that everyone in the group may benefit from everyone else's experiences and insights [25]. Sharing specialized information is made easier through conversation, which is vital for leveraging the group's knowledge.

Fundamental Elements of a Transaction Memory System

- Having specialized knowledge among members
- Common mental representation of specialized information
- Work obligations requiring specialist expertise
- Members who have faith in one another's expertise
- Participants who are willing to impart their expertise

Potential Benefits for Macro-tasking: TMS enables crowd workers to implicitly coordinate with one another. It's a more efficient alternative to direct speech. Task management systems can also be used to allocate responsibilities. Workers in the crowd are assigned tasks based on their areas of expertise as they become available.

Potential Drawbacks for Macro-tasking: Either a common background in the project's development is necessary for a crowd to create a TMS, or there must be some way for members of the crowd to convey their respective areas of expertise to one another. Crowd employees may not have the time necessary to develop a TMS. It is possible to construct platforms that reveal the experts among a group. However, when people join and depart the crowd, it may be challenging for those already working there to keep track of who knows what.

2.2.2 Coordination Theory

What is it? One of the most well-known schools of thought on the topic of coordination is called "coordination theory" [26]. Coordination, according to this theory, requires controlling what are called "dependencies between activities" [27]. The innovative use of coordination theory in the context of creating collaborative systems is the use of coordinating mechanisms based on the character of links between tasks [28]. In their analysis of coordination, Malone and Crowston established a framework based on the three axes of "actors," "interdependent tasks," and "resources" (1994). After conducting this analysis, proper coordination mechanisms can be put in place to address the coordination issues that develop due to interdependencies among tasks, actors, and resources.

How does it work? Coordination theory differs from other coordination theories in some key respects. It emphasizes the interdependencies between tasks, rather than between people or departments [29]. It is the relationships between activities, rather than between workers, that this concept of coordination emphasizes. The second thing it does is classify different kinds of interdependence between tasks. This makes it easier to understand the potential outcomes of particular dependence. One further benefit of this theory is that it can be used to describe coordination, which improves our ability to see how different job allocations and reassignments affect overall productivity. This raises consciousness of the impact that individual differences have on a team's cohesiveness. However, recent studies have revealed that coordination theory has limitations when it comes to coordinating collective actions [30, 31].

Essential Elements of Coordination Theory

- Identify tasks
- Identify and classify interdependencies between activities
- Utilize the Method that is Most Appropriate for Each Dependency Type

Potential Benefits for Macro-tasking: Using the tools provided by coordination theory, we may pinpoint and eliminate any obstacles standing in the way of completing a task by a large group. Plans for completing tasks based on coordination theory provide not only the direction workers need but also act as a means of communication that helps everyone in the crowd be on the same page.

Potential Drawbacks for Macro-tasking: Coordination theory is less helpful when job requirements are unclear or when task needs are emergent and change over time since it requires an individual or group to pre-plan the action. This limitation is especially noticeable when crowdsourcing complex adaptive tasks, as demonstrated by at least one study [32].

2.2.3 Coordination Theory based on Role

In a role-based structure, the duties and obligations associated with a certain position are clearly defined and consistently carried out. The duties and obligations related to a person's role are part and parcel of their definition. For a long time, roles have been considered the fundamental building block of organizational coordination [33, 34].

Some research has shown that role-based coordination can be helpful for complex and interdependent crowd action involving several participants without necessitating the engagement of any one person to carry out the essential duties [35, 36].

Coordination theories like role-based concepts emphasise on assigning defined responsibilities to team members and keeping them accountable for carrying those responsibilities through to completion. One way to conceptualize structure is as a hierarchical system, whereas another way is as a flat, decentralized network. Role-based coordinating concepts are used to get things done by determining who needs to do what, assigning that task to them, and then making sure their roles are set up in the best way possible to facilitate the work that must be done.

2.2.4 Interactive Coordination Theory

According to the relational coordination hypothesis, productivity greatly increases when workers have positive relationships with one another [37]. Trust and open dialogue are the bedrock of harmonious relationships between individuals. Coordination, it stands to reason, requires not just the interdependencies between tasks but also the interactions between the persons performing those tasks. Therefore, improved coordination and performance are likely to result from interactions among the people performing these jobs that are of higher quality [38, 39]. Relational coordination theorists claim that high-quality connections are especially crucial for achieving enhanced performance when labour is complicated, interdependent, and time-constrained [40]. Both research and common sense show that a company's success is linked to the degree to which its personnel can work together effectively [41].

According to the theory of relational coordination, coordination is "a process of communication and connection construction that serves to combine previously independent responsibilities". According to the notion of relational coordination, the key ingredients in a successful partnership are shared values, mutual respect, and mutual effort. Four dimensions of communication are theorized under the framework [42].

When people work together "through high-quality communication, underpinned by relationships of shared goals, shared knowledge, and mutual respect," this is called relational coordination [43]. This suggests that organizations will have an easier time coordinating their activities and accomplishing their goals if they communicate with one another on a consistent, accurate, timely, and problem-solving basis.

Essential Elements of the Theory of Coordinated Relationships

- Shared goals
- Relationships
- Communication
- Timely
- Shared knowledge
- Accurate
- Mutual respect
- Problem-solving focus
- Frequent

2.2.5 Stigmergic Coordination Theory

Cooperation in which each member of a group works to improve the final product is known as "stigmergic coordination". The term "stigmergy" was developed by entomologists after studying social insects. Traces (such as pheromones) left behind by insects like ants and termites encourage other insects to follow suit. The stigmergic coordination used by stingless ants to locate the quickest path to food is also used by termites in their construction and maintenance of neotropical anthills. The idea of stigmergy has influenced the development of many other types of collaborative activity, including multi-agent systems, free open-source software creation, and collective robotics. The stigmergic coordination method has been developed to meet the challenge of coordination in such fluid and emergent environments whereby workers and agents are not in constant communication with one another.

In addition to doing tasks, members of a specific collective also leave behind evidence of those efforts. This necessitates making those traces readily apparent to other participants. They use this information in conjunction with their foresight to identify what tasks must be completed next. Last but not least, while they go about their tasks, they leave behind evidence for the rest of the group to follow. Coordination occurs in a stigmergic fashion when numerous tasks are performed by many people. This finding suggests that stigmergic coordination can emerge in the absence of overt communication between group members and situated action provides the inspiration for the concept of "stigmergic coordination," which appears to function through the establishment of norms and practices for doing work that is shared across professionals in a certain field.

2.2.6 Coordination Framework

Among the goals of Okhuysen and Bechky's (2009) integrative framework was to speed up the identification of coordinating mechanisms. After reviewing the literature on coordination (predictability, accountability, and a common understanding), they identified three fundamental needs and five distinct coordination mechanisms (including but not limited to routines, roles, objects, representations, plans and rules, and proximity). In their comprehensive approach, Okhuysen and Bechky (2009) identify five types of coordination mechanisms, each of which helps with coordination by bolstering one of the three requirements. In other words, their framework can identify the conditions that encourage various forms of coordination.

How does it work? The integrative framework's ability to zero in on the precise mechanisms that will achieve the trick encourages coordination. Assuming the importance of transparency, predictability, and agreement, one may pick mechanisms that would help with all three. Similarly, other processes may be used to reinforce a certain condition if coordination was still an issue. For instance, the integrative framework might be used to locate a mechanism for enhancing accountability in cases when groups are having trouble holding themselves accountable.

Coordination Mechanisms

Plans and rules: A set of guidelines for organizing tasks, groups, and sections is often referred to as "plans and rules." Plans and guidelines help with things like this by outlining who is responsible for what. Coordinated decisions regarding what (further) activities should be done and which choices should be made among the possibilities to fulfil tasks are made possible by having a plan and guidelines to follow.

Objects and representations: When used properly, objects, representations, and technologies provide information that aids in task coordination and leads to more efficient results (direct information-sharing). For teams to effectively communicate problems, ideas, and actions, boundary objects (such as data spreadsheets) must be used. Scaffolding, a graphical representation of roles and responsibilities, provides a framework for keeping everyone on the same page about what must get done, who is responsible for what, how the various pieces of the puzzle fit together, and how far along they actually are (acknowledging and aligning work).

Roles: There are two ways in which roles can be used as a kind of coordination. At the same time as they represent a set of duties and activities performed by an actor in a certain position, roles also make it possible for responsibilities to be reshaped in response to changes in the workplace (monitoring and updating). By establishing clear divisions of labor, a shared viewpoint can be developed. If everyone is on the same page, switching roles is simple.

Routines: Traditional organizational settings define routines as "regular patterns of conduct that are governed by rules and norms". But there has been a growing consensus in the scholarly literature that "routines" incorporate "social relevance and social interaction... embedded inside them".

Proximity: The term "proximity" is used to describe a method of cooperation that is not necessarily tied to actual physical proximity. The degree of exposure and familiarity are two such characteristics. The term "visibility," meaning the capacity to observe the actions of others, is frequently used in conjunction with collocation but is not essential for it to exist. Assisting in the coordination of actions is facilitated by one's "familiarity," or their reliance on their past relationships with other people. Once again, though, it is not always the case that familiarity is a prerequisite for collocation.

Conditions

Accountability: Who is responsible for certain parts of a larger work is described by its "accountability." The expectation is that coordinated actions in a group will result from making explicit and apparent who is responsible for what tasks. In the integrative framework, accountability extends to the means developed through unplanned and impromptu exchanges, such as dialogues among team members. This scaffolding might take the form of plans, rules, or physical objects, and it can connect tasks to the people who are responsible for them. In addition to facilitating constant monitoring, updating, and handoffs among workers, roles, routines, and transparency all contribute to these processes.

Predictability: Workers' knowledge of which smaller jobs make up larger tasks and in what order, as well as the specific actions needed to complete each task, can be explained by their experience with predictable work environments. Coordination relies on predictability since it allows workers to modify their job to the activity of others and perform it accordingly, highlighting the anticipation of later activities and associated actions of others. The coordinating mechanisms that produce predictability are plans and objects, which specify what actions must be taken. Understanding how other employees value the work is another way in which familiarity and routines increase predictability.

Common understanding: A common understanding exists when all employees have the same picture in their heads of what the finished product looks like, including the work's purpose, objectives, and methods of completion. More efficient coordination is achieved through the use of plans and regulations, which foster a shared comprehension of the complete dependently dependent task and process. Workers benefit from routines and familiarity because they learn how the pieces fit together to form the whole. Sharing and learning new ways to do the same tasks also helps objects and roles acquire a shared perspective.

Elements of a Framework for Coordinated Integration
- Coordination mechanisms
- Objects and representations
- Plans and rules
- Roles
- Routines
- Proximity
- Accountability
- Conditions
- Predictability

Local Training: Local training is the process by which each participating user or device trains the model using its data. This allows for the preservation of data privacy, as the sensitive information never leaves the local device. In the context of crowd coordination, local training enables participants to contribute their unique insights and expertise to the shared model without exposing their data, thus promoting collaboration while maintaining privacy. This is particularly important when tackling complex macro-tasks that require a diverse set of skills and knowledge, as individuals may be more willing to contribute if their privacy is assured.

Model Aggregation: Once local training is completed, each participant shares only their model updates, such as gradients, weights, or parameters, with a central server or aggregator. The central server is responsible for aggregating these updates from all participating devices to update the global model. This process allows for the consolidation of individual contributions into a coherent, shared model that represents the collective knowledge of the crowd. By using model aggregation, crowd coordination efforts can harness the power of the crowd while minimizing the exposure of sensitive data, as only model updates, rather than raw data, are shared.

Global Model: The global model represents the combined knowledge and expertise of all participants, synthesized through the aggregation of local model updates. After the central server updates the global model, it is shared back with the participating devices for further local training. This iterative process continues until the global model reaches a desired level of accuracy or convergence. In the context of crowd coordination, the global model serves as the common goal that participants work towards. As the model improves, it becomes increasingly capable of addressing the complex macro-tasks that are the focus of the crowd coordination effort.

Crowd coordination strategies play a critical role in organizing large groups of people to work together towards a common goal, particularly when addressing complex macro-tasks. Integrating federated learning into these strategies can help to overcome several challenges, such as model and data heterogeneity, scalability, and efficiency. In this section, we will explore how federated learning can enhance crowd coordination strategies by addressing these challenges.

Efficiency: Efficient communication and computation are crucial for the success of crowd coordination efforts. Federated learning can improve efficiency by minimizing data transfers between participating devices and a central server, as only model updates are shared during the aggregation process. This reduces bandwidth requirements and communication overhead, ensuring that crowd coordination efforts remain efficient even when dealing with large volumes of data and numerous participants. Additionally, various optimization techniques can be employed to further enhance the efficiency of federated learning systems, such as model compression, sparsification, and quantization.

By addressing the challenges of model and data heterogeneity, scalability, and efficiency, federated learning can significantly improve crowd coordination strategies. The integration of federated learning into existing crowd coordination methodologies enables the harnessing of diverse talents and knowledge from large groups of people, while respecting privacy and data ownership. This, in turn, can lead to more effective and efficient solutions for addressing challenging macro-tasks through the collective efforts of large groups of people.

3. Recent Research on Coordinated Macro task Crowdsourcing

3.1 Search Methods

We started by researching "microtask," "coordination," and "crowdsourcing" in Google Scholar to assess recent research on coordination in macro-tasking. In August 2018, we searched, and 60 items were returned. After reviewing abstracts, we chose which articles would be included based on whether they (1) dealt with coordination issues for macro-tasks or (2) proposed and empirically tested ideas or designs for crowdsourcing macro-tasks. Researchers excluded works published in languages other than English, patent applications, patent review papers, review articles, and works that were not originally written in English. As a result of the initial search, eight studies were found to completely meet the criteria. We went all the way back to the beginning to trace the original leads in the hunt. That's because we found certain studies that the keyword search had missed, despite their significance to the field of macro-task coordination. Despite their frequent mention in discussions of studies that explore macro-task coordination issues. Ten publications were chosen to include in the literature evaluation since they matched all the criteria.

3.2 *Coordinate Crowdsourcing Methods for Macro-tasks*

To determine which of the five coordinating mechanisms had been put into practice, researchers conducted a comprehensive literature review. The authors did this by first understanding the fundamental concepts and presumptions of all the existing coordination theories. Next discussed our readings of each study, trying to determine which of the five mechanisms of coordination best described the approach to coordination.

The coordination theory method seems to best fit most investigations. Typical research in this area has been on how best to detect and control the many interdependencies that exist across jobs and between workers. To identify and manage reliance, researchers in this study employed numerous strategies and tools. Systems provide workers with blueprints for the work, detailing the sequence and structure of the job in units of subtasks, helping them to better comprehend relationships at the task level. The organization of labor according to workers' level of ability and competence, with the goal of better coordinating available capable personnel.

According to the authors, familiarity among workers is a necessary element for executing jobs for dispersed populations. When forming teams, they took into account not only availability, but also participants' prior experience working together. They also offered a means of instantaneous communication and a venue for group authoring. Researchers found that employees who worked with people they already knew on the job performed better because they were able to capitalize on each member's talents and understand the team's workflow more easily. This research did not take place in the same setting as the face-to-face organizational engagement that has been the focus of prior studies on relational coordination. This study, however, successfully evaluated the value of relational communication by bringing together people who already knew each other and making the most of their shared knowledge using appropriate communication technologies.

In conclusion, it seems that academics rely heavily on coordination theory to investigate macro-task management in crowdsourcing. The role-based theory of coordination and the relational theory of coordination comes in much behind the first. The use of TMS or stigmergic methods was absent from all the included research. However, the literature base is still fairly young, with only two studies published before 2015 and more than half published in 2017 or 2018.

3.3 Coordination Mechanisms for Macro-task Crowdsourcing

3.3.1 Rules and Plans

Tasks that need to be done and the personnel responsible for doing them have been determined with the use of plans and rules. In the context of macro-tasking, in particular, plans and rules for crowd workers should modify and adapt as the job develops. Turkomatic, a platform for editing workflows in real time that relied on crowdsourcing. The inspiration behind Turkomatic came from a desire to equip workers with a resource that would make it easier for them to tackle complex tasks. To help large groups of people with difficult tasks like story writing, a reflect-and-revise strategy. Their success in completing narrative writing tasks based on top-down goals exemplifies the value of such objectives for work that is both complex and indeterminate. While the goals did their job of accommodating the outputs of many crowd workers, a unique aspect of this approach was that the goals weren't hardcoded into the writing system, but rather selected from among the workers who had already completed their stages. In this way, by cycling through the phases of introspection and aim-setting, workers were able to provide better solutions to problems.

3.3.2 Routines

Routines as a tool for coordination were mentioned in only one of the articles reviewed. Routines are most useful when there is little room for error and little complexity to navigate. They noted that routines can help workers learn from one another and become more familiar with how to complete jobs in compliance with policies and procedures. Better teamwork was shown to result from increased familiarity among workers, which routines would facilitate.

3.3.3 Proximity

Authors found used physical closeness as a stand-in for familiarity but couldn't find any that did the same for visibility in the context of coordinating massive undertakings. Potentially, this is because the investigations looked at were motivated by the need to solve issues unique to online crowdsourcing, a trend characterized by the dispersed and asocial nature of its workforce.

3.4 Summary

Importantly, our study team discovered that few coordination options have been studied in the context of crowdsourcing for macro-tasks. Specifically, authors discovered that most of the research into macro-level task coordination has centered on planning and establishing guidelines (80%), which define an overall objective and a set of sub-objectives. Research into constructing objects and representations came next (50%). Several researchers also employed role-based approaches to coordinate large-scale undertakings (40%). One analysis focused on the relationship between routine and proximity.

4. Results and Discussions

The authors performed a rapid literature review of the several coordination theories utilized in macro-tasking and discovered that while concepts of proximity (visibility and familiarity) and routines have been examined extensively, stigmergic and relational coordination concepts have gotten less attention. The greatest promise for the crowdsourcing of large projects appears to lie in these ideas and approaches. To begin with, these frameworks are dependent on interactive social processes and adjusting to emergent conditions. They don't put nearly as much weight on identifying job interdependence or on assigning roles in advance to members of a crowd. Initial work definition approaches will almost certainly always rely significantly on requestors. However, informal coordination is established by proximity (visibility and familiarity), routines, and stigmergic coordination, allowing for more spontaneous coordination of tasks. As crowdsourcing evolves to focus more on large-scale projects than small ones, authors anticipate that these kinds of informal approaches to coordination will prove to be the most efficient. The "shared knowledge" concept in relational coordination also incorporates many TMS ideas.

Here, the authors present and discuss several intriguing research ideas that could help us learn more about stigmergic and relational coordination. The stigmergic and relational coordination display in our designs is also highlighted. The propositions of a design are the overarching statements that explain how all the parts of the design work together. The design hypotheses discussed in this chapter are broad generalizations about the link between system design and coordinating methods.

4.1 Stigmergic Coordination

Stigmergic coordination is communication between people that takes place mostly through clues rather than overt words. Stigmergic mechanisms may be useful for the coordination of macro-tasks since they do not necessitate communication between workers and allow for coordination to be done by observation of the environment and the traces left by previous workers. The process of stigmergic cooperation, for instance, does not include instituting frameworks of rules and regulations. As a result, more people in the crowd might get to work faster and more easily adapt their actions to the current situation. With this in mind, the authors propose research issues that, with stigmergic methods, could improve macro-task coordination.

RQ 1: "How can authors support the traces of prior work in the crowdsourcing of macro-tasks?"

It has been shown that traces of stigmergic coordination mediate between the activities of separate employees.

Workers can use traces to learn about their accomplishments and the tasks still ahead of them. Thus, technologies that promote the efficient leaving of traces could be created to aid stigmergic collaboration for crowdsourcing macro-tasks. It's possible, for instance, that crowdsourcing platforms will provide tools for users to add comments and notes to their contributions. Features that monitor and highlight the work's progress could be built into these systems.

RQ 2: "How can authors promote the shared interpretations of traces in the crowdsourcing of macro-tasks?"

These traces are used by employees engaging in stigmergic coordination to deduce what has been done and what should be done next. Because they are part of a cohesive group with a common history and experience, workers can coordinate without consciously trying to do so. Through exposure to similar circumstances, individuals of a community are more likely to develop shared practices and habits for completing tasks. This is what makes it possible for employees to use traces as a means of engaging in implicit cooperation. The authors next go through three strategies for using stigmergic coordination with crowdsourcing for large-scale tasks.

One strategy is to hire members of the crowd who already share a culture, set of values, and work habits. Given enough time, the crowd workers would develop stigmergic collaboration strategies like those used by members of established online communities. For instance, a group of GitHub users could be contracted to complete a massive project. This workforce would already share a common history, culture, and practices. Users will benefit most from the crowdsourcing platform if it is organized similarly to the popular code repository GitHub. The new crowdsourcing platform on GitHub should enable stigmergic cooperation among crowd workers to accomplish macro-tasks, as it fosters a communal workspace where employees may establish and maintain common practices and conventions.

One alternative is to organise a group of people in a virtual space to find people to work in the crowd. Two benefits can be gained by using this strategy. The first benefit is that it would make it easier for workers in a crowd to establish common ground and establish routines. Given enough time, the crowd workers would develop stigmergic collaboration strategies similar to those used by members of established online communities. Second, this approach would make it possible to create a new online community for a topic or interest that does not yet have its own online home. Consider the ramifications of a scenario in which workers performing macrotasks needed expertise in a particular programming language, such as the common business-oriented language (COBOL). This language is not commonly taught, but it is still used by many main frames that run older applications. The recruiting of COBOL-needing macrotasks would benefit from the establishment of an online community of COBOL programmers.

The third strategy involves vetting prospective macro-tasking crowd employees by looking at their track record in a given online community. To be considered for selection for macro-tasking, prospective workers would be encouraged to join a designated online community. This would provide crowd workers access to a preexisting online community to pick up ground rules and acquire fundamental expertise. The team would get closer over time, establishing conventions and habits that would allow them to take on larger-scale projects and accomplish more together.

The three stigmergic coordination design hypotheses are summarized in Table 1. Research questions 1 and 2 served as inspiration for potential design solutions.

Table 1. Stigmergic coordination design propositions.

Design propositions of Stigmergic coordination	
Design proposition 1	Crowd workers can be more efficient at completing macro-tasks if crowdsourcing platforms allow for stigmergic cooperation.
Design proposition 1a	Crowdsourcing systems that aim to foster stigmergic collaboration will make it easier for users to leave and make public the traces of previous effort.
Design proposition 1b	Stimulating stigmergic cooperation requires crowdsourcing platforms to make it easy for participants to agree on how to interpret the results of previous efforts.
Design proposition 1c	Crowdsourcing platforms that aim to foster stigmergic coordination should allow for the utilisation of established workplace practices and procedures.

4.2 Relational Coordination

There are three types of connectivity (based on the quality of the relationships established) and four dimensions (based on the frequency, accuracy, timeliness, and emphasis placed on problem-solving) by which they can be evaluated (shared knowledge, shared goals, and mutual respect). The strength of interpersonal bonds within a society is both an indicator of and a driver of such qualities. Workers can coordinate complicated tasks in uncertain contexts, which is one of the many advantages of relational coordination. To do this, people are given the opportunity to coordinate their efforts while solving challenges collectively. A group of people can work together to create art on a shared canvas thanks to some sort of relational coordination. Collaborative groups can efficiently coordinate their efforts by capitalizing on aspects of the relationships they have fostered. Relational coordination is distinguished from stigmergic coordination by its emphasis on connection quality.

After that, the authors propose research questions that, if answered, would deepen our knowledge of crowdsourcing's potential for coordinating large-scale projects.

RQ 3a: How can shared knowledge be promoted in the crowd sourcing of macro-tasks?

According to the relational coordination concept, team members are more likely to get along and value one another when they have a firm grasp of their specific duties. Knowledge of this kind aids in the development of clear and precise expression. There are two major obstacles to overcome before crowdsourcing can achieve a high enough degree of shared knowledge. One, crowdsourcing typically involves workers who are not permanent and who have not had much practice cooperating with one another. That means they don't know a whole lot about each other, to begin with. Second, crowd workers may not have enough time to acquire common knowledge, which can increase productivity and efficiency, depending on the length of time needed to finish the activity. Both problems severely weaken the coordinating mechanism of crowd workers relying on previously acquired information.

Crowdsourcing systems can be designed in several ways, all of which have the potential to increase the spread of knowledge. To begin, technology could aid crowd workers in determining who has certain knowledge. Having a system that shows off employee profiles could help with this. Information about the employee's education and employment history could be included in this profile. The viewing of an employee's profile is restricted unless the employee gives explicit permission, and the information displayed on the profile may change depending on the viewer's identity. In some cases, the macro task force may have more detailed information about each employee than the public. Second, it's important to create systems that make as much of the knowledge held by each crowd worker as feasible available to everyone. To accomplish this, the team could be encouraged to increase the flow of information by sharing, applying, and ultimately integrating its collective expertise. For crowdsourcing platforms to be effective, they will need to have not only synchronous and asynchronous communication options, but also a slew of other useful functions. Consider the following as evidence of the usefulness of such a system: a search of the communications archive will yield results for any visual aids used in conversations, including sketches, photographs, whiteboards, papers, links, and templates. Furthermore, these platforms should support in-the-moment revisions and comments, enabling users to provide additional context for their work and respond to questions from coworkers in real time.

RQ 3b: "*How can shared goals be leveraged in the crowdsourcing of macro-tasks?*"

Coordination in relationships relies significantly on common objectives, which could be a stumbling block when crowdsourcing massive projects. When employees are working toward the same goals, they are more likely to communicate effectively with one another. This directs employees to prioritize communication that helps them solve problems above chatter that serves no useful purpose. When it comes to crowdsourcing larger projects, encouraging everyone to work toward the same goals should be a breeze. Each member of the crowd workforce has been recruited for a certain macro-task. This massive undertaking is the true common objective. Conversely, it may be challenging for crowd workers to keep a unified perspective on the success or failure of their collective efforts to achieve the goals they have set. More so than in more static micro-tasking contexts, this can become troublesome in more dynamic macro-tasking settings.

Authors use boundary objects to facilitate a common understanding of objectives in the crowdsourcing of large-scale projects. Boundary objects are a technique for coordinating things and representations. Boundary objects facilitate the sharing of information about issues, ideas, and actions between groups. The primary value of boundary objects is that they provide a framework for situating the unique perspective of an individual within the greater framework of a group's experience. Therefore, employees can use boundary objects to convey the collective's condition without needing to completely comprehend the circumstances of each member. By facilitating the pursuit of individual interests while still contributing to the collective's end aim, boundary objects may be useful in the context of crowdsourcing macro-tasks. Which border items are appropriate is ambiguous. One solution could be to raise people's awareness of their immediate environment.

RQ 3c: "How can mutual respect be promoted in the crowdsourcing of macro-tasks?"

When individuals are treated with dignity and consideration, they are more likely to listen to one another, which in turn increases the potential for mutual understanding and the development of creative approaches to problems. One way in which crowd sourcing and mutual regard are similar is that both require work to reach a common understanding of the issue at hand. Due to the ad hoc nature of crowd sourcing, it sometimes necessitates the coordination of a diverse collection of individuals who have never worked together before to complete a certain task. There are also some unique difficulties, such as the possibility that crowd employees will come to look down on one another. All these dangers make it more difficult for the crowd to rely on mutual respect as a means of maintaining order.

Finally, solutions should be developed to foster mutual respect when needed. The authors suggest some interesting approaches, but more research is required to establish the optimal path of action. There is a wealth of literature to draw from in the study of conflict and its resolution. Relationship conflicts, process conflicts, and task conflicts are just a few examples that have emerged. Personal disputes between team members are what cause relationship conflict, while arguments about tasks and processes are work-related but not interpersonal. Studies have demonstrated that interpersonal conflicts always harm productivity, but conflicts over tasks and procedures might have a positive effect. Conflict resolution systems need to identify the nature of the conflict at hand.

Methods for resolving group conflict have been identified in the literature on the topic. Relationships between people can take many different forms, including those characterized by accommodation, avoidance, collaboration, competition, and compromise. It is evident that there are benefits and drawbacks to each approach, as well as opportunities for system interventions, but doing so would go beyond the scope of this chapter. A research program could investigate the effectiveness of various approaches to crowdsourcing macro-tasks, as well as the optimal design of technologies to facilitate various approaches.

The four relational coordination design hypotheses are summarized in Table 2. Propositions for the design were generated from the answers to research questions 3a, b, c, and e.

Table 2. Relational coordination design propositions.

Design propositions for relational coordination	
Design proposition 2	Relational coordination-enabled crowdsourcing technologies will aid crowd workers in efficiently completing macro-tasks.
Design proposition 2a	Crowdsourcing systems should encourage relational coordination by easing the production and dissemination of shared expertise.
Design proposition 2b	Crowdsourcing solutions that aim to improve relational coordination should make it easy for people to set and discuss shared objectives.
Design proposition 2c	Crowdsourcing platforms that aim to improve relational coordination should encourage the maturation of mutual respect.
Design proposition 2d	Crowdsourcing systems that aim to improve relational coordination ought, therefore, to prioritize clear and open lines of communication.

4.3 Limitations

Theories of coordination are shown to share or overlap in some key conceptual areas, as discussed in this chapter. However, while debating their benefits and drawbacks, authors generally viewed them as individual entities. Authors may have been too hasty or too rigid in dividing up the various coordinating theories. Academics looking into crowdsourcing coordination problems should think about using a hybrid strategy that combines different parts of each theory. Adding role-based coordination to the already-existing stigmergic coordination is just one example. To accomplish this, it may be necessary to introduce new people into the workplace who are not familiar with the established procedures and policies. Disruption to the workplace caused by their unfamiliarity with traces can be minimized by clearly outlining their responsibilities. The authors also recognize that the literature on each theory is vast and deep, far exceeding the scope of this one chapter. Each theory is briefly described in this chapter. It's often difficult to determine where clarity and completeness begin, and brevity and confusion begin. Therefore, this chapter's objective was to highlight problems associated with macro-task coordination in crowdsourcing settings. Our suggestions should be seen as that, and readers are encouraged to do their own research on the topics. The authors conclude by recommending some design concepts that bridge the gap between theoretical and practical considerations. All our hypotheses are generalizations, as are all hypotheses. Design hypotheses should be derived from design propositions and then tested empirically. Our expectation is that in the future, a growing number of scholars will rise to this challenge.

5. Conclusion

The integration of federated learning into crowd coordination efforts holds significant promise for enhancing the effectiveness of large groups of people working together to address complex macro-tasks. However, several areas of research need to be further explored to realize the full potential of this approach. In this section, we will discuss two key future research directions: incentive mechanisms for motivating user participation and the development of secure, efficient communication protocols to support large-scale crowd coordination.

One of the challenges in federated learning-based crowd coordination is motivating users to participate in the process. Effective incentive mechanisms are crucial to ensuring that users are willing to share their resources, time, and expertise for the collective benefit of the crowd. Future research should explore various incentive mechanisms, including monetary rewards, social recognition, and gamification, to encourage user participation in federated learning-based crowd coordination efforts. Moreover, the development of fair and transparent reward systems that account for the varying levels of user contributions, the quality of their data, and the computation resources provided are essential for fostering continued engagement and collaboration.

Federated learning relies heavily on the communication between participating devices and a central server for model aggregation. As crowd coordination efforts scale to accommodate larger groups of participants and more complex tasks, the need for secure and efficient communication protocols becomes increasingly important. Future research should focus on the development of communication protocols that ensure data privacy, maintain the integrity of model updates, and minimize the potential for adversarial attacks. This includes exploring techniques such as secure multi-party computation, homomorphic encryption, and differential privacy. Additionally, optimizing communication efficiency through

330 Handbook on Federated Learning: Advances, Applications and Opportunities

techniques like model compression, scarification, and quantization can help to reduce the overhead associated with data transfers and improve the overall performance of federated learning-based crowd coordination efforts.

By addressing these future research directions, we can continue to advance our understanding of how federated learning can be effectively integrated into crowd coordination methodologies and enhance the success of such efforts. Exploring innovative incentive mechanisms and developing secure, efficient communication protocols will be essential for unlocking the full potential of federated learning-based crowd coordination, enabling large groups of people to tackle complex macro-tasks while maintaining privacy and data ownership.

References

[1] Thomas, N., Balagopal, A. and Varghese, S.M. 2019. Green route: An ecofriendly route suggestion and description based on congestion and air quality. Int. J. Innov. Technol. Explor. Eng., 8(5): 414–419.

[2] Musil, J., Musil, A. and Biffl, S. 2013. Elements of software ecosystem early-stage design for collective intelligence systems. In 2013 1st International Workshop on Software Ecosystem Architectures, WEA 2013 - Proceedings, pp. 21–25. doi: 10.1145/2501585.2501590.

[3] Wang, L., Yang, Y. and Wang, Y. 2019. Do higher incentives lead to better performance?—An exploratory study on software crowdsourcing. In International Symposium on Empirical Software Engineering and Measurement, vol. 2019-Septe. doi: 10.1109/ESEM.2019.8870175.

[4] Maheswari, R., Vignesh, S., Kumar, R., Venkatesh, T.M. and Sundar, R. 2022. Voice control and eyeball movement operated wheelchair. In 2022 International Conference on Edge Computing and Applications (ICECAA), pp. 805–809.

[5] Yang, Y. and Saremi, R. 2015. Award vs. worker behaviors in competitive crowdsourcing tasks. In International Symposium on Empirical Software Engineering and Measurement, pp. 68–77. doi: 10.1109/ESEM.2015.7321192.

[6] Ponmalar, A., Renukadevi, B., Anand, J., Kamalisri, G., Dharshini, N. and Sobana, N. 2022. Automatic forensic analysis of criminal navigation system using machine learning. In 2022 1st International Conference on Computational Science and Technology (ICCST), pp. 1–5.

[7] Lykourentzou, I., Vergados, D.J., Papadaki, K. and Naudet, Y. 2013. Guided crowdsourcing for collective work coordination in corporate environments. Lecture Notes in Computer Science (including subseries Lecture Notes in Artificial Intelligence and Lecture Notes in Bioinformatics), vol. 8083 LNAI, pp. 90–99. doi: 10.1007/978-3-642-40495-5_10.

[8] Sundaram, S.G., Ponmalar, A., Deeba, S., Aarthi, S., Shwetha, S. and Anand, J. 2022. Plane detection and product trail using augmented reality. In 2022 1st International Conference on Computational Science and Technology (ICCST), pp. 887–892.

[9] Sims, R. 2023. Minimal cognition and stigmergic coordination: An everyday tale of building and bacteria. Cogn. Syst. Res., 79: 156–164. doi: 10.1016/j.cogsys.2023.01.008.

[10] Xu, Z., Wang, G. and Guo, X. 2023. Event-driven daily activity recognition with enhanced emergent modeling. Pattern Recognit., 135. doi: 10.1016/j.patcog.2022.109149.

[11] Sims, R. and Yilmaz, Ö. 2023. Stigmergic coordination and minimal cognition in plants. Adapt. Behav. doi: 10.1177/10597123221150817.

[12] Ribino, P., Islam, S., Ciampi, M. and Papastergiou, S. 2023. Swarm intelligence based multi-agent communication model for securing healthcare ecosystem. Lecture Notes in Networks and Systems, vol. 594 LNNS, pp. 50–61. doi: 10.1007/978-3-031-21333-5_5.

[13] Xu, X., Li, R., Zhao, Z. and Zhang, H. 2022. Stigmergic independent reinforcement learning for multiagent collaboration. IEEE Trans. Neural Networks Learn Syst., 33(9): 4285–4299. doi: 10.1109/TNNLS.2021.3056418.

[14] Alajami, A.A., Moreno, G. and Pous, R. 2022. Design of a UAV for autonomous RFID-based dynamic inventories using stigmergy for mapless indoor environments. Drones, 6(8). doi: 10.3390/drones6080208.

[15] Tamari, R., Friedman, D., Fischer, W., Hebert, L. and Shahaf, D. 2022. From users to (Sense)Makers: On the pivotal role of stigmergic social annotation in the quest for collective sensemaking. In HT 2022: 33rd ACM Conference on Hypertext and Social Media - Co-located with ACM WebSci 2022 and ACM UMAP 2022, pp. 236–239. doi: 10.1145/3511095.3536361.

[16] Gurcan, O. 2022. Proof of work is a stigmergic consensus algorithm: Unlocking its potential. IEEE Robot. Autom. Mag., 29(2): 21–32. doi: 10.1109/MRA.2022.3165745.

[17] Schubotz, R., Spieldenner, T. and Chelli, M. 2022. stigLD: Stigmergic coordination in linked systems. Mathematics, 10(7). doi: 10.3390/math10071041.

[18] Shaw, S., Wenzel, E., Walker, A. and Sartoretti, G. 2022. ForMIC: Foraging via multiagent RL with implicit communication. IEEE Robot. Autom. Lett., 7(2): 4877–4884. doi: 10.1109/LRA.2022.3152688.

[19] Xu, X., Li, R., Zhao, Z. and Zhang, H. 2022. Trustable policy collaboration scheme for multi-agent stigmergic reinforcement learning. IEEE Commun. Lett., 26(4): 823–827. doi: 10.1109/LCOMM.2022.3144451.

[20] Tran, J., Fule, P. and Szabo, C. 2022. Effects of information sharing on swarm based communication in dynamic environments. In Proceedings - Winter Simulation Conference, vol. 2022-Decem, pp. 461–472. doi: 10.1109/WSC57314.2022.10015493.

[21] Dounas, T., Voeller, E., Prokop, S. and Vele, J. 2022. The architecture decentralised autonomous organisation - a stigmergic exploration in architectural collaboration. In Proceedings of the International Conference on Education and Research in Computer Aided Architectural Design in Europe, 1: 567–576.

[22] Casalnuovo, G. and Erioli, A. 2022. Deep trails: Coupling of structural optimization and self-organization processes for the computational design of composite surface tectonics. In Proceedings of the International Conference on Education and Research in Computer Aided Architectural Design in Europe, 2: 85–94.

[23] Alajami, A.A., Moreno, G. and Pous, R. 2022. A ROS Gazebo Plugin design to simulate RFID systems. IEEE Access, 10: 93921–93932. doi: 10.1109/ACCESS.2022.3204122.

[24] Spieldenner, T. and Chelli, M. 2022. Linked data as medium for stigmergy-based optimization and coordination. Communications in Computer and Information Science, vol. 1622 CCIS, pp. 1–23. doi: 10.1007/978-3-031-11513-4_1.

[25] Gürcan, Ö. 2022. Proof-of-work as a stigmergic consensus algorithm. In Proceedings of the International Joint Conference on Autonomous Agents and Multiagent Systems, AAMAS, 3: 1613–1615.

[26] Schubotz, R., Spieldenner, T. and Chelli, M. 2022. stigLD: Stigmergic coordination of linked data agents. Communications in Computer and Information Science, vol. 1566 CCIS, pp. 174–190. doi: 10.1007/978-981-19-1253-5_13.

[27] Kopetz, H. 2022. Data in communication. Springer Briefs in Computer Science, pp. 27–34. doi: 10.1007/978-3-030-96329-3_6.

[28] Faulk, D.K., Frey, T.L. and Holkem, L. 2022. Ad-hoc Stigmergic Load Shaping. doi: 10.2514/6.2022-1761.

[29] Chen, J., Liu, Y., Pacheck, A., Kress-Gazit, H., Napp, N. and Petersen, K. 2022. Errors in collective robotic construction. In Springer Proceedings in Advanced Robotics, vol. 22 SPAR, pp. 269–281. doi: 10.1007/978-3-030-92790-5_21.

[30] Bolshakov, V., Alfimtsev, A., Sakulin, S. and Bykov, N. 2022. Deep reinforcement ant colony optimization for swarm learning. Studies in Computational Intelligence, vol. 1008 SCI, pp. 9–15. doi: 10.1007/978-3-030-91581-0_2.

[31] Palade, A. and Clarke, S. 2022. Collaborative agent communities for resilient service composition in mobile environments. IEEE Trans. Serv. Comput., 15(2): 876–890. doi: 10.1109/TSC.2020.2964753.

[32] Batra, S., Dey, A.K., Singh, R. and Chaudhuri, M. 2023. Influence of transactive memory systems and strategic orientations on the performance of hospitality firms. J. Hosp. Tour. Insights, 6(1): 131–150. doi: 10.1108/JHTI-03-2021-0071.

[33] Zhang, X., Zhou, X., Wang, Q., Wu, Z. and Sui, Y. 2023. Political skills matter: The role of academic entrepreneurs in team innovation. Eur. J. Innov. Manag. doi: 10.1108/EJIM-08-2022-0456.

[34] Yan, Y., Xin, S. and Zha, X. 2023. Understanding TMS and knowledge transfer in the social media mobile App context. Aslib. J. Inf. Manag. doi: 10.1108/AJIM-08-2022-0366.

[35] Chen, Y. and Yi, Y. 2023. TMT transactive memory system and business model design: The moderating effect of strategic orientation. J. Knowl. Manag. doi: 10.1108/JKM-07-2022-0546.

[36] Huan Zhang, Yibei Liu, Xin Wang, Ziqian Cui, Haiman Wan, Xiping Liu, Ullrich Wagner and Gerald Echterhoff. 2023. Benefits of collaborative remembering in older and younger couples: The role of conversation dynamics and gender. Memory. doi: 10.1080/09658211.2023.2166963.

[37] Tuan, L.T. 2023. Fostering green product innovation through green entrepreneurial orientation: The roles of employee green creativity, green role identity, and organizational transactive memory system. Bus Strateg. Environ., 32(1): 639–653. doi: 10.1002/bse.3165.

[38] Scheibe, K.P., Mukandwal, P.S. and Grawe, S.J. 2022. The effect of transactive memory systems on supply chain network collaboration. Int. J. Phys. Distrib. Logist. Manag., 52(9-10): 791–812. doi: 10.1108/IJPDLM-07-2021-0288.

[39] González-Ponce, I., Díaz-García, J., Ponce-Bordón, J.C., Jiménez-Castuera, R. and López-Gajardo, M.A. 2022. Using the conceptual framework for examining sport teams to understand group dynamics in professional soccer. Int. J. Environ. Res. Public Health, 19(23). doi: 10.3390/ijerph192315782.

[40] Xu, N., Ghahremani, H., Lemoine, G.J. and Tesluk, P.E. 2022. Emergence of shared leadership networks in teams: An adaptive process perspective. Leadersh Q, 33(6). doi: 10.1016/j.leaqua.2021.101588.

[41] Xiao, Y., Cen, J. and Hao, J. 2022. Transactive memory system and green innovation: A cross-level mediation of social network. Ind. Manag. Data Syst., 122(12): 2737–2761. doi: 10.1108/IMDS-04-2021-0254.

[42] Okada, M. and Shirahada, K. 2022. Organizational learning for sustainable semiconductor supply chain operation: A case study of a japanese company in cross border M&A. Sustain., 14(22). doi: 10.3390/su142215316.

[43] Palermos, S.O. 2022. Responsibility in epistemic collaborations: Is it me, is it the group or are we all to blame? Philos. Issues, 32(1): 335–350. doi: 10.1111/phis.12230.

Chapter **15**

Stock Market Prediction via Twitter Sentiment Analysis using BERT

A Federated Learning Approach

M Rajeev Kumar,[1,*] *S Ramkumar,*[2] *S Saravanan,*[3]
R Balakrishnan[4] and *M Swathi*[5]

1. Introduction

Federated Learning (FL) has had high popularity in working out collaborative models with no disclosing sensitive data over the previous

[1] Professor, Department of Computer Science and Engineering, Vel Tech Rangarajan Dr. Sagunthala R&D Institute of Science and Technology, Avadi, Chennai – 600 062.

[2] Associate Professor, Department of Electronics and Communication Engineering, Sri Eshwar College of Engineering, Coimbatore – 641 202, Tamil Nadu, India.

[3] Associate Professor, Department of AI&DS, Karpagam Academy of Higher Education, Faculty of Engineering, Coimbatore – 641 021, Tamil Nadu, India.

[4] Assistant Professor, Department of Robotics And Automation Engineering, Vel Tech Multi Tech Dr. Rangarajan Dr. Sakunthala Engineering College, Chennai – 600 062, Tamil Nadu, India.

[5] Assistant Professor, Department of Science and Humanities, Sri Sairam College of Engineering, Anekal – 562 106, Bengaluru, India.

Emails: sram0829@gmail.com; ssaravananme@gmail.com; balakrishnan007r@gmail.com; swathimanjunath.56@gmail.com

* Corresponding author: smrajeevkumar@gmail.com

decade. Since its inception, the most popular approach in literature has been centralized FL (CFL), in which a single organization builds global models. On the other side, a centralised system has the drawbacks of a server node bottleneck, a single point of failure, and trust requirements. Decentralised Federated Learning (DFL), which rejects the usage of centralised infrastructures and embraces the ideas of data sharing minimization and decentralised model aggregation, was created to overcome these problems. Despite the work that has been done in DFL, the literature has not (i) examined the fundamental principles that set DFL apart from CFL; (ii) reviewed application circumstances and solutions using DFL; and (iii) analysed DFL frameworks to build and evaluate new solutions.

This study identifies and examines the most crucial DFL fundamentals in terms of federation architectures, topologies, communication strategies, security precautions, and key performance indicators to achieve that goal. The current research also investigates methods for enhancing crucial DFL fundamentals. The most popular DFL application scenarios and solutions are then analysed and contrasted considering the aforementioned fundamentals. Then, the key elements of the present DFL frameworks are compared and evaluated. Lastly, the progress of offered DFL systems is examined to identify patterns, lessons learned, and unresolved difficulties. Federated learning is a machine learning strategy in which numerous clients collaborate to train a model under the regulation of a central server while maintaining the data on the clients. It enables the preservation of data privacy while also increasing learning generalization capacity. Furthermore, the global model of federated learning must adapt to the unique characteristics of the client's data. We propose fine-tuning as a customization strategy for modifying a global federated learning model to fresh client input.

Data scientists can utilise the FL machine learning framework to train statistical models using user data. It is a scattered training method in which testing and training are carried out at the edges. The meta-data is transmitted to a centralized server, which integrates numerous model updations (from various clients) into a novel model (Fig. 1).

Since stock price prediction can provide considerable gains, it has been a hot topic for years. Because a stock time series exhibits behaviour that is nearly identical to a random walk, the stock market is unpredictable, and forecasting it is challenging [5, 7]. Stock quotes can shift dramatically in a couple of seconds when breaking news emerges. Stock price fluctuations can be caused by a variety of factors. Mostly stock prices vary based on supply and demand. When a big number of people want to acquire a stock rather than sell it, or when demand exceeds supply, the price of the

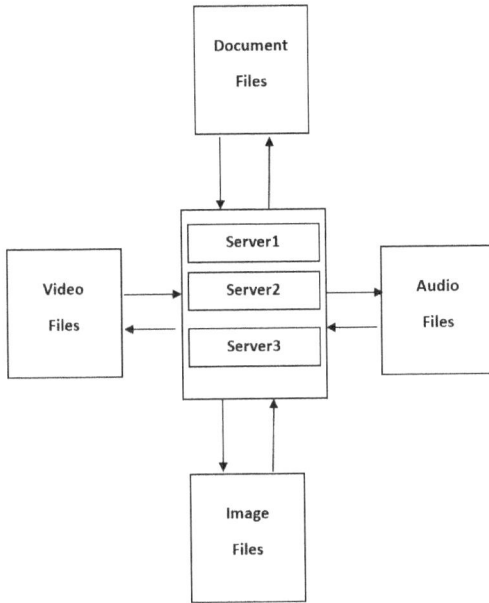

Figure 1. Without sending confidential user data to servers.

stock rises. The price of a stock falls when supply exceeds demand. The company's performance [2] has an impact on its stock price. Aside from fake news, human emotions such as fear and greed, financial results, and other socio-economic elements such as country's political situations and economic conditions all have an impact on the stock market [34]. Human analysis of breaking news might take several minutes, and financial market investors must make choices quickly. Previously, investors would go to a stock market [12] such as the BSE or NSE. Online trading has simplified the process. As a result, many investors enter the capital market. During the COVID-19 pandemic, the number of new investors in the stock market increased significantly. The stock market appears to have attracted new investors throughout the epidemic due to online trading and digitization.

As previously stated, an impact on the stock market was created by psychological factors such as human sentiment and emotions [1, 3, 14]. Twitter is a popular forum to follow stock market news. The BERT and transformers models are utilised to analyse sentiment in this paper. For sentiment prediction, altered BERT is used to train the model using tweets [6, 27]. The BERT classifier is used to make bullish and bearish predictions. The tweets are then pre-processed using the classified results [8, 13, 29] which may subsequently be used to forecast real-time sentiment in tweets.

1.1 Aim and Scope

To train the Machine Learning model and find tweets about stocks and to build a stock market prediction system based on Twitter by evaluating tweets and predicting sentiment. To use Twitter as a source of stock market information, such as Moving Average Convergence and Divergence (MACD), Moving Averages, and other stock indicators. The scope of this study is to decrease the risk of investing in the stock market using twitter sentiment analysis and to aid stockholders in making prudent investments that provide good returns [25].

2. Literature Survey

We've all seen how useful crowd sourcing can be. Customers can report all kinds of driving circumstances on mobile apps like Waze, including cars parked on the side of the road, traffic jams, car accidents, police speed traps, and a lot more. As a result, other users can benefit from this cooperation and make wiser driving choices. As a straightforward illustration, Waze is capable of selecting some alternate routes to reach the destination if there is a major traffic jam on a particular route. Similarly, while looking for a hotel to reserve for a family holiday, we usually check multiple customer reviews. On websites like TripAdvisor or Booking. com, we can learn about almost anything, including the cleanliness of the location, the staff's customer service skills, and much more. Finally, utilising crowdsourcing platforms like Amazon Mechanical Turk to speed up machine learning development is rather prevalent. The concept is to outsource a certain process using a highly dispersed platform. This method is divided into discrete, easy activities that novices may do in a short period. These tasks could involve data annotation, validation, or straightforward data cleaning techniques like deduplication.

Federated learning's underlying concept. Large corporations and hospitals are seeking to figure out how to improve their services by utilising massive volumes of sensitive data. Imagine, for example, if we could train machine learning systems using all the data that is currently available from hospitals around the world. This collaboration would surely benefit applications such as melanoma and breast cancer diagnostics. The same logic applies to insurance and banking companies that may utilise private information to construct more accurate predictive models. However, another crucial issue – privacy — is what is really standing in the way of us working together to improve healthcare. Concern about how businesses around the world might utilise our data grows as predictive algorithms become more commonplace in our daily lives. Knowing that businesses can utilise the data we produce to develop commercial machine learning algorithms without getting our permission does not feel appropriate. Furthermore, it is inconceivable (privacy perspective) for a hospital to give the photos from your most recent examination to a foreign firm that will develop predictive models. To make sure that client data is kept secure at any cost, laws like HIPAA and GDPR were created. Federated learning tries to make use of numerous participants who can each separately contribute to a larger job. Participants in this game can include a wide range of gadgets, including smartphones, tablets, healthcare information stored in hospitals, self-driving cars, and any internet-connected computing device with enough processing power and user interaction data.

The data on each client device has unique characteristics. It is personal, to begin with. Think about a smartphone app for a virtual keyboard. Every day, users connect with coworkers, fill out forms, and communicate with friends using mobile keyboards. Users create passwords, send voice messages, divulge personal information about themselves or their employers, confer with their doctors, and other tasks during this process. However, the data is severely uneven across participants. It is clear that people utilise the keyboard at various speeds for different purposes. While one user may fix all her bank concerns using the mobile app, another user may choose to visit the bank's physical facility.

Think about using a navigation app on your phone, like Waze or Google Maps. Once more, the kind of data produced by such apps is very delicate. It can be used to find out where someone has been and if they are or were riding a bike, in a car, on a plane, or out for a stroll in the park. Statistical models that can learn anyone's movement patterns can also be trained using them. These models could be used to predict the location of his or her trip for tomorrow or the following weekend. A severe infringement of privacy has occurred.

A GAN network (Generative Adversarial Network) based on a Hybrid Prediction Algorithm (HPA) has been proposed [23] to attain the Stock GAN structure by using a descriptive approach. combined information from two different systems. The first is historical data from the Yahoo Finance API, and the second is Twitter data from the Kaggle website. Then, after pre-processing tweet data, BERT and Word2Vec models were trained to get the expected outcomes. Because BERT is stronger at classifying tweet sentiment and can attain lower Mean Square Error, it outperforms Word2Vec in predicting stock price (MSE).

The research analysis [35] offered a novel technique for increasing the 'Fears' that uses search term embedding to better convey investor emotion in their work. On a weekly dataset, the studies reveal that semantic information aids in stock return prediction, and that a trade-off between price and search volume data can assist enhance performance [17]. Employ Support Vector Machine (SVM) and SS kernels to extract correlation between inputs and use this new method to predict stock price trends using financial news [25, 31]. The accuracy is 73% when forecasting price changes with a two-day lag.

A comparison technique [10] uses SVM and a personalised sentiment lexicon for stock price prediction. –5 to +5, with –5 signifying the highest amount of negativity, +5 signifying the largest amount of positivity, and 0 signifying a stop word [28]. SVM has an accuracy rate of 59.1734%. CNN (Convolutional Neural Network) is used as a base learner for sentiment index by [11]. EMD is based on attention LSTM (Layer enhances Long Short-Term Memory). While predicting the stock market, the proposed method achieves the highest accuracy, the shortest time offset, and the closest predictive value.

The Radial Basis Function (RBF) and Support Vector Machine (SVM) are utilized to predict the stock market [24]. Numerical results back up the extraordinary efficiency. To anticipate stock movements, Weng et al., used ANN, SVM, and decision Tree machine learning techniques [33]. The proposed approach achieves an accuracy of approximately 85%.

To calculate the sentiment score [16], suggested using SVM combined with the Harvard IV-4 Sentiment Dictionary (HVD) and Loughran McDonald Financial Sentiment Dictionary (LMD), both of which have accuracy values of LMD = 0.5527 and HVD = 0.5460. To predict stock prices using financial technical indicators, Hegazy et al., devised a machine learning model [9] that combines the particle swarm optimization (PSO) method and the Least Square Support Vector Machine (LS-SVM). The most effective algorithm is the LS-SVM-PSO [4].

3. Materials and Methodologies

3.1 Existing System

There is already a prediction system in place that uses the Long Short-Term Memory (LSTM) algorithm. The processing device known as the memory cell proposed by LSTM [15] takes the place of conventional artificial neurons. Networks are the best choice for dynamically recording data structure throughout time with a high predictive limit because they can link memory and remote input in time with efficiency. Artificial neural networks (ANNs) and genetic algorithms (GAs) are other prediction methods [18]; however, they do not account for long-term temporal correlations between stock prices. Fundamental analysis, which considers a stock's past performance and the company's overall credibility, and statistical analysis, which focuses solely on number crunching and identifying patterns in stock price variation, are traditional methods for stock market analysis and stock price prediction.

Moving averages, MACD, Relative Strength Indicator (RSI), and Channel Commodity Index (CCI) are some of the technical indicators used to anticipate the stock market. Moving Average and MACD [30] are technical indicators that identify stock trends based on prices above or below the moving average. Using the MACD buy or sell signal, investors can make informed stock market investments. Some indicators, such as the RSI and CCI, suggest overbought and oversold conditions by comparing current prices to past values. Technical indicators determine whether stock prices are bearish or bullish [32]. Pattern-based signals are the end results.

3.1.1 Limitations of the Existing System

As analyzed, the existing system has the following shortcomings:

Technical indicators employ a traditional method to make investing decisions. A stock can be oversold or overbought for an extended length of time.

To predict price, technical analysis tools use short intervals. As a result, inaccurate stock price indicators emerge, potentially leading to a disastrous outcome. When using MACD, it is discovered that this technical indicator does not perform well in choppy markets with sideways movement [20–22, 26]. The RSI compares bullish and bearish price momentum in a price chart oscillator and displays the results. Its signals, like those of other technical indicators, are most credible when they correspond to the long-term trend. To anticipate the value for the next day, LSTM just utilises a value that is very close to the previous day's closing price, which may or may not offer an accurate prediction result [19]. When the weights of a big network get too big or too small, respectively, it causes the bursting/disappearing gradient problem, which significantly reduces the convergence to the best value.

3.2 Proposed System

Federated Learning is a strategy for discreetly preserving on-device data while training large machine learning models. Federated Learning, described in the 2015 paper "Communication-Efficient Learning of Deep Networks from Decentralised Data" by Google researchers, is a distributed method for training a centralised server on decentralised data. One of Federated Learning's most important benefits is that it frees machine learning training from the need to retain a dataset on a central server. In Federated Learning, the user's device never loses access to the data. Utilising the local computing power and data of the device, machine learning models are trained. The locally trained model's training meta-information (weights and biases) is sent to a central server. Only the locally trained model's training meta-information (weights and biases) is sent to a central server. We must first investigate how Federated Learning differs from conventional machine learning development to fully understand it. Indeed, a mentality change is necessary to achieve federated learning. When developing machine learning models, it is a common standard to mix training and validation data in a centralised location, such as a server. With all the data in one place, data exploration, visualisation, validation, and batch training procedures (needed for deep neural networks) may be performed fast and reliably.

Most machine learning algorithms are based on the premise that data is IID (Identically Distributed and Independently). Federated Learning violates this rule. In this situation, each member only keeps consumption statistics. We cannot thus assume that each data point (from each client device) is indicative of the entire population. In Federated Learning, model development, training, and evaluation are conducted without direct access to raw data. Costs and dependability of communication, however, provide considerable obstacles. This is how it works. Based on various criteria, the algorithm selects a selection of eligible participants from a pool of candidates.

If we limit our example to mobile devices (such as smartphones and tablets), the devices that were completely charged, had a specific hardware configuration, were connected to a reliable, free WiFi network, and were not in use would be eligible. The fact that not all devices join the federation is a critical element of the Federated Learning architecture. Only qualifying devices are provided with the training model. The purpose is to mitigate any negative consequences that local training may have on the user experience. Even if we believe that local model training will just take a few minutes, no one will be thrilled to experience a lagging device (induced by local model training) when using the device to talk to a friend or get essential information. Then, utilising the local data, each device begins the training model's local fine-tuning phase. Following training, the changed parameters of each local model are sent to the central server for a global update.

The attribute of this paper is to create a stock prediction method that will help stock market investors and inspire newcomers to the industry. We discovered that prediction algorithms such as LSTM and GAN-HPA are utilised for stock price prediction utilising Yahoo finance data or other datasets based on a literature survey and research publications. As a result, our primary goal is to create a prediction system based on sentiment analysis of stock-related tweets. They proposed a method for predicting stock prices using the BERT algorithm and sentiment analysis of tweets with stock market content. The tweets and essential datasets are acquired via Twitter API and other sources because this paper is now focusing on stock price prediction from Twitter data. BERT is a semi-supervised machine learning algorithm, and it is used for stock price prediction from Twitter data.

This Algorithm tries to improve the classification of tweets by categorizing them into two sections, Positive and Negative are reduced by using a feature determination and extraction method called word embedding, in which words with similar meanings have comparable representations. This approach is shown in Fig. 2.

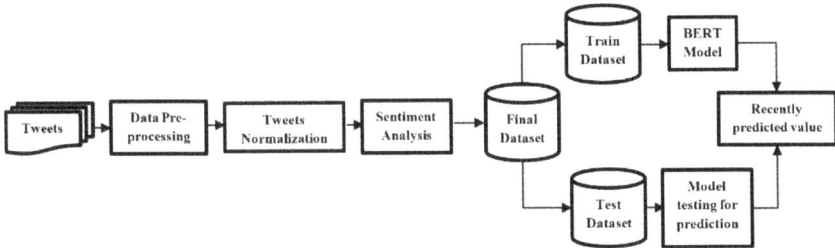

Figure 2. Enlarged block architecture.

The detailed architecture operation of the tweeter sentiment analysis is depicted in Fig. 2. The tweets are the raw data that is processed further. The design contains a mechanism for classifying sentiments on a Twitter platform by normalizing the text and recognizing the positive/negative duality of the tweets. The completed dataset is then sent to be trained and tested on. For stock prediction, the training dataset is transmitted to a pre-trained BERT model. The model is tested using the dataset that was tested. Essentially, BERT is a Transformer's bi-directional Encoder Representation. It's a Machine Learning Model that's semi-supervised ([13, 14]). The Bi-directional Encoder is a Natural Language Processing (NLP) tool. The tweets are retrieved and transformed to a vector format of 0 and 1, with 1 denoting 'Positive' and 0 denoting 'Negative.' In tweets, human emotions are expressed using a specific language, which is then systematically transformed into mathematical form. Finally, we may use a combination of BERT and NLP to forecast if a particular stock will perform well in the future.

Sentiment analysis is a type of machine learning activity in which we try to figure out how people feel about a piece of text. To extract certain elements from the provided text and categories them as positive or negative, we'll employ machine learning and natural language processing techniques. This is beneficial since we can get a general sense of whether to purchase or sell the stock. We attempted to categories tweets from Twitter as favorable or bad in this research. Twitter is a social networking platform that allows users to express themselves rapidly and spontaneously under a limit of 140 characters. Twitter is an excellent source for understanding current overall impressions about anything due to its widespread use. The suggested system primarily consists of three key steps: data acquisition, data pre-processing, and data classification.

Server executes:

    initialize ω_0

    for each round t = 1,2,...do

        $m \leftarrow max(C.K,1)$

        $S_t \leftarrow$ *(random set of m clients)*

        for each client $k \in S_t$ *in parellel do*

            $\omega_{t+1}^k \leftarrow$ *ClientUpdate*(k, ω_t)

$$w_{t+1}^k \leftarrow \sum_{k=1}^{K} \frac{n_k}{n} w_{t+1}^k$$

ClientUpdate(k, ω): *//Run on client k*

        $\mathcal{B} \leftarrow$ *(split* $\wp_k$ *into batches of size B)*

    for each localepochi from 1 to E do

        for batch $b \in \mathcal{B}$ *do*

            $\omega \leftarrow \omega - \eta \nabla l(\omega; b)$

    return ω *to server*

The Federated Averaging algorithm is depicted in the pseudo-algorithm above. It is divided into two halves, the first of which is executed by the server and the second by the clients. The server initially initialises the model parameters — normally with random values. The server coordinates the many execution rounds. The server selects a selection of clients (eligible devices) at random for each round and provides a copy of the training model in parallel. Each client utilises their data to do a series of gradient descent steps to fine-tune the training model's copy. Following training, each client sends the weights and biases of the local model to the server. The server gathers all client changes and initiates a new cycle.

Realising that Federated Learning is not totally private is crucial. An attacker (or even the server) may acquire access to the raw updates and undertake some reverse engineering to study information about each client's data because each device's unique model changes are sent to a coordinating server. End-to-end encryption is therefore included in Federated Learning in its standard form. The training metadata is, therefore, encrypted throughout its transfer from client to server, further securing the procedure. Secure aggregation, a method that prevents the server access to the raw updates, enables the server to compile encrypted models from various participants and only decode the aggregated training results.

As a result, the server never sees the training results from a specific device. For added security, we may combine Federated Learning with Differential Privacy. Testing is another key distinction between Federated and traditional machine learning. When validating machine learning models, we should use data that is as like actual production data as possible. The combined model cannot be evaluated by the server after it has been updated with client contributions because it lacks access to the training data. As a result, testing and training are conducted on consumers' devices. It is imperative to emphasise that distributed testing restores the benefits of testing the new model on user-facing devices, which is where it matters the most.

The novel approach of federated learning, in conclusion, has great promise. It is a decentralised, collaborative approach that enables researchers to train machine learning models with sensitive data while upholding privacy standards. Federated Training may be employed in more applications because of current improvements in 5G technology, which will need us to rely on stronger and faster internet connections. Although federated learning is still in its infancy, it has already found success in several applications, including Apple Siri and the Google virtual keyboard (GBoard).

3.2.1 Data Acquisition

Twitter API: There are numerous options for gathering data. To collect tweets, the Twitter API v2 version is suggested. The following steps can be used to gather tweets using the Twitter API:

i. A developer account with access level control is required to send requests over the Twitter API.

ii. Log in to your developer account once you've created one. Create a new article and name it.

iii. Once everything is in place, you will be given a set of four keys that will be used to access the Twitter database.

Collection: The Tweepy library, which is available in Python, to collect tweets via the Twitter API shall be used here.

- Use the pip installs Tweepy command to install Tweepy.
- Use your keys and tokens to verify your identity.

- Create a search query—A search query is a string that tells Twitter API what kind of tweets you're looking for. For example, if you want to locate tweets regarding Tesla, simply search for Tesla or @Tesla on Twitter.

- To get tweets using Twitter API, we'll use the Tweepy cursor. The response is then recorded in a list that can be iterated to obtain the needed information.

Dataset: The Panda library from Python is recommended for this purpose. A data frame can be made from the response's desired properties.

3.2.2 Pre-processing

It is essential to pre-process the tweets because this has a direct impact on the classifier's performance. The tweets will be pre-processed using the 'Regex' library. For this objective, the following procedures will be carried out:

- Conversion: The text is converted to lowercase. Tokenization is the method of separating raw text into words or sentences known as tokens. For further processing, the tweet will be broken down into words.

- Stop words Removal: These are the words that appear frequently in all tweets and may not express any emotions usually.

- Emoticons Removal: We're also eliminating all emojis from the tweet because we're only interested in the text.

- Hashtags and URLs Removal: Tweets can include several hashtags and URLs. Because these are useless for sentiment analysis, they must be eliminated.

3.2.3 Classification

Researchers developed an algorithm to automatically identify tweets as negative or positive in previous articles. To train the model, some people used data that was already public. We manually classified tweets as negative or positive in this work. Our model is then trained using manually categorized data.

3.2.4 Advantages of the Proposed System

The advantages of the proposed system are as follows:
   To pre-train BERT, a large amount of data was used. Over 2500 million words have been taught. This means it can work with small datasets and still get good results.

Previously, no matter how a word was used, it would return the same vector; now, depending on the words around it, BERT returns multiple vectors for the same word.

It is said that this paradigm is bidirectional. Contrary to directed models, BERT can read a text at any time and in any direction. Because of its bidirectionality, this model can understand the meaning of each word based on the context to the right and the left of the word; this represents a significant advancement in the field of context learning. Masked Level Modelling (MLM) and Next Sentence Prediction (NSP) are combined in BERT. It is currently the best NLP strategy for comprehending texts with a lot of contexts. Unlike other algorithms, BERT concentrates on individual words rather than an entire text. BERT also recognises each word and anticipates the word's left and right counterparts.

3.3 Transformers

Because BERT is based on the Transformers franchise, it's crucial to comprehend the Transformers model. Transformers models usually have two layers process as shown in Fig. 3.

Encoder: This part of the model examines the input text for key phrases and creates embeddings for each word based on how those words relate to one another.

Decoder: The decoder transforms the embedding-based output of the encoder into text output, which is the translated form of the input text.

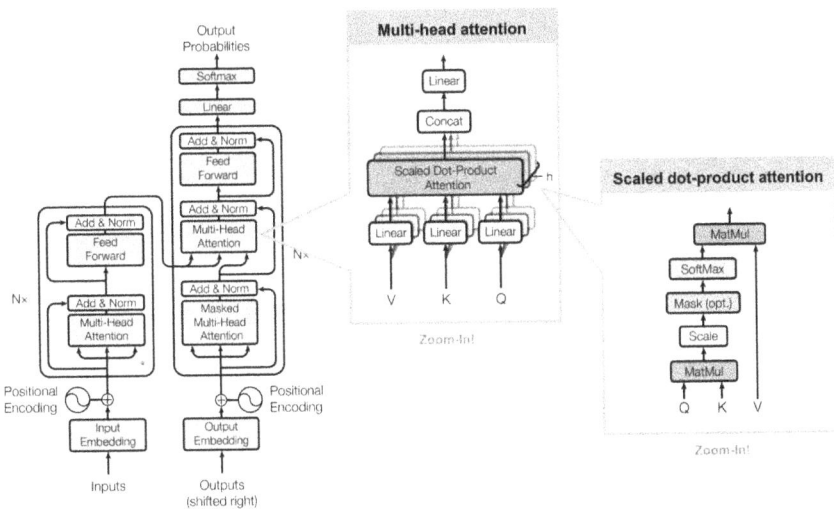

Figure 3. Transformers architecture.

The most essential part of the study is the layers used to make the encoder and decoder as depicted in Fig. 3. Contrary to conventional RNNs, neither the encoder nor the decoder employed recurrence or looping. Instead, they made use of "attention" levels, which enable a straight flow of information. The Transformer transmits the input across several attention layers rather than looping it many times.

3.3.1 BERT

BERT is a Transformers-based machine learning model. Before, text input could only be read from right to left or from left to right, but not at once. BERT's ability to be bi-directional allows it to complete both tasks simultaneously.

3.3.2 How BERT Works?

The primary goal of NLP is to comprehend human language. Models are trained using a big resource to achieve this objective. Only unlabelled, plain text corpus was used to train BERT. Unsupervised learning from unlabelled text can help it improve. Pre-training BERT can serve as a foundation of "knowledge" on which to build. It can be further trained based on the developer's requirements. BERT is the first NLP system that entirely relies on the self-attention mechanism. Because of its bidirectionality, this is conceivable. BERT may concentrate on each word rather than the entire speech. The word in focus becomes increasingly unclear the more words there are in a sentence or phrase.

3.3.3 Training BERT

BERT is trained using two different methods:

- Masked language modelling (MLM) is used to randomly substitute 15% of the words in the sequence before feeding it to BERT. Using the non-masked words in the sequence, the model then attempts to predict the original value of the masked words. The BERT loss function does not take non-masked words into account; only masked word prediction is. As a result, the model converges later than directed models.

- Predicting the Next Sentence: The model is trained on a pair of sentences and asked to predict whether the second sentence in the source text will come next. In input pairings used for training, the second sentence is found in 50% of cases in the original text and 50%

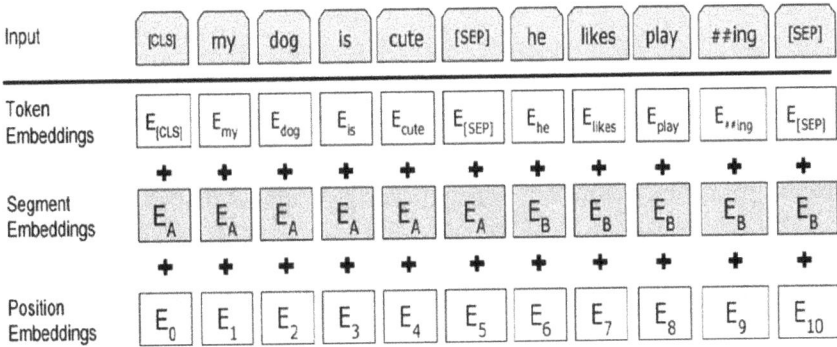

Figure 4. Token embeddings.

of cases in a random sentence. To discriminate between two sentences during training, the input is handled as shown in Fig. 4:

- A [CLS] token is added to the beginning of the first sentence, and a [SEP] token is added at the end of each sentence. Each token has an embedded sentence that denotes Sentences A and B. Each token is given a position embedded to identify its place in the sequence.

To minimise the combined loss function of the two strategies, Masked LM and Next Sentence Prediction are merged when training the BERT model. Finally, as illustrated in Fig. 3, the model can be adjusted to the user's requirements.

3.4 Testing and Validation

The stock prediction input and output designs are thoroughly analysed and tested using the testing methods discussed in this section.

3.4.1 Input Design

This usually entails developing a set of valid and invalid inputs to check if the prototype can accurately detect invalid conditions and throw an exception or error. Currently, there is a large amount of stock available. Take a specific stock, such as Tesla, to make a valid input. Then we'll obtain all the tweets on that stock. The sentiment analysis uses those tweets as inputs. Because the BERT technique is being used, the tweets will be translated into mathematical form, i.e., vectors, which will then be used as inputs for sentiment analysis.

3.4.2 Output Design

When our algorithm processes raw tweets, it looks for stop words, emoticons, and links, among other things. We need to filter the tweets for our model to run efficiently. The cleaned and filtered tweets using regular expressions are one of the results (regex). The feelings of those cleaned tweets are then divided into two categories: good and negative.

We get the average positive Tweet after sentiment analysis on cleaned tweets, and we may infer the average positive Tweet to produce the desired output: BETTER NOT TO BUY, NO HEAVY PROFITS, and CAN BE BOUGHT.

3.4.3 Testing

Modules are tested against comprehensive designs using unit testing. The process's inputs are typically compiled modules from the coding process. During the unit testing process, all the modules are usually brought together and combined into a larger unit.

The following processes presented obstacles to us during the implementation:

- Removal of Irrelevant tweets: Several irrelevant tweets, stop words, emoticons, and links were captured while obtaining tweets via Twitter API. To solve this problem, we had to filter the tweets using a certain term and a regular expression in the model.

- Updates of Tweets: Due to the massive volume of trending tweets created each second concerning the stock market, we had some difficulty retrieving recent stock-related tweets. So, to obtain the requisite amount of data, we tried multiple times to generate tweets using the Twitter API. We also attempted to expand our system. We experimented with increasing the size of the data frame. In our model, re-tweets are insignificant. The amount of data is significantly decreased after the re-tweets are removed.

4. Results and Discussion

The efficiency of a proposed model was investigated using a range of evaluation criteria to evaluate how effectively it could discriminate positive from negative emotion.

Because the RMSE value is always less than 10, the system is extremely accurate (Equation 1).

$$Prediction = \frac{No. of\ Positive}{No. of\ Negative} \tag{1}$$

If the prediction is greater than 1, the stock will rise, and if it is less than 1, the stock will fall. When comparing values predicted by a model to values observed, the root-mean-square deviation, also known as root-mean-square error, is commonly determined (Equation 2).

The following is the general formula for calculating the Root Mean Square Error (RMSE) value:

$$RMSE = \sqrt{\sum_{i=1}^{N} \frac{(predicted_i - actual_i)^2}{N}} \qquad (2)$$

where N = Total Prediction, i = variable.

The first stage would involve a thorough assessment of the literature regarding data classification techniques and stock prediction algorithms. To build a reliable system, we will be able to strengthen the structure of the proposed security solution design by using a literature study to address the shortcomings of existing schemes while keeping in mind their advantages. Figure 5 depicts the proposed solution's implementation together with all relevant input and output parameters.

```
[ ]  #Get the root mean squared error (RMSE)
     rmse=np.sqrt(np.mean( predictions - y_test)**2)
     rmse

5.25987907409668
```

Figure 5. Sample implementation outcome of RMSE.

The comparisons of the proposed method's accuracy based on RMSE with other methods are provided in Table 1.

Table 1. Performance comparison.

Methodology	RMSE
Proposed Method	< 10
Linear regression [Van, 2022]	Between 10 and 15
Structural similarity [Long et al., 2019]	Between 12 and 15
Granger causality analysis [Mittal and Goel, 2012]	Between 16 and 20

Figure 6 represents the stock movement on timeline, and colour notation was used to differentiate between the actual and system predicted

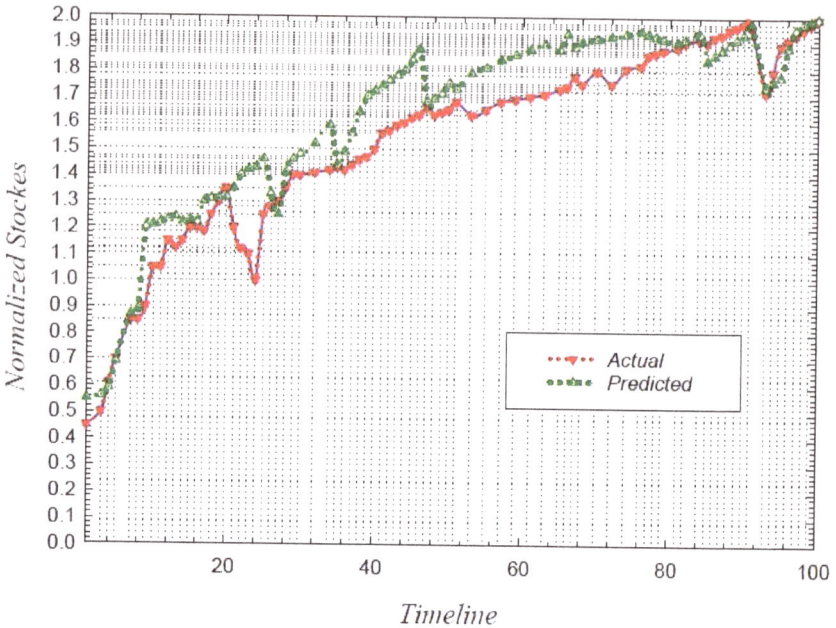

Figure 6. Graphical representation between actual stock and system predicted stock.

stock movement. Through this graph stock prediction the actual stock movement can be identified.

5. Conclusion

The goal is to use the proposed technology to the advantage of stock market investors. This technique is focused on forecasting stock prices using sentimental analysis of tweets. This model uses an NLP pipeline to train an updated BERT model with the federated learning approach along with Twitter tweets to predict their bullish and negative stock sentiments. BERT is a cutting-edge NLP model that has shown results in a variety of disciplines. BERT was able to increase accuracy on several Natural Language Processing and Language Modeling tasks (or F1-score).

This work's main contribution is that it makes it possible to apply semi-supervised learning for a variety of NLP tasks, facilitating transfer learning in NLP. These qualities led us to use BERT to increase the accuracy of the current model. Our model is fairly clear and makes use of a simple Twitter and Machine Learning interface. As a result, it can be conclude that this model is closer to human inference in terms of financial comprehension. It has also simplified the process by including different Python and machine learning libraries. The model can directly advise each investor on whether to invest, which is something that most existing models lack. There will be more open-source articles in the future. Tech giants will be attracted to it. As more money is invested in this, a system that can accurately anticipate stock prices will emerge. An algorithm for stock categorization and prediction shall be constructed in future work to solve the shortcomings of the existing system and create a new and robust stock prediction system. The stock market prediction accuracy can also be improved by combining trending factor data with online source stock data.

References

[1] Anjaria, M. and Guddeti, R.M.R. 2014. Influence factor-based opinion mining of Twitter data using supervised learning. In 2014 Sixth International Conference on Communication Systems and Networks (COMSNETS), pp. 1–8.

[2] Barbosa, L. and Feng, J. 2010. Robust sentiment detection on twitter from biased and noisy data. In COLING '10: Proceedings of the 23rd International Conference on Computational Linguistics: Posters, pp. 36–44.

[3] Barhan, A. and Shakhomirov, A. 2012. Methods for sentiment analysis of twitter messages. In Proceedings of the XXth Conference of Open Innovations Association FRUCT., 325(12): 215–222.

[4] Boiy, E. and Moens, M.F. 2009. A machine learning approach to sentiment analysis in multilingual Web texts. Information Retrieval, 12(5): 526–558.

[5] Cai, R., Qin, B., Chen, Y., Zhang, L., Yang, R., Chen, S. and Wang, W. 2020. Sentiment analysis about investors and consumers in energy market based on BERT-BiLSTM. IEEE Access, 8: 171408–171415.

[6] Chen, L.S., Liu, C.H. and Chiu, H.J. 2011. A neural network-based approach for sentiment classification in the blogosphere. Journal of Informetrics, 5(2): 313–322.

[7] Giachanou, A. and Crestani, F. 2016. Like it or not: A survey of twitter sentiment analysis methods. ACM Computing Surveys (CSUR), 49(2): 1–41.

[8] Go, A., Bhayani, R. and Huang, L. 2009. Twitter sentiment classification using distant supervision. CS224N Article Report, Stanford, 1(12): 2009–2018.

[9] Hegazy, O., Soliman, O.S. and Salam, M.A. 2014. A machine learning model for stock market prediction. International Journal of Computer Science and Telecommunications, 4(12): 17–23.

[10] Jiang, L., Yu, M., Zhou, M., Liu, X. and Zhao, T. 2011. Target-dependent twitter sentiment classification. In Proceedings of the 49th Annual Meeting of the Association for Computational Linguistics: Human Language Technologies, 1: 151–160.

[11] Jin, Z., Yang, Y. and Liu, Y. 2020. Stock closing price prediction based on sentiment analysis and LSTM. Neural Computing and Applications, 32(13): 9713–9729.

[12] Kharde, V. and Sonawane, P. 2016. Sentiment analysis of twitter data: A survey of techniques. International Journal of Computer Applications, 139(11): 5–15.

[13] Koppel, M. and Schler, J. 2006. The importance of neutral examples for learning sentiment. Computational Intelligence, 22(2): 100–109.

[14] Kumar, H.P., Prashanth, K.B., Nirmala, T.V. and Patil, S.B. 2012. Neuro fuzzy based techniques for predicting stock trends. International Journal of Computer Science Issues (IJCSI), 9(4): 385–394.

[15] Li, M., Chen, L., Zhao, J. and Li, Q. 2021. Sentiment analysis of Chinese stock reviews based on BERT model. Applied Intelligence, 51(7): 5016–5024.

[16] Li, X., Xie, H., Chen, L., Wang, J. and Deng, X. 2014. News impact on stock price return via sentiment analysis. Knowledge-Based Systems, 69(14-23).

[17] Long, W., Song, L. and Tian, Y. 2019. A new graphic kernel method of stock price trend prediction based on financial news semantic and structural similarity. Expert Systems with Applications, 118: 411–424.

[18] Mittal, A. and Goel, A. 2012. Stock prediction using twitter sentiment analysis. Stanford University, CS229, pp. 1–5.

[19] Mullen, T. and Collier, N. 2004. Sentiment analysis using support vector machines with diverse information sources. In Proceedings of the 2004 Conference on Empirical Methods in Natural Language Processing, pp. 412–418.

[20] Pang, B. and Lee, L. 2004. A sentimental education: Sentiment analysis using subjectivity summarization based on minimum cuts. In Proceedings of the 42nd ACL, pp. 271–278.

[21] Pang, B. and Lee, L. 2008. Opinion mining and sentiment analysis. Foundations and Trends in Information Retrieval, 2(1-2): 1–135.

[22] Pang, B., Lee, L. and Vaithyanathan, S. 2002. Thumbs up? Sentiment classification using machine learning techniques. In Proceedings of the 2002 Conference on Empirical Methods in Natural Language Processing (EMNLP), pp. 79–86.

[23] Polamuri, S.R., Srinivas, K. and Mohan, A.K. 2021. Multi-Model Generative Adversarial Network Hybrid Prediction Algorithm (MMGAN-HPA) for stock market prices prediction. Journal of King Saud University-Computer and Information Sciences, 34(9): 7433–7444.

[24] Reddy, V.K.S. 2018. Stock market prediction using machine learning. International Research Journal of Engineering and Technology (IRJET), 5(10): 1033–1035.

[25] Romero, D.M., Meeder, B. and Kleinberg, 2011. Differences in the mechanics of information diffusion across topics: Idioms, political hashtags, and complex contagion on twitter. In Proceedings of the 20th International Conference on World Wide Web, 695–704.

[26] Saif, H., He, Y. and Alani, H. 2012. Semantic sentiment analysis of twitter. The Semantic Web – ISWC 2012. ISWC 2012. Lecture Notes in Computer Science, 7649. Springer, Berlin, Heidelberg.

[27] Sanh, V., Debut, L., Chaumond, J. and Wolf, T. 2020. DistilBERT, a distilled version of BERT: Smaller, faster, cheaper and lighter. Computational Language.

[28] Sebastian, W. and Isa, S.M. 2020. Stock price prediction using BERT and Word2Vec sentiment analysis. International Journal of Emerging Trends in Engineering Research, 8(9): 5430–5433.

[29] Tang, H., Tan, S. and Cheng, X. 2009. A survey on sentiment detection of reviews. Expert Systems with Aplications, 36(7): 10760–10773.

[30] Tomihira, T., Otsuka, A., Yamashita, A. and Satoh, T. 2020. Multilingual emoji prediction using BERT for sentiment analysis. International Journal of Web Information Systems, 16(3): 265–280.

Index

A

A3C 177, 184–187, 189, 191, 192
Advantage function 185
Alzheimer's disease 264–267, 269–272, 278, 279
Architecture 61, 71
Artificial Intelligence (AI) 202–206, 211–213, 216, 221, 222, 223
Autonomous self-aware intelligent agents 32, 54

B

BERT 333, 335, 337, 340, 341, 344–347, 350
Blockchain 283–286, 288, 289, 292–294, 301–305

C

Centralized 106, 107, 110, 112, 113, 115–117, 119, 120, 123, 128
Cloud environment 155, 171
Cognitive science 33, 37, 38, 50, 52, 54
Collaborative learning 203, 204
Computation offloading 186
Convolution neural network 245, 251, 256
Crowd coordination 308, 309, 318, 319, 329, 330
Crowdsourcing 308–310, 314, 319–329
Customer satisfaction 283

D

Data loss 155, 171
Data-driven 202, 216
Deep learning 1, 20, 27, 84, 228, 230–232, 238, 239, 246, 250, 251, 265–267, 269, 270–272, 279
Deep Neural Network (DNN) 132–136, 138, 142, 147, 148, 150

Deep reinforcement learning 174–176, 179, 198, 199
Differential privacy (DP) 83–87, 89, 99, 102
Distributed machine learning 151
Distributed machine learning 7, 9, 10, 24, 26, 27

E

Edge devices 134, 141, 142, 147, 148, 150, 152
Encoder 341, 345, 346
Encrypted model 231, 238

F

Federated learning 1, 3, 4, 11–17, 20–24, 26, 27, 80–83, 85, 88, 90, 91, 97–99, 102, 155–159, 161–171, 175, 186, 188, 191, 202–211, 213–217, 219–224, 245, 261, 264–266, 269, 270, 274, 275, 283, 286, 287, 289, 291–293, 296, 297, 301, 304, 305, 333, 334, 336, 337, 339, 340, 342, 343, 350
Federation learning 107, 132–134, 136, 139, 140, 145, 150, 151

G

General data protection regulation 111, 112, 117, 119–121, 124, 125, 127

H

Healthcare 203–205, 210–216, 219, 221–224

I

Internet of Things (IoT) 131–139, 142, 145, 147, 148, 150, 151

L

LSTM 246, 251, 256–261

M

Machine learning 1, 4, 5–10, 16, 24, 26,
27, 33, 34, 36, 40, 52, 54, 84, 106–110,
112, 113, 115, 116, 120, 122, 123, 126,
127, 202–207, 209–211, 213–216, 219,
222–224, 226–228, 232, 235, 236, 238,
240, 247, 249, 250, 267, 268, 270, 271,
274, 277–279
Macro-tasks 308, 310, 311, 318–330
MRI images 265–269, 271, 272, 280

N

Neural network models 6, 20

P

Preservation 108, 110, 117, 118
Privacy 61, 63–77, 106–108, 110–113,
116–119, 121–128, 155–162, 164, 167,
168, 170, 171

R

Residence time 177, 181, 186, 189, 193–195
Resource allocation 157

S

Security 61, 63–67, 70, 74, 76, 77
Security monitoring 131–133, 136, 139, 152
Sensor device 298
Sentiment analysis 333, 336, 340, 341, 344,
347, 348
Stochastic gradient descent (SGD) 84–86

T

Transactive Memory Systems (TMS)
310–312, 321, 323
Transformers 335, 341, 345, 346
Twitter API 341, 343, 344, 348

For Product Safety Concerns and Information please contact our EU
representative GPSR@taylorandfrancis.com
Taylor & Francis Verlag GmbH, Kaufingerstraße 24, 80331 München, Germany

www.ingramcontent.com/pod-product-compliance
Lightning Source LLC
Chambersburg PA
CBHW052011230326
41598CB00078B/2509